Cuatro Ciénegas Basin: An Endangered Hyperdiverse Oasis

Series Editors

Valeria Souza
Ecology Institute
Universidad Nacional Autónoma de México
Mexico City, Distrito Federal, Mexico

Luis E. Eguiarte
Ecology Institute
Universidad Nacional Autónoma de México
Mexico City, Distrito Federal, Mexico

This book series describes the diversity, ecology, evolution, anthropology, archeology and geology of an unusually diverse site in the desert that is paradoxically one of the most phosphorus-poor sites that we know of. The aim of each book is to promote critical thinking and not only explore the natural history, ecology, evolution and conservation of the oasis, but also consider various scenarios to unravel the mystery of why this site is the only one of its kind on the planet, how it evolved, and how it has survived for so long.

More information about this series at http://www.springer.com/series/15841

Maria C. Mandujano • Irene Pisanty
Luis E. Eguiarte
Editors

Plant Diversity and Ecology in the Chihuahuan Desert

Emphasis on the Cuatro Ciénegas Basin

Springer

Editors
Maria C. Mandujano
Departamento de Ecología de la Biodiversidad, Instituto de Ecología
Universidad Nacional Autónoma de México, UNAM
Mexico City, Mexico

Irene Pisanty
Departmento Ecología y Recursos Naturales, Facultad de Ciencias
Universidad Nacional Autónoma de México, UNAM
Mexico City, Mexico

Luis E. Eguiarte
Departamento de Ecología Evolutiva, Instituto de Ecología
Universidad Nacional Autónoma de México, UNAM
Mexico City, Mexico

Cuatro Ciénegas Basin: An Endangered Hyperdiverse Oasis
ISSN 2523-7284 ISSN 2523-7292 (electronic)
ISBN 978-3-030-44962-9 ISBN 978-3-030-44963-6 (eBook)
https://doi.org/10.1007/978-3-030-44963-6

This Springer imprint is published by the registered company Springer Nature Switzerland AG
The registered company address is: Gewerbestrasse 11, 6330 Cham, Switzerland

Foreword

Perhaps no landscape in North America has ecological contrasts and visual juxtapositions as dramatic and awe-inspiring as the Cuatro Ciénegas Basin in the Chihuahuan Desert. After miles of travel through hyperarid landscapes dominated by the low-lying xerophilous vegetation known as desert scrub, you arrive at an archipelago of oases unmatched by any habitat for 500 km in any direction. Cool blue waters with an abundance of aquatic life provide a stark contrast to pale gypsum-derived dunes, with cacti such as *Grusonia* which literally crawl across the sands!

You feel as if you have stumbled upon an ecological and evolutionary puzzle as complex as that which Darwin discovered in the Galapagos. That is because the levels of biodiversity in the basin are unusually high compared to any other desert, but its levels of endemism outdistance those of any other habitat complex found elsewhere in the arid regions of the Americas. Diversity and endemism are often negatively correlated, but in Cuatro Ciénegas, this is not the case.

If that puzzle alone were not enough to keep your mind active and your senses attentive, the landscapes and waterscapes of Cuatro Ciénegas face daunting challenges with regard to "whole ecosystem conservation." We find it difficult—if not disturbing—to see the integrity of this habit complex reduced to a few fragments, for the synergies between the geohydrology, the flora, and the fauna form an indivisible, "unified field" for inquiry and appreciation.

We are now fortunate enough to have some of the finest and most forward-thinking ecologists, biogeographers, and biosystematists in Mexico to help us solve these puzzles. While scientific collaborations have advanced the knowledge of the Basin's biogeography for over a quarter-century, never have we had such a remarkable synthesis of a single landscape in arid America. I am grateful to the editors and contributors of this volume for their detailed treatment of biogeographic patterns that ripple out from the Basin to the Chihuahuan Desert as a whole. But more than that, their collective lifework on the ecology and evolution of desert and aquatic organisms found in the Cuatro Ciénegas stand as some of the most exciting field research ever accomplished in any of the deserts of North America.

You cannot spend time "beyond the rainbow's end" in Cuatro Ciénegas without feeling that your own life and sense of what a desert can be have been forever changed and enriched. While many of us have felt this shift in perception, until now, few of us have had the full scientific context which may explain why we so strongly feel that way.

I am humbled and gratified by the great work that the editors have brought together for this unprecedented volume. The editors were already among my list of heroes before they joined forces for this book; but now they deserve to be honored as "true desert saints." May the power of their synthesis convince both local communities and national policy-makers that the ecosystem-level conservation and restoration of Cuatro Ciénegas habitat complexes deserve a public investment commensurate with the distinctiveness of this unforgettable landscape.

Sonora, Mexico
March 2019

Gary Paul Nabhan

Preface

Plants are great storytellers, despite their innate silent life. After "Once upon a time…," they can tell us about the movement of continental masses, the behavior of water, and the vicissitudes of climate through really long-term time-lapses. They can also tell stories of more recent events and even tell us what happened a few days ago. If you "listen" to them, you will hear such stories, and it will be hard to differentiate between fantasy and reality, for plants can tell awesome things about themselves and about the places they live in, as they are such sensitive, green, picky species.

In this book, we recapitulate some stories accumulated by plants that inhabit the Chihuahuan Desert, in Northern Mexico and Southwestern USA, and more specifically of those living in the Cuatro Ciénegas Basin. Stories based on carefully gathered scientific information are presented in 19 chapters in this compendium. These chapters describe aspects of the life history of plants, vegetation distribution, origin and affinities with landforms, and biologically obligated interactions such as water dependence or plant pollination.

Deserts, whose allure deserves to be sensed, have environmental conditions that can be identified and understood simply by looking at plant adaptations such as their reduction of leaves to spines, their different structures that retain water, their cuticles that diminish evapotranspiration, their different photosynthetic paths, physiologies, and complex life cycles and morphologies, including different shapes and number of flowers, fruits, and seeds as well as contrasting reproductive strategies. In the unbelievably beautiful Cuatro Ciénegas Basin—which is at the heart of the Chihuahuan Desert—plant species and vegetation types tell us about a patchy environment resulting from ancient conditions that have been changing for millions of years. In each patch, plants deal with lack of nutrients, high salinity, gypsum soils, or even flooded conditions. No one could expect less from a place located within the Chihuahuan Desert with an elevation of ca. 700 m above sea level and surrounded by very high and steep mountains that were originally covered by an ancient ocean. In the actual basin, complex underground hydrological systems still emerge and create the pools (locally called *pozas*), lakes, creeks, and rivers that make this place unique not only because of its beauty but also because of its biological history and its ecological characteristics and importance.

Environmental and specific diversity in the Chihuahuan Desert in general, and in the Cuatro Ciénegas Basin in particular, has long been recognized as outstanding. A global ecological overview and in-depth studies of specific processes are reviewed in this book. The Chihuahuan Desert is the largest hot desert in North America and has a complex geologic, climatic, and biogeographical history, which affects todays' distribution of vegetation and plants and generates complex phylogeographic patterns. The knowledge of the changing climatic conditions allows the reconstruction of paleogeographic vegetation. The high number of endemic species is related to this complex set of environmental traits. The modern distribution of environments, including aquatic and subaquatic systems, riparian environments, gypsum dunes and gypsum-rich soils, low content of phosphorus and organic matter, and high salinity, combined with an extreme climate demands a set of adaptations to respond to different combinations of these conditions. Plants are distributed in a patchy pattern according to punctual variations, and many of them respond with a considerable morphological plasticity to different resources and conditions. Physiological, morphological, and ecological variability allows the identification of different adaptations to different environments that can be shown in different manners in species and individuals.

Plants tell us their stories through morphology, physiology, all sorts of ecological traits, and, last but not least, their genetics. The authors of the different chapters proved to be nosy enough to ask interesting questions in very different fields and scales, and the stories that plants had to tell are embodied, with a deep scientific insight, in each of them. That is enough to make this book valuable. However, the aim of this book is not only to show what our favorite storytellers tell us about their lives, but also to gather in a single volume information that is relevant to all stakeholders—not only researchers and enlightened public—but specifically decision-makers at all levels. Not even the very adaptable plants of the desert are able to neglect the intense disturbance the Cuatro Ciénegas Basin has been undergoing for more than a decade already, putting plants, as well as all domains of life, in jeopardy. The effects of this disturbance go beyond the scientific interest that Cuatro Ciénegas represents. These effects involve local and regional aspects, including the livelihood and well-being of the people that live in the Chihuahuan Desert.

The consequences of the disturbance of desert ecosystems and in particular of the Cuatro Ciénegas Basin include the loss of water bodies with the concomitant extinction of endemic species, ranging from plants and fishes to bacteria and the stromatolites they form, which are only found in this and few other places in Mexico and in the world. Ecosystem services are being lost, and not even plants, with all their adaptive baggage, can ignore how serious the loss of water can be in an arid zone like this. Aquatic, subaquatic, and riparian plants are quickly disappearing in the Chihuahuan Desert, and punctual dramatic vegetation changes can be observed, and they are described in this book.

As we said, plants are great storytellers and have great stories to tell. We hope the information gathered in this book can help reverse the worst story plants from

Cuatro Ciénegas can tell nowadays: the irreversible disturbance and eventual disappearance of water in one of the most relevant arid lands of the world. Once all ecosystem services are lost, little will remain to be told. We hope this book will help both to understand plant life in Cuatro Ciénegas as much as to reverse the damage that has already been done and to prevent further mismanagement.

This book comprises a foreword by Professor Gary Nabham and 19 peer-reviewed chapters written by expert scientists from Mexico and the USA. The book is organized into three broad sections. The first section includes four chapters that represent a general overview of the origin, evolution, diversity, and floristic composition of the Chihuahuan Desert (Chaps. 1, 2, 3, and 4). The next section comprises nine chapters that provide fascinating examples of the ecology of different plant groups and landforms that are emblematic of Cuatro Ciénegas Basin, as well as of different desert species subject to management and use (Chaps. 5, 6, 7, 8, 9, 10, 11, 12, and 13). For instance, deserts are frequently viewed as not having any resources for human societies, because agriculture and cattle raising are limited. However, local dwellers have found ways to use a rich diversity of wild species, as it happens with *candelilla* from which wax to prevent the early maturation of citric, among other uses, is extracted. Wild species are an important source of resources for the rural inhabitants of arid and semiarid regions, and Cuatro Ciénegas is not an exception. The last section includes six chapters that describe the environmental problems that are risking the preservation of riparian vegetation in Cuatro Ciénegas such as invasive species and water overexploitation and about scientific strategies that may support or lead to conservation (Chaps. 14, 15, 16, 17, 18, and 19). Along the reading of the chapters in this last section, you will find why the Chihuahuan Desert is so special, and how cacti were identified as the most important group in specific environments like *bajadas*, characterized by high diversity values, while gypsophytes and gypsovagues of different phylogenies including species with restricted distribution and endemics proved to constitute one of the most diverse gypsophylous floras in the world. Sexual reproduction, clonality, floral biology, life history, and germination were studied for several species of which very little was known, providing useful information for the recovery of the Cuatro Ciénegas populations. Riparian species, that form a discrete landscape that indicates the presence of water, are losing their habitat as the water bodies desiccate, but proved to be good colonizers of newly opened habitats formed by the drastic disturbance of underground water system. Thus, they are found colonizing numerous sinkholes formed due to the subsurface water flow of water, as well as the dry beds of the now dry water bodies. Germination patterns and morphological and ecological variations of these species, as well as the interactions among them, can help explain their colonizing ability in an unusual succession process. Genetic analysis of native and invasive species helped understand their distribution and differentiation as well as their invasive potential.

This book is the fourth in a series of books on the ecology, natural history, biodiversity, evolution, conservation, and geology of Cuatro Ciénegas and the surrounding Chihuahuan Desert. The first book, edited by Valeria Souza, Gabriela Olmedo-Álvarez, and Luis E. Eguiarte (2018), described the general natural and physical settings of Cuatro Ciénegas, including 9 chapters dealing with general aspects of the ecology, natural history, geological history, climate, and the microbiology of the area, including a brief description of the different scientific research programs that have been conducted in the valley. In the second book in the series, edited by Felipe García-Oliva, James Elser, and Valeria Souza, and also published in 2018, the ecosystem ecology and geochemistry of Cuatro Ciénegas were carefully described in 12 chapters, again with emphasis, but not exclusively, on microorganism and the aquatic systems. The third book of the series analyzed the rich animal diversity in Cuatro Ciénegas and contiguous Chihuahuan Desert, and was carefully edited by Fernando Alvarez and Margarita Ojeda and published in 2019. In 14 chapters, the animal diversity and biogeographical patterns of diversity were described and analyzed, ranging in studies from the helminth parasites and soil microarthropods to birds and mammals. The future two books in this series will deal with the astrobiological relevance of Cuatro Ciénegas and the conservation perspectives of this beautiful but endangered valley.

We are sure that you will enjoy reading the series and will discover all sorts of interesting and amazing details from our favorite silent storytellers.

Mexico City, Mexico

María C. Mandujano
Irene Pisanty
Luis E. Eguiarte

Contents

Chapter 1
Diversity and Uniqueness at Its Best: Vegetation of the Chihuahuan Desert

José Alejandro Zavala-Hurtado and Monserrat Jiménez

Abstract About 10,500 years ago at the beginning of the Holocene, the first humans in the north of Mexico found themselves in the middle of an aridification process that culminated about 4000 years ago in the modern Chihuahuan Desert, which is the largest desert in North America and the second most diverse on Earth with about 3400 plant species, including cacti that reach their greatest diversity in this region. Nearly 25% of the species are endemic, most notably in the Cuatro Ciénegas Basin with 86 plant taxa. The climate is hot-dry with summer rain. Most of the region has calcareous soils, although there are important outcrops of gypsum in patchy arrangements throughout the region. These environmental pressures have generated a variety of adaptive strategies among the organisms evolving here, resulting in great species' richness. Creosote bush (*Larrea tridentata*) dominates in the driest sites, frequently accompanied by lechuguilla (*Agave lechuguilla*) and tar-bush (*Flourensia cernua*); in the *bajadas* (lowlands) we find the less drought-tolerant plants such as yuccas (*Yucca* spp.) and sotol (*Dasylirion wheeleri*). Grasslands integrate grass and shrub mosaics, with species such as bush muhly (*Muhlenbergia porteri*), bluegrass (*Bouteloua gracilis*), and purple three-awn (*Aristida purpurea*). In gypsum outcrops a diverse flora with several endemics is found. The vegetation is highly variable, responding to water availability, physical–chemical and biological dynamics of soils and fires, among other factors. Currently, there are serious pressures modifying the natural dynamics of the desert, such as the indiscriminate extraction of water, agricultural and livestock practices, water and soil contamination, and invasive species that threaten this unique place.

Keywords Arid lands · Cuatro Ciénegas Basin · Desert plants communities · Cacti · Endemic plants

---

to Carlos Montaña and Francisco González-Medrano, in memoriam

---

J. A. Zavala-Hurtado (✉) · M. Jiménez
Departamento de Biología, Área de Ecología, Universidad Autónoma Metropolitana-Iztapalapa, Mexico City, Mexico
e-mail: jazh@xanum.uam.mx

M. C. Mandujano et al. (eds.), *Plant Diversity and Ecology in the Chihuahuan Desert*, Cuatro Ciénegas Basin: An Endangered Hyperdiverse Oasis,
https://doi.org/10.1007/978-3-030-44963-6_1

In the Beginning

The groups of hunters that inhabited what is now the north of Mexico and the Southeastern United States some 10,500 years ago, in the Early Holocene (Epstein 1980), had the sight of vast wooded areas. In these woods a transition was already taking place from the mesic subalpine forest, dominated by (Pinaceae) *Picea* and limber pine (*Pinus flexilis*), to a more xeric forest with *Pinus edulis*, *Pseudotsuga menziesii*, and (Fagaceae) *Quercus gambelii* (Van Devender et al. 1977). In this forest matrix, elements characteristic of the current Chihuahuan Desert, such as (Asparagaceae) *Agave lechuguilla*, *Dasylirion* sp., *Nolina* sp., *Yucca rostrata*, *Y. torreyi*, (Cactaceae) *Echinocereus dasyacanthus*, *Opuntia macrocentra*, (Leguminosae) *Acacia roemeriana*, and (Ephedraceae) *Ephedra aspera*, among others, were already there (Wells 1966, 1977).

A constant decrease in humidity, despite some intervals of increasing humidity in the Late Holocene, generated the expansion of the desert, and it is estimated that the Chihuahuan Desert acquired its current characteristics only about 4000 years ago (Polyak and Asmerom 2001). Thus, it is one of the youngest deserts on Earth.

The Physical Scenario

The Chihuahuan Desert is the largest in North America and, together with the Great Sandy-Tanami Desert of Australia and the Namib-Karoo of Southern Africa (Olson and Dinerstein 1998) is among the most diverse deserts worldwide. Its extension is shared by the USA (New Mexico and Texas) and Mexico (Chihuahua, Coahuila, Durango, Zacatecas, San Luis Potosí, and Nuevo León), ignoring political borders and intended walls. As happens with almost any natural system, the precise definition of its limits is practically impossible (Gleason 1926); different criteria (climate, soil, vegetation, vertebrates) have been used to demarcate its boundaries (Granados-Sánchez et al. 2011), but there is not a consensus regarding the extension of the Chihuahuan Desert. Area estimations go from 350,000 km^2, using climatic criteria based on the De Martonne aridity index (Schmidt Jr. 1979), to 507,000 km^2 from the distribution of its flora (Goettsch and Hernández 2006), and up to 629,000 km^2 when the adjacent Central Plateau is included as part of this ecoregion in a conservation proposal (Dinerstein et al. 2001). For this reason, the biodiversity evaluations of the Chihuahuan Desert can vary widely, although in this chapter an extension of 507,000 km^2 is taken as reference.

Chihuahuan Desert covers an altitudinal range of 1000–2000 m asl, although within the region there are mountain ranges of up to 3050 m. The largest area of the desert is composed of calcareous soil on limestone, although some parts of the mountains are of igneous origin (Ferrusquía-Villafranca 1993). There is an important presence of alluvial fans with deep soil and localities with gypsum outcrops (Ochoterena et al. 2020, this volume). It is also common to find endorheic basins or "bolsones" (González-Medrano 2012). The climate is dry with very hot summers and cold winters. The average annual rainfall is 235 mm, most of which occurs during the summer.

Diversity of Plant Communities

Although a first visual impression of the Chihuahuan Desert could consider it a somewhat monotonous ecosystem with a high predominance of only a few species, it harbors an extraordinary richness of plant species. This richness is clearly related to high environmental heterogeneity. There have been 3382 species reported (Henrickson and Johnston 2007), of which almost a quarter (826 species) are endemic to the region (Villarreal-Quintanilla et al. 2017). This plant richness has been systematized into different types of vegetation considering the patterns of coexistence of groups of species in different patches with particular environments (Table 1.1). These habitats are defined fundamentally by the type of substrate, geomorphology, and microclimatic variations of temperature and precipitation. This characterization of vegetation types varies with different authors; up to 17 different types of vegetation have been proposed in the Chihuahuan Desert (Johnston 1977). However, Rzedowski (1965, 1978) and González-Medrano (2012), authors with greater experience and recognized expertise in the definition of vegetation types, at a general level and specifically of arid ecosystems, recognized three general types of desert plant communities for the Chihuahuan Desert. Difference in criteria and conceptions of the vegetation is reflected in the fact that the three types of vegetation recognized by these authors are not fully equivalent. Considering both proposals, the three general vegetation types are:

1. Microphyllous desert scrub (Fig. 1.1) is found in alluvial fans and foothill depressions, on calcareous sedimentary material derived from limestone or, in some cases, igneous rock. The substrates are coarse-textured loams on gravelly plains and slopes (Schulz and Muldavin 2015). It is also known as Chihuahuan Desert scrub (Brown 1982). This scrub is the most conspicuous and extensive along the Chihuahuan Desert and is dominated by shrubby species with small leaves that may have thorns. The most conspicuous and abundant component is the creosote bush (Zigophyllaceae) *Larrea tridentata* and, to a lesser extent, the tarbush (Compositae) *Flourensia cernua*. Other important elements of this scrub are: (Fouquieriaceae) *Fouquieria splendens*, (Compositae) *Zinnia acerosa*, *Parthenium incanum*, (Euphorbiaceae) *Jatropha dioica*, (Koeberliniaceae) *Koeberlinia spinosa*, (Leguminosae) *Prosopis glandulosa*, *Mimosa aculeaticarpa*, *M. zygophylla*, *Acacia berlandieri*, *A. farnesiana*, *A. tortuosa*, *Eysenhardtia polystachya*, (Asparagaceae) *Agave lechuguilla*, *A. scabra*, *Yucca carnerosana*, and *Y. filifera*, among others (Granados-Sánchez et al. 2011, Ezcurra et al. 2020, this volume; Flores Vázquez et al. 2020, this volume). Also noteworthy are the cacti with several species of *Ariocarpus*, *Astrophytum*,

Table 1.1 Main traits of vegetation types in the Chihuahuan Desert

Type of vegetation	Physiognomy	Representative species	Distribution in the Chihuahuan Desert[a]	Main threats [b]
Microphyllous desert scrub	Shrubby	*Larrea tridentata, Flourensia cernua, Fouquieria splendens, Zinnia acerosa, Astrophytum spp., Coryphantha spp.*	Is the vegetation matrix of the Chihuahuan Desert in plains, alluvial fans, basins, piedmont	OG, CC, LUC, HS, P, IPC, HF
Rosetophyllous desert scrub	Arboreal rosette	*Agave spp., Yucca spp.*	Bolsón de Mapimí, Durango; trans-Pecos, Texas; Cuatro Ciénegas and mountain range, Coahuila	OG, CC, LUC, HS, P, IPC, HF
Lechuguillal	Arboreal rosette	*Agave lechuguilla, Hechtia glomerata, Agave striata, Yucca carnerosana, Agave asperrima*	Coahuila; South of Zacatecas; North of San Luis Potosí	OG, CC, LUC, HS, P, IPC, HF
Izotal	Arboreal rosette	*Yucca carnerosana, Yucca filifera, Yucca rigida, Larrea tridentata, Cylindropuntia imbricata, Cylindropuntia leptocaulis*	Potosino-Zacatecano plateau	OG, CC, LUC, HS, P, IPC, HF
Crassicaulous desert scrub	Thorny tall succulent	*Myrtillocactus geometrizans, Cylindropuntia imbricata, Cylindropuntia tunicata, Opuntia leucotricha, Opuntia streptacantha, Stenocereus griseus*	South and Southeast of San Luis Potosí; South and Southeast of Zacatecas	OG, CC, LUC, HS, P, IPC, HF
Grassland	Perennial grasses	*Bouteloua gracilis, Bouteloua curtipendula, Bouteloua eriopoda, Bouteloua barbata, Aristida adscensionis, Aristida curvifolia*	North of Chihuahua; Southwest Arizona; South New Mexico and Texas; Margin of the Sierra Madre Oriental; North of Coahuila	OG, CC, F, WM, LUC, SE, HS, P, HF
Halophytic vegetation	Herbaceous and shrubby	*Acacia greggii, Allenrolfea occidentalis, Atriplex acanthocarpa, Atriplex canescens, Clappia suaedifolia*	Bolsón de Mapimí, Chihuahua; Cuatro Ciénegas Coahuila; Trans-Pecos, Brewster Co., Texas; Vanegas, San Luis Potosí	CC, F, WM, LUC, HS, M, P, CP, HF

(continued)

Table 1.1 (continued)

Type of vegetation	Physiognomy	Representative species	Distribution in the Chihuahuan Desert[a]	Main threats [b]
Gypsophilous vegetation	Herbaceous	*Tiquilia hispidissima*, *Atriplex canescens*, *Calylophus hartwegii*, *Ephedra torreyana*, *Frankenia jamesii*, *Bouteloua breviseta*, *Mentzelia perennis*	Cuatro Ciénegas, Coahuila; Mazapil, Zacatecas; Los Cerritos, San Luis Potosí; Río Nazas, Durango; Chihuahua center; Southwest and Northwest of Nuevo León; North of San Luis Potosí; Bolsón de Mapimí, Durango; Río Grande, Texas	CC, F, WM, LUC, HS, M, P, CP, HF
Chaparral	Shrubby	*Quercus intricata*, *Quercus mohriana*, *Quercus vaseyana*, *Quercus laceyi*, *Quercus hypoxantha*, *Quercus pringlei*, *Fraxinus greggii*, *Arctostaphylos pungens*	Sierra El Carmen, Coahuila; Brewster Co., Trans-Pecos, Texas; Río Grande, Chihuahua	CC, F, LUC, HS, P, CP, HF
Montane woodlands				
Pynion-Juniper low-stature woodlands	Arboreal	*Pinus cembroides*, *Juniperus monticola*, *Juniperus deppeana*, *Juniperus flaccida*, *Arbutus xalapensis*, *Pinus greggii*	Southeast of Arizona and Southwest New Mexico; Sierra de Parras, Sierra El Carmen and South of Coahuila	F, CC, LUC, HS, P, HF
Coniferous forests	Arboreal	*Cupressus arizonica*, *Juniperus pachyphlaea*, *Pinus arizonica*, *Pinus ayacahuite*, *Populus tremuloides*, *Pinus teocote*, *Acer brachypterum*, *Quercus gravesii*	Sierra El Carmen, Coahuila; Bolsón de Mapimí, Santa Elena canyon, Chihuahua	F, CC, LUC, HS, P, HF
Abies forests	Arboreal	*Abies* sp., *Pinus ayacahuite*, *Populus tremuloides*, *Pseudotsuga menziesii*, *Abies coahuilensis*, *Pinus arizonica*	Low part of Coahuila; South of Zacatecas; North of San Luis Potosí	F, CC, LUC, HS, P, HF

(continued)

Table 1.1 (continued)

Type of vegetation	Physiognomy	Representative species	Distribution in the Chihuahuan Desert[a]	Main threats[b]
Riparian vegetation	Arboreal	*Populus nigra, Salix babylonica, Prosopis glandulosa, Cercidium texanum, Chilopsis linearis*	Río Grande and tributaries, Texas	CC, WM, LUC, HS, P, HF

[a]Because of the amplitude of distribution of most vegetation types, only some representative sites are shown

[b] Main threats: *OG* Overgrazing; *CC* Climatic change; *F* Fires; *WM* Water management; *LUC* Land use change; *SE* Shrub encroachment; *HS* Human settlements; *M* Mining; *P* Air, water, and soil pollution; *IPC* Illegal plant collection; *HF* Habitat fragmentation

Fig. 1.1 Microphyllous desert scrub of *Larrea tridentata* (creosote bush) near Van Horn, Texas. Photo by Leaflet, licensed under the Creative Commons Attribution-Share Alike 3.0 Unported (https://creativecommons.org/licenses/by-sa/3.0/deed.en) license

Coryphantha, *Echinocactus*, *Echinocereus*, *Escobaria*, *Ferocactus*, *Lophophora*, *Mammillaria*, *Opuntia*, and *Turbinicarpus* (Granados-Sánchez et al. 2011; Flores Vázquez et al. 2020, this volume). Creosote bush scrub has increased its extension, particularly in the last 150 years, by invading large areas previously occupied by grasslands, mainly in the north of the Chihuahuan Desert (Alvarez et al. 2011).

2. Rosetophyllous desert scrub (Fig. 1.2) is found in hills formed of sedimentary material, mainly limestone, although it is also found on rocky substrates of igneous origin. This scrub is dominated by plants with succulent leaves arranged in

Fig. 1.2 Rosetophyllous desert scrub. Izotal of *Yucca carnerosana* near San Luis Potosí. Photo by Tomás Castelazo, licensed under the Creative Commons Attribution-Share Alike 2.5 Generic license

the form of a rosette, in individuals with arboreal physiognomy (e.g., *Yucca* spp.) or without an apparent stem, with leaves emerging from the base of the plant (e.g., *Agave* spp.). They form high density patches of individuals on hills with limestone substrates. Within this type of vegetation, several different types of communities have been described:

- Lechuguillal, a desert scrub established in lower parts of the limestone slopes, usually below 1400 m. The dominant species is (Asparagaceae) *Agave lechuguilla*, frequently accompanied mainly by *Agave striata*, *Yucca carnerosana*, and (Bromeliaceae) *Hechtia glomerata* (González-Medrano 2012). Other relevant species are (Asparagaceae) *A. asperrima*, *Dasylirion cedrosanum*, *Nolina parviflora*, *Y. rigida*, (Scrophulariaceae) *Buddleja marrubiifolia*, (Euphorbiaceae) *Euphorbia antisyphilitica*, *Jatropha dioica*, (Fouquieriaceae) *Fouquieria splendens*, (Cactaceae) *Opuntia stenopetala*, and (Compositae) *Parthenium argentatum*, among others (Granados-Sánchez et al. 2011). This scrub is also rich in herbaceous species such as (Compositae) *Eupatorium calophyllum*, *Verbesina pedunculosa*, *Zaluzania triloba*, *Ageratum corymbosum*, *Zinnia acerosa*, (Poaceae) *Bouteloua curtipendula*, (Polemoniaceae) *Loeselia coerulea*, (Brassicaceae) *Lesquerella fendleri*, and (Linaceae) *Linum scabrellum*, among others (González-Medrano 2012).
- Izotal, a type of arboreal community dominated physiognomically by the *Yucca* genus that is generally established in relatively deep soil or caliche on alluvial fans.

3. Crassicaulous desert scrub (Fig. 1.3) is found mainly in the south and center of the Chihuahuan Desert, on substrate of igneous material (rhyolites or basalt). Physiognomically, they are thorny, tall succulent plants with a predominance of cacti. Among the dominant species are (Cactaceae) *Myrtillocactus geometrizans*, *Cylindropuntia imbricata*, *Cylindropuntia tunicata*, *Opuntia leucotricha*, *Opuntia streptacantha*, *Stenocereus griseus*, *Echinocactus platyacanthus*, *Stenocactus multicostatus*, (Euphorbiaceae) *Jatropha dioica*, *Euphorbia maculata*, (Asparagaceae) *Yucca carnerosana*, *Agave lechuguilla*, (Compositae) *Verbesina oreopola*, *Parthenium incanum*, *Gymnosperma glutinosum*, (Leguminosae) *Mimosa zygophylla*, *Eysenhardtia polystachya*, (Poaceae) *Bouteloua curtipendula*, *Bouteloua gracilis*, *Stipa eminens*, (Convolvulaceae) *Dichondra argentea*, (Anacampserotaceae) *Talinopsis frutescens*, (Oleaceae) *Menodora coulteri*, and (Nyctaginaceae) *Boerhavia intermedia*, among others (González-Medrano 2012).

In the Chihuahuan Desert we find other types of vegetation that occupy smaller areas than the three mentioned above, but which are important for their contribution to plant diversity and constitute peculiar environmental scenarios for the evolution of different lineages of desert plants (Pinkava 1984; Meyer 1986; Granados-Sánchez et al. 2011; Moore and Jansen 2007):

4. Grasslands (Fig. 1.4a) are dominated by perennial grasses, "*zacates*," in flat lowlands with relatively more developed soil and higher humidity conditions (Granados-Sánchez et al. 2011). They occupy about 20% of the surface of the Chihuahuan Desert, although it is estimated that between 25 and 50% of the cur-

Fig. 1.3 Crassicaulous desert scrub with *Opuntia* sp., *Cylindropuntia imbricata*, *Myrtillocactus geometrizans*, *Agave* sp., and *Prosopis glandulosa* near San Luis Potosí. Photo by Evelyn M. Rosas-García

Fig. 1.4 Gypsophilous vegetation with *Dasylirion cedrosanum* in Cuatro Ciénegas, Coahuila. Photo by Eduardo Casas Hernández

rent desert scrubs were grasslands in the past (Dinerstein et al. 2001). Among the dominant grass species are (Poaceae) *Bouteloua gracilis*, *B. curtipendula*, *B. eriopoda*, *B. barbata*, *Aristida adscensionis*, *A. curvifolia*, *A. purpurea*, *Hilaria mutica*, *Eragrostis lehmanniana*, *Sporobolus airoides*, and *S. palmeri*, among others (Granados-Sánchez et al. 2011). Currently, stands where the grassland species are interspersed with individuals of microphyllous species are frequently found, particularly *Larrea tridentata*, forming ecotones between the two types of vegetation. These transition zones are evidence of the process of invasion of the creosote bush over the grassland (Dinerstein et al. 2001; Gibbens et al. 2005; Alvarez et al. 2011; Moreno de las Heras et al. 2016).

5. Halophytic vegetation is distributed throughout the Chihuahuan Desert, frequently around dry lakes, beaches, or in salt flats at the bottom of basins with internal drainage. Various salts, derived from weathering and filtration of mineral material or salty sediment, accumulate in the soil (Hendrickson 1977). Halophytic vegetation accounts for the lowest number of species in the Chihuahuan Desert; they are primarily herbaceous plants although some shrubs can be found. The dominant species include (Leguminosae) *Acacia greggii*, *Prosopis glandulosa*, (Amaranthaceae) *Allenrolfea occidentalis*, *Atriplex acanthocarpa*, *A. canescens*, *Salsola tragus*, *Suaeda mexicana*, (Compositae) *Clappia suaedifolia*, (Poaceae) *Cynodon dactylon*, *Distichlis spicata*, *Sporobolus airoides*, (Solanaceae) *Lycium berlandieri*, and (Aizoaceae) *Sesuvium verrucosum*, among others (Granados-Sánchez et al. 2011). The majority of these species present some physiological

mechanism to face conditions of high salinity. Some accumulate salts in their tissues, while others reduce the concentration of salts by increasing their intake of water, which may imply an increase in the succulence or the deep vertical extension of their roots in order to exploit water tables (Hendrickson 1977).

6. Gypsophilous vegetation (Fig. 1.4) is found in outcrops of gypsum-rich substrate that sustain a community of endemic plants with a grassland appearance (Rzedowski 1955; Luévano 2009). Gypsum deposits on the Chihuahuan Desert are widely dispersed but well localized, ranging from less than one hectare to several km^2 in extension, and they have developed a diverse endemic flora that presumably has evolved through several million years, resulting in at least 200 species and plant varieties restricted to gypsum substrates on the Chihuahuan Desert (Moore and Jansen 2007; Ochoterena et al. 2020, this volume). Among the gypsophilic species found in these communities we can mention (Boraginaceae) *Tiquilia hispidissima*, (Amaranthaceae) *Atriplex canescens*, (Onagraceae) *Oenothera hartwegii*, (Ephedraceae) *Ephedra torreyana*, (Frankeniaceae) *Frankenia jamesii*, (Poaceae) *Bouteloua breviseta*, *Sporobolus nealleyi*, *Sporobolus airoides*, (Loasaceae) *Mentzelia perennis*, (Boraginaceae) *Nama carnosa*, (Nyctaginaceae) *Acleisanthes lanceolata*, and (Compositae) *Sartwellia flaveriae*.
7. Chaparral consists of evergreen scrub characterized by plants with sclerophyllous leaves that have settled at mid elevations of the mountains (ca. 2000 m), often between grasslands and pine forests. These communities are dominated by oak species, such as (Fagaceae) *Quercus intricata*, *Q. mohriana*, *Q. vaseyana*, *Q. laceyi*, *Q. hypoxantha* and *Q. pringlei*, as well as other species such as (Oleaceae) *Fraxinus greggii*, (Ericaceae) *Arctostaphylos pungens*, (Berberidaceae) *Berberis trifoliolata*, (Rhamnaceae) *Ceanothus pauciflorus*, *Condalia ericoides*, (Rosaceae) *Cowania plicata*, *Cercocarpus montanus*, (Asparagaceae) *Dasylirion* sp., *Nolina erumpens*, *Yucca carnerosana*, (Garryaceae) *Garrya ovata*, (Rhamnaceae) *Condalia ericoides*, (Anacardiaceae) *Rhus microphylla*, *R. trilobata*, and *R. virens* (Muldavin et al. 2004; Granados-Sánchez et al. 2011).
8. Montane woodlands: within the surface of the Chihuahuan Desert there are several mountain ranges above 1500 m with temperature, humidity, and soil conditions that allowed the establishment of wooded communities with a temperate affinity. Depending on altitude, slope orientation and inclination, as well as on the substrate, different types of forest communities can be recognized (Muldavin et al. 2004; Granados-Sánchez et al. 2011):

   - In the relatively lower parts with gentle slopes and shallow soil, still under xeric conditions, there are pinyon-juniper low-stature woodlands dominated by the pinyon pine, *Pinus cembroides*, which can form associations with (Cupressaceae) *Juniperus monticola*, *J. deppeana*, *J. flaccida* and different tree species such as (Ericaceae) *Arbutus xalapensis*, (Pinaceae) *Pinus greggii*, *P. edulis*, (Fagaceae) *Quercus chihuahuensis*, *Q. deserticola*, *Q. emoryi*, *Q. laeta* and (Asparagaceae) *Yucca carnerosana*.

- At higher altitudes up to 2300 m, on steep slopes and mountain tops, on igneous and karstic substrates, with climate patterns from temperate to cold, and with higher humidity, coniferous forests are established. The characteristic species of one of the modalities of these forests at an altitude of 2000 m are (Cupressaceae) *Hesperocyparis arizonica*, *Juniperus deppeana* var. *deppeana*, (Pinaceae) *Pinus arizonica*, *P. ayacahuite*, and (Salicaceae) *Populus tremuloides*. At higher altitudes, between 2300 and 3000 m, there are pine-oak forests, characterized by *Pinus arizonica*, *P. teocote*, (Sapindaceae) *Acer saccharum* var. *sinuosum*, *Cupressus* sp., *Juniperus* sp., (Fagaceae) *Quercus gravesii*, *Q. hypoleucoides*, and *Q. muehlenbergii*, among other species.
- In the highest mountain ranges, above 2500 m, under a temperate-humid climate regime, high and closed fir *(Abies)* forests are established. The characteristic arboreal species are (Salicaceae) *Populus tremuloides*, (Pinaceae) *Abies* sp., *Pinus ayacahuite*, and *Pseudotsuga menziesii*, being also *Abies coahuilensis* and *Pinus arizonica*.

9. Riparian vegetation: arboreal plant communities established on the banks of streams and rivers are among the least studied communities in the Chihuahuan Desert (Soykan et al. 2012), despite being considered sites with high diversity. The dominant tree species are (Salicaceae) *Populus nigra*, *Salix babylonica*, (Leguminosae) *Prosopis glandulosa*, *Cercidium texanum*, and (Bignoniaceae) *Chilopsis linearis*. The invasive (Tamaricaceae) *Tamarix ramosissima* is also frequently found. Among the shrub species, (Compositae) *Baccharis glutinosa*, *Chloracantha spinosa*, (Leguminosae) *Prosopis juliflora*, and (Cannabaceae) *Celtis pallida* can be recognized (Cornell et al. 2008). Herbaceous ensemble is represented by many endemic species such as (Cyperaceae) *Carex potosina*, (Poaceae) *Bouteloua kayi*, (Potamogetonaceae) *Potamogeton clystocarpus*, (Compositae) *Helianthus neglectus*, *Chromolaena bigelovii*, (Caryophyllaceae) *Drymaria pachyphylla*, (Convolvulaceae) *Bonamia ovalifolia*, (Phyllanthaceae) *Phyllanthus ericoides*, (Gentianaceae) *Eustoma barkleyi*, (Lythraceae) *Ammannia grayi*, (Onagraceae) *Oenothera arida*, (Papaveraceae) *Argemone turnerae*, and (Solanaceae) *Solanum davisense* (Villarreal-Quintanilla et al. 2017). The diversity of these communities increases with the integration of elements of nearby associations, such as grasslands and scrubs (Soykan et al. 2012).

The Dynamic Nature of Chihuahuan Desert Communities

The Chihuahuan Desert communities present a high variability in their spatial and temporal dimensions, which contributes to their high biological diversity. This variability is driven primarily by the input of water, which is scarce and has unpredictable fluctuations at different time scales (Noy-Meir 1973). In addition, plants respond in different ways, depending on their life history characteristics and in different circumstances of their environmental envelope, mainly related to the availability of water and nitrogen, which has been reported as a limiting element in arid ecosystems (Ladwig 2014).

The variety of landforms present in the region is also a determining factor of diversity. A study conducted in the south of the Chihuahuan Desert (Montaña 1990) shows the relationship between landforms and diversity of species and life forms in different types of vegetation. He found significant differences in species richness between landforms. Lowlands, hills, and mountains were the most diverse landforms, and the beaches and inter-dunes plains, the poorest. Foothills and dunes showed intermediate values. Although no significant differences in the composition of life forms between landforms were found, landforms actually differed in the relative cover of life forms. These differences would be a consequence of patterns of spatial and temporal distribution of species, regulated by the hydrological and radiation dynamics in the different landforms throughout the Chihuahuan Desert landscape. The author suggests that the lack of differentiation in the life forms' spectra between landforms would be an expression of the relatively young age of the Chihuahuan Desert.

One of the most notorious processes in the dynamics of Chihuahuan Desert communities is the continuous and accelerating invasion of microphyll shrub species, mainly the creosote bush, *Larrea tridentata*, and the honey mesquite, *Prosopis glandulosa*, over grassland communities. This process has been documented since 1858 (Gibbens et al. 2005). It has caused changes in the patterns of primary production, reduction of biodiversity, and an increase in wind and water erosion (Moreno de las Heras et al. 2016). It has been proposed that these changes, which some authors (Van Auken 2000; Ladwig 2014; Caracciolo et al. 2016) consider irreversible, are due to factors such as temperature increase, overgrazing, and poor management practices (Ladwig 2014). This would have caused alterations in soil characteristics and the dynamics of the seed bank that are reflected in the exacerbation of the asymmetry in the competitive capacity of grasses and shrubs (Montaña et al. 1995).

The Uniqueness of Cacti

The Chihuahuan Desert has the greatest richness of cacti species on Earth. Currently, 329 species have been reported (including five hybrids), of which 229 are endemic to the region (Hernández et al. 2004). This extraordinary diversity is not apparent to the naked eye, since the populations of most species are made up of small individuals that are difficult to observe. This has made completing an exhaustive inventory of the species actually present in the region problematic. Most of the registered species belong to the genera *Mammillaria* (79 species), *Opuntia* (46), *Coryphantha* (36), and *Echinocereus* (30) (Hernández et al. 2004).

The great diversity of cacti in the patches of vegetation that make up the types of the Chihuahuan Desert communities described above is inevitably related to the high environmental heterogeneity that characterizes the region, in geomorphology, solar radiation regime, water and nutrient availability, and soil (Goettsch and

Hernández 2006), as well as conditions generated by coverage of other plants. A large part of cactus species shows specific and restrictive habitat preferences, in addition to limited dispersal ability, which limits their distribution to only certain patches. However, there are more generalist species, such as opuntias, which have a wide distribution in the Chihuahuan Desert.

The adequate conditions for the establishment of several species of cacti in the harsh environment of the desert are generated by the coverage of other plants, the nurse plants, which favor microenvironments with higher humidity, less extreme temperatures, nutrient accumulation, and protection against predators, compared to the microenvironment found in unsheltered areas (Pérez-Sánchez et al. 2015). Although the importance of facilitating interactions in the determination of patterns of coexistence in desert communities has been documented for several species, there is still a long way to go to cover the extraordinary diversity of cacti in the Chihuahuan Desert (Muro-Pérez et al. 2011).

A Glimpse of the Uniqueness of the Cuatro Ciénegas Basin

The Cuatro Ciénegas basin (CCB) is a unique and extraordinary place in the Chihuahuan Desert, as it is highlighted in this volume (Ezcurra et al. 2020; Flores Vázquez et al. 2020; Ochoterena et al. 2020) and in companion books. This basin of only about 2000 km^2 has the highest number of endemic plant taxa (86), of a total of 902 plant species throughout the Chihuahuan Desert (Villarreal-Quintanilla et al. 2017). An area of 840 km^2 is under the Mexican regime of Flora and Fauna Protection Area (INE 1999).

CCB has an intricate water network with underground interconnections. At a superficial level, these systems form ponds, rivers, and lakes, building up a great diversity of terrestrial, aquatic, and semi-aquatic environments (Meyer 1973; Wolaver et al. 2006; Souza et al. 2012; Pisanty et al. 2013). The substrate is limestone, and the topography is very rough.

Pinkava (1984) recognizes eight major vegetation types for the CCB: (1) sacaton grasslands on the saline basin floor, (2) aquatic and semi-aquatic habitats, (3) gypsum dunes, (4) transition zone, (5) desert scrub on the *bajadas*, (6) chaparral, (7) oak-pine and oak woodlands, and (8) montane forests. Although these groups roughly correspond to those described previously in this chapter for the entire Chihuahuan Desert, they are not fully equivalent. Pinkava's desert scrub (5) encompasses microphyllous and rosetophyllous desert scrub, and his transition zone (4) is a kind of transition between the microphyllous desert scrub and the herbaceous vegetation of the halophytic communities. Grasslands (1), chaparral (6), woodlands (7), and montane forests (8) do correspond with those described for the Chihuahuan Desert. The gypsophylous vegetation (3) found mainly in the surroundings of the Churince water system is noteworthy. In addition, associated with bodies of water, aquatic and semi-aquatic vegetation (2) can be found, including species such as (Nymphaeaceae) *Nymphaea ampla*, (Characeae) *Chara* spp., (Typhaceae) *Typha domingensis*, (Cyperaceae) *Eleocharis* sp., and (Juncaceae) *Juncus torreyi* (INE 1999).

Diversity and Uniqueness Under Threat

The Chihuahuan Desert has a long history of human intervention, with increasing intensity and extension, that has necessarily modified the spatial and temporal patterns of natural communities. Perhaps the most notorious disturbance agents so far are changes in regimes from agriculture, grazing, and induced fires, as well as the overexploitation of aquifers in the region (Dinerstein et al. 2001). In addition, there is strong pressure from plant collecting, mainly cacti and other species such as *Fouquieria splendens* (CONABIO 2014).

Likewise, air and water pollution, mining, human settlement development, introduction of invasive species (Colin and Eguiarte 2020, this volume), and of course climate change are generating new scenarios for the coexistence and evolution of the species that inhabit the Chihuahuan Desert (Hultine et al. 2016).

The challenge of maintaining ecological and evolutionary processes in a matrix of inevitable pressures of human societies is huge. There is no proposal for an integrated solution. Here we only make a call to try to incorporate the knowledge of patterns and processes of the natural systems of the Chihuahuan Desert into whatever we intend to do with this unique and diverse ecosystem.

Acknowledgements We thank Evelyn Rosas-García and Eduardo Casas who kindly provided the pictures included in this chapter. We also thank Luisa Granados, Irene Pisanty, and Meli (María C.) Mandujano for their detailed and constructive review of the draft of this chapter.

References

Alvarez LJ, Epstein HE, Li J et al (2011) Spatial patterns of grasses and shrubs in an arid grassland environment. Ecosphere 2(9):1–30

Brown DE (1982) Chihuahuan desert scrub. Desert Plants 4:169–179

Caracciolo D, Istanbulluoglu E, Noto LV et al (2016) Mechanisms of shrub encroachment into Northern Chihuahuan Desert grasslands and impacts of climate change investigated using a cellular automata model. Adv Water Resour 91:46–62

Colin R, Eguiarte LE (2020) Genetic and ecological characterization of the invasive wetland grasses *Arundo donax* and *Phragmites australis* in the Cuatro Ciénegas Basin. In: Mandujano MC, Pisanty I, Eguiarte LE (eds) Plant diversity and Ecology in the Chihuahuan desert. Springer International, Cham, Switzerland

CONABIO (2014) La biodiversidad en Chihuahua: Estudio de Estado. Comisión Nacional para el Conocimiento y Uso de la Biodiversidad, México, DF, Mexico

Cornell JE, Gutiérrez M, Wait DA et al (2008) Ecological characterization of a riparian corridor along the Río Conchos, Chihuahua, Mexico. Southwest Nat 53:96–100

Dinerstein E, Olson D, Atchley J, et al (2001) Ecoregion-Based Conservation in the Chihuahuan Desert. A Biological Assessment. 2nd Printing with Corrections. World Wildlife Fund,

Comisión Nacional para el Conocimiento y Uso de la Biodiversidad (CONABIO), The Nature Conservancy, PRONATURA Noreste, and the Instituto Tecnológico y de Estudios Superiores de Monterrey (ITESM). Washington, DC, USA

Epstein JF (1980) Some reflections on the nature of the Northeast Mexico. Lithic tradition and the problem of its origin. In: Epstein JF, Hester TR, Graves C (eds) Papers on the prehistory of Northeastern Mexico and adjacent Texas. Center for archaeological Research. The University of Texas at San Antonio, TX, USA. Special Report No. 9, pp 81–92

Ezcurra E, Martínez-Berdeja A, Villanueva-Almanza L (2020) The evolution of North American deserts and the uniqueness of Cuatro Ciénegas. In: Mandujano MC, Pisanty I, Eguiarte LE (eds) Plant diversity and Ecology in the Chihuahuan desert. Springer International, Cham, Switzerland

Flores Vázquez JC, Rosa Barrera MD, Golubov J, et al (2020) Ecological importance of Bajadas in the Chihuahuan Desert. In: Mandujano MC, Pisanty I, Eguiarte LE (eds) Plant diversity and Ecology in the Chihuahuan desert. Springer International, Cham, Switzerland

Ferrusquía-Villafranca I (1993) Geología de México: Una Sinopsis. In: Ramamoorthy TP, Bye R, Lot A, Fa J (eds) Diversidad Biológica de México. Orígenes y Distribución. Instituto de Biología, UNAM, México, DF, Mexico, pp 3–108

Gibbens RP, McNeely RP, Havstad KM et al (2005) Vegetation changes in the Jornada Basin from 1858 to 1998. J Arid Environ 61:651–668

Gleason HA (1926) The individualistic concept of the plant association. Bull Torrey Bot Club 55:7–26

Goettsch B, Hernández HM (2006) Beta diversity and similarity among cactus assemblages in the Chihuahuan Desert. J Arid Environ 65:513–528

González-Medrano F (2012) Las zonas áridas de Mexico y su vegetación. Secretaría del Medio Ambiente y Recursos Naturales, México, DF, Mexico

Granados-Sánchez D, Sánchez-González A, Granados Victorino RL et al (2011) Ecología de la vegetación del Desierto Chihuahuense. Revista Chapingo Serie Ciencias Forestales y del Ambiente 17:111–130

Hendrickson J (1977) Saline habitats and halophytic vegetation of the Chihuahuan Desert region. In: Wauer RH, Riskind DH (eds) Transactions of the symposium on the biological resources of the Chihuahuan desert region. US Department of the Interior, National Park Service Transactions and Proceedings Series 3. Washington, DC, USA, pp 289–314

Henrickson J, Johnston MC (2007) A flora of the Chihuahuan Desert region. Published by J Henrickson, Los Angeles, CA, USA

Hernández HM, Gómez-Hinostrosa C, Goettsch B (2004) Checklist of Chihuahuan Desert Cactaceae. Harv Pap Bot 9:51–68

Hultine KR, Majure LC, Nixon VS et al (2016) The role of botanical gardens in the conservation of Cactaceae. Bioscience 66:1057–1065

INE (1999) Programa de Manejo del Área de Protección de Flora y Fauna Cuatrociénagas. Instituto Nacional de Ecología, Secretaría de Medio Ambiente, Recursos Naturales y Pesca (SEMARNAP), México, DF, Mexico

Johnston MC (1977) Brief resume of botanical, including vegetational, features of the Chihuahuan Desert region with special emphasis on their uniqueness. In: Wauer RH, Riskind DH (eds) Transactions of the symposium on the biological resources of the Chihuahuan desert region. US Department of the Interior, National Park Service Transactions and Proceedings Series 3. Washington, DC, USA, pp 335–359

Ladwig LM (2014) Abiotic drivers of Chihuahuan Desert plant communities. PhD Dissertation. The University of New Mexico, Alburquerque, NM, USA

Luévano J (2009) Aspcctos ccológicos y dinámica de un mosaico de pastizal gipsófilo y matorral micrófilo en el sur del Desierto Chihuahuense. PhD Dissertation. IPICYT, San Luis Potosí, Mexico.

Meyer ER (1973) Late-quaternary paleoecology of the Cuatro Cienegas Basin, Coahuila, Mexico. Ecology 54:982–995

Meyer SE (1986) The ecology of gypsophile endemism in the eastern Mojave Desert. Ecology 67:1303–1313
Montaña C (1990) A floristic-structural gradient related to land forms in the Southern Chihuahuan Desert. J Veg Sci 1:69–674
Montaña C, Cavagnaro B, Briones O (1995) Soil water use by co-existing shrubs and grasses in the Southern Chihuahuan Desert, Mexico. J Arid Environ 31:1–13
Moore MJ, Jansen RK (2007) Origins and biogeography of Gypsophily in the Chihuahuan Desert plant group *Tiquilia* subg. *Eddya* (Boraginaceae). Syst Bot 32:392–414
Moreno-de las Heras M, Turnbull L, Wainwright J (2016) Seed-bank structure and plant-recruitment conditions regulate the dynamics of a grassland-shrubland Chihuahuan ecotone. Ecology 97:2303–2318
Muldavin EH, Harper G, Neville P, et al. (2004) A vegetation classification of the sierra del Carmen, U.S.A. and Mexico. In: Hoyt C.A. and Karges J. (editors). Proceedings of the sixth symposium on the natural resources of the Chihuahuan Desert region October 14–17, 2004. p 117–150
Muro-Pérez G, Jurado E, Flores J et al (2011) Positive effects of native shrubs on three specially protected cacti species in Durango, Mexico. Plant Spec Biol 27:53–58
Noy-Meir I (1973) Desert ecosystems: Environment and producers. Annu Rev Ecol Syst 4:25–51
Ochoterena H, Flores-Olvera H, Gómez-Hinostrosa C, et al (2020) Gypsum and plant species: a marvel of Cuatro Ciénegas and the Chihuahuan Desert. In: Mandujano MC, Pisanty I, Eguiarte LE (eds) Plant diversity and Ecology in the Chihuahuan desert. Springer International, Cham, Switzerland
Olson DM, Dinerstein E (1998) The global 200: A representation approach to conserving the Earth's most biologically valuable ecoregions. Conserv Biol 3:502–512
Pérez-Sánchez R, Flores J, Jurado E et al (2015) Growth and ecophysiology of succulent seedlings under the protection of nurse plants in the Southern Chihuahuan Desert. Ecosphere 6(3):1–21
Pinkava DJ (1984) Vegetation and Flora of the Bolson of Cuatro Cienegas region, Coahuila, Mexico: IV. Summary, endemism and corrected catalogue. J Ariz Nev Acad Sci 19:23–47
Pisanty I, Pérez y Sosa C, Gálvez G (2013) Agriculture, water mismanagement and ecosystem transformations in the Cuatrociénegas Valley in the Chihuahuan Desert, Mexico. In: Schwabe K, Albiac J, Connor JD et al (eds) Drought in arid and semiarid regions. Springer, Dordrecht, Netherlands, pp 199–216
Polyak VJ, Asmerom Y (2001) Late Holocene climate and cultural changes in the Southwestern United States. Science 294:148–152
Rzedowski J (1955) Notas sobre la flora y la vegetación del estado de San Luis Potosí. 2. Estudio de diferencias florísticas y ecológicas condicionadas por ciertos tipos de sustrato ecológico. Ciencia (Mexico) 15:141–158
Rzedowski J (1965) Vegetación del Estado de San Luis Potosí. Acta Científica Potosina 1(2):5–291
Rzedowski J (1978) Vegetación de Mexico. Limusa, México, DF, Mexico
Schmidt RH Jr (1979) A climatic delineation of the 'real' Chihuahuan Desert. J Arid Environ 2:243–250
Schulz, KA, Muldavin E (2015) G288 *Larrea tridentata – Flourensia cernua – Acacia neovernicosa* Chihuahuan Mixed Desert Scrub Group. The U.S. National Vegetation Classification. Group Detail report G288. https://www1.usgs.gov/csas/nvcs/nvcsGetUnitDetails?elementGloballd=837711
Souza V, Siefert JL, Escalante AE et al (2012) The Cuatro Ciénegas Basin in Coahuila, Mexico: An Astrobiological Precambrian Park. Astrobiology 12:641–647
Soykan CU, Brand LA, Ries L et al (2012) Multitaxonomic diversity patterns along a desert riparian–upland gradient. PLoS One 7(1):e28235. https://doi.org/10.1371/journal.pone.0028235
Van Auken OW (2000) Shrub invasions of North American semiarid grasslands. Annu Rev Ecol Syst 31:197–215
Van Devender TR, Martin PS, Phillips AM III (1977) Late Pleistocene biotic communities from the Guadalupe Mountains, Culberson County, Texas. In: Wauer RH, Riskind DH (eds) Transactions of the symposium on the biological resources of the Chihuahuan desert region.

US Department of the Interior, National Park Service Transactions and Proceedings Series 3. Washington, DC, USA, pp 85–106

Villarreal-Quintanilla JA, Bartolomé-Hernández JA, Estrada-Castillón E et al (2017) El elemento endémico de la flora vascular del Desierto Chihuahuense. Acta Botanica Mexicana 118:65–96

Wells PV (1966) Late Pleistocene vegetation and degree of pluvial climatic change in the Chihuahuan Desert. Science 153:970–975

Wells PV (1977) Post-glacial origin of the present Chihuahuan Desert less than 1500 years ago. In: Wauer RH, Riskind DH (eds) Transactions of the symposium on the biological resources of the Chihuahuan desert region. US Department of the Interior, National Park Service Transactions and Proceedings Series 3. Washington, DC, USA, pp 67–83

Wolaver BD, Sharp JM Jr, Rodriguez JM (2006) Gravity geophysical analysis of spring locations in a karstic desert basin, Cuatro Cienegas Basin, Coahuila, Mexico. Gulf Coast Assoc Geol Soc T 56:885–897

Chapter 2
Phylogeography of the Chihuahuan Desert: Diversification and Evolution Over the Pleistocene

Enrique Scheinvar, Niza Gámez, Alejandra Moreno-Letelier, Erika Aguirre-Planter, and Luis E. Eguiarte

Abstract The Chihuahuan desert is the largest and most diverse arid desert in North America. Geological and climatic events of the Miocene–Pliocene, as well as the climatic cyclical changes of the Pleistocene, had an important effect on the diversity patterns of the species in this desert. Several areas of the Chihuahuan desert have been identified as "refuges" in which the arid biota survived or thrived during the complex Pleistocene climate changes, including Cuatro Ciénegas, and Mapimian valleys, among other areas. We analyzed bibliographic genetic structure information from plant and animal species of the Chihuahuan desert, using the data to explore for general phylogeographic patterns. Our hypothesis was that the recent history of the genetic diversity currently observed could be interpreted in terms of the effects of large-scale geological and climatic events that occurred during Miocene/Pliocene. We analyzed 24 studies, 9 involving plants (*Agave lechuguilla, A. striata, A. stricta, A. victoria-reginae, Astrophytum* spp., *Berberis trifoliate, Ephedra compacta, Leucophyllum* spp., *Larrea tridentata*), and the rest involving

E. Scheinvar
Departamento de Ecología Evolutiva, Instituto de Ecología, Universidad Nacional Autónoma de México, UNAM, Mexico City, Mexico

Dirección General de Informática y Telecomunicaciones, Secretaría de Medio Ambiente y Recursos Naturales, Mexico City, Mexico

N. Gámez
Facultad de Estudios Superiores-Zaragoza, Universidad Nacional Autónoma de México, UNAM, Mexico City, Mexico

A. Moreno-Letelier
Jardín Botánico, Instituto de Biología, Universidad Nacional Autónoma de México, UNAM, Mexico City, Mexico

E. Aguirre-Planter · L. E. Eguiarte (✉)
Departamento de Ecología Evolutiva, Instituto de Ecología, Universidad Nacional Autónoma de México, UNAM, Mexico City, Mexico
e-mail: fruns@unam.mx

M. C. Mandujano et al. (eds.), *Plant Diversity and Ecology in the Chihuahuan Desert*, Cuatro Ciénegas Basin: An Endangered Hyperdiverse Oasis,
https://doi.org/10.1007/978-3-030-44963-6_2

animals. We detected three main patterns: (1) An ancient differentiation of the North of the Chihuahuan desert in the Cochise filter barrier area, within the Late Miocene (~11.6–5.3 MY), supported by three species; (2) A North/South differentiation from the Chihuahuan desert into the Altiplano Norte and Altiplano Sur, broadly congruent with the Pliocene (~5.3–2.5 MY) supported by seven species; (3) A divergence congruent with the Pleistocene (~2.5 MY), involving recent events within lineages. Another commonly detected pattern is an East/West differentiation. The current genetic differentiation patterns of the Chihuahuan desert species can be explained, at least in part, in terms of Pleistocene climate dynamics along an altitudinal and latitudinal gradient of locally adapted populations that underwent cyclic processes of contraction, isolation, and divergence, followed by expansion and secondary contact.

Keywords *Agave* · Climatic change · Desert animals · Desert plants · Genetic structure · Geographic distribution · Morafka biogeographic classification · Pleistocene refuges

Introduction

The study of the geographical dynamics of the biota has remained a central element in modern evolutionary biology. In particular, the comparative study of spatial genetic variation, i.e., the phylogeography of co-distributed taxa, provides important insights into the effects of historical events in the evolutionary process (Morrone 2009). Population dynamics, such as demography (growth or decrease of populations), isolation, gene flow, and natural selection leave a signal in the distributional patterns of the genetic variation in species; its study allows understanding the historical processes that have modeled their populations. The comparative analyses of these phylogeographic patterns are a powerful tool to explore the generality and relevance of these historical processes in modeling the ecology and diversity of an area.

Climatic changes and geological events, such as tectonism, global temperature fluctuations, glaciation, rainfall, and sea-level variation are critical factors in fragmenting populations, changing the patterns of genetic diversity, and driving evolutionary diversification (Brooks and McLennan 2012). Specifically, climatic changes are one of the most important drivers of genetic diversification and subsequent speciation, because variable environments can allow different specialized genotypes to evolve and coexist (Ackermann and Doebeli 2004; Aguirre-Liguori et al. 2014).

When the environmental conditions are not adequate, populations may move, tracking the geographic extent of the environmental variables where they can survive—i.e., their fundamental niche (Holt 1990; Peterson et al. 2011). When populations move, they also change their population sizes, age structures, and reproductive patterns. In this dispersal process, populations can sometimes colonize new areas or experience local extinction or genetic change, either by adaptation or by genetic drift (Holt 1990; Peterson et al. 2011). In species with greater geographical variability, effective population sizes may decrease, and in such cases, evolution would be dominated by genetic drift, causing genetic variation within populations to decrease, and genetic differentiation among populations to increase. If isolation occurs for a long period, speciation may occur. Alternatively, if environmental conditions change allowing for population expansion and an increase in population connectivity and gene flow (secondary contact), this could result in an increase in the total genetic variation within populations and a decrease in differentiation among populations, which in turn could inhibit a speciation process.

Mexico is an extremely diverse country, where both geology and climate are variable, dynamic, and complex. The territory of Mexico has a complex topography, with a rugged relief dominated by mountain ranges, wide elevation variation, and intense seismic and volcanic activity that result in 11 morphotectonic provinces (Ferrusquía-Villafranca 1993). This rugged and sinuous relief translates into an immense diversity of environments and climates that generate an impressive biological diversity resulting from complex historical processes and local adaptation (Aguirre-Liguori et al. 2019).

Multiple geological and climatic events in the history of Mexico have created geographic and ecological barriers that have affected their populations through isolation, diversification, and speciation events, followed by gene flow, secondary contact, and hybridization, which in turn have generated complex scenarios for the formation of biodiversity and exceptional levels of richness and diversity in the Mexican territory. For instance, the inventory of vascular plant species in Mexico records a total of 23,314 species (Villaseñor 2016). Arid zones of Mexico are very diverse, containing ca. 25% of the total flora of Mexico. Among these arid environments, the Chihuahuan desert has a high level of endemism, with 560 species distributed only in the Chihuahuan desert (strict endemics), 165 quasi-endemic, and 176 species endemics to Mexico (Rzedowski 1993; Toledo and Ordoñez 1993; Villarreal-Quintanilla et al. 2017). Among the remarkable plant groups inhabiting this desert, it harbors the highest worldwide total diversity of Cactaceae (Hernández and Godínez 1994; Hernández and Barcenas 1995) and of *Agave* species (Gentry 2004).

The Chihuahuan desert is the largest and most diverse arid desert in North America (Laity 2008). Depending on its delimitation (see below), it can have an approximate area of 518,000 km^2 (Hernández et al. 2001; Olson and Dinerstein 2002; Zavala-Hurtado and Jiménez 2020, this volume). It is located on an immense plateau in southern North America, in the Rio Grande basin of New Mexico and east of Texas (USA) and in the states of Chihuahua, Coahuila, Durango, Zacatecas, San Luis Potosí, from 35° to 22° North latitude.

Several studies have found that the geological and climatic events of the Miocene–Pliocene (Raymo and Ruddiman 1992; Wilson and Pitts 2010), as well as the climatic cyclical changes of the Pleistocene, had an important effect on the diversity patterns of the species currently found in the Chihuahuan desert (Van Devender 1977; Van Devender and Spaulding 1979; Lanner and Van Devender 1981; Van Devender and Burguess 1985; Castoe et al. 2007; Gándara and Sosa

2014; Castellanos Morales 2016a, b; Angulo et al. 2017; Gámez et al. 2017; Loera et al. 2017; Scheinvar et al. 2017). Over the Neogene, atmospheric CO_2 concentrations and air temperature decreased and the North American Cordillera rose, doubling in size over the last 15 million years, more rapidly from 7 to 5 million years ago (MY hereafter) (Raymo and Ruddiman 1992). Over this period, new geographic barriers were generated by the climate change that made North America dryer. These climatic and geographic changes also formed the great arid zones found in North America (Raymo and Ruddiman 1992; Axelrod 1985).

Different authors have studied the distribution patterns of plant and animal species from the Chihuahuan desert. A detailed review of the biogeographic patterns can be found in Morrone (2005), who reviewed biogeographic information from birds, reptiles, plants, and mammals, and identified the Chihuahuan desert as a biogeographic province called the Mexican Altiplano, which is a great plain located between the Eastern and Western Sierra Madre and Trans-Mexican Volcanic Belt, and its northern limit is found in New Mexico and southern Texas in the USA (Morrone 2005). This same territory was subdivided by Arriaga et al. (1997) in two regions named Altiplano Norte and Altiplano Sur, divided by the Sierra de Saltillo-Parras, on the Sierra Plegada, at 25° of latitude North.

Another regionalization was proposed by Morafka (1977), who divided the Chihuahuan desert into four subprovinces (Trans Pecos, Mapimian, Saladan, and Rio Pánuco) and two transition zones (Cochise filter barrier, Saladan filter barrier in Fig. 2.1a). Each area defined by Morafka has very contrasting conditions in elevation, temperature, and precipitation (Fig. 2.1b–d). The limits established by Morafka are similar to those used in other studies such as Rzedowski (1978), which unifies this area as a floristic province called Altiplanicie; Halffter (1987), using a biogeographical analysis of entomofauna defined it as the Altiplano pattern; Zink et al. (2000) defined the group as Desierto de Chihuahua using bird cladograms; and Morrone (2001a, b, 2004, 2005, 2009) performed various panbiogeographic and biogeographic cladistic analyses of different animal and plant taxa and defined the area as a biogeographic province called the Mexican Altiplano.

Several areas of the Chihuahuan desert have been identified as sites in which the arid biota of North America survived or thrived during the complex Pleistocene climate changes. These areas are sometimes called "refuges," from where the biota later recolonized to the current distribution when environmental conditions were more appropriate, as is the case for Cuatro Ciénegas, Mapimian, and Zimapán Metztitlán valleys, among other (Van Devender 1977; Van Devender and Spaulding 1979; Lanner and Van Devender 1981; Van Devender and Burguess 1985; Castoe et al. 2007; Gándara and Sosa 2014; Castellanos Morales 2016a, b; Angulo et al. 2017; Gámez et al. 2017; Loera et al. 2017; Scheinvar et al. 2017). Nearly 50% of the endemisms of the Chihuahuan desert can be found in the Santa Elena Canyon in the Big Bend National Park, in the nearby area of Maderas del Carmen and in the Cuatro Ciénegas and Mapimí valleys, which indicates that these areas could represent Pleistocene refuges (Villaseñor 2016; Villarreal-Quintanilla et al. 2017).

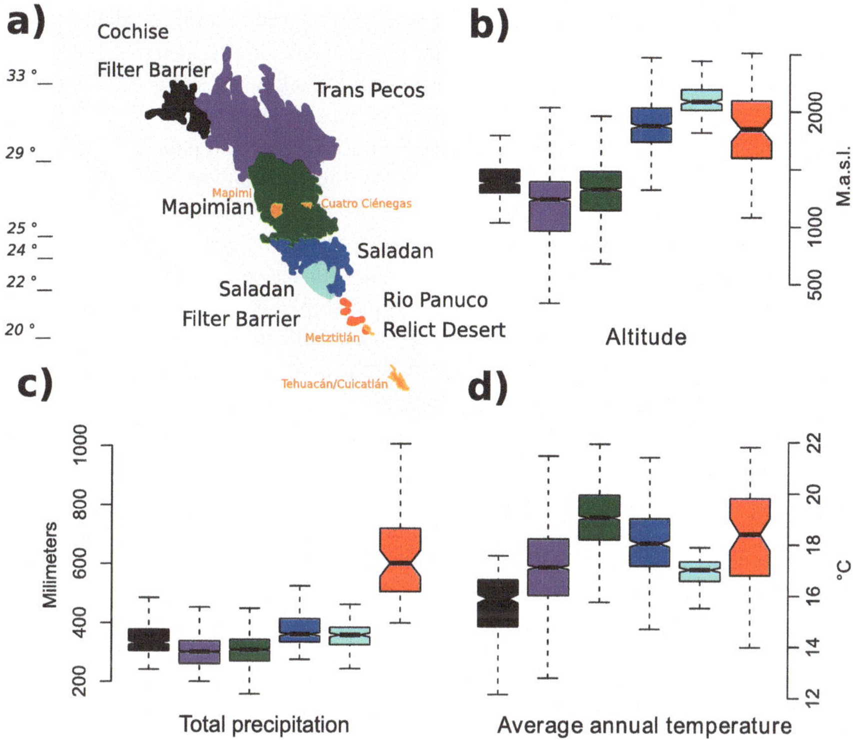

Fig. 2.1 (**a**) Morafka regionalization (1977) in colors: Cochise filter barrier in black, Trans Pecos in purple, Mapimian in green, Saladan in dark blue, Saladan filter barrier in blue, and Río Pánuco Relict Desert in red, and 4 Mexican Biosphere Reserves as geographical reference in orange. (**b**) Altitude, (**c**) total precipitation, and (**d**) average annual temperature observed in each region. The colors as in figure (**a**). Environmental data were obtained sampling from the 10% of the pixels of each area from the observed current climate data layers of World Clim (V. 1.4, Hijmans et al. 2005)

In this chapter, we carried out a bibliographic review of different evolutionary studies from species distributed in the Chihuahuan desert based on the use of different molecular markers, including studies that described genetic structure and explored general phylogeographic patterns of this desert. While the focus of this book is on plants of the Chihuahuan desert, we consider that it is critical to also include animal studies, as they share the same climatic and geological changes, and geographic and ecological barriers, providing important information in order to detect common biogeographic patterns. Our hypothesis was that the recent history of genetic diversity currently observed in the Chihuahuan desert can be interpreted in terms of the effects of large-scale geological and climatic events that occurred during Miocene/Pliocene, as well as by events of isolation and diversification that occurred during the climatic changes of the Pleistocene and that these events left perceptible marks in the patterns of genetic diversity and structure observed today.

Methods

We conducted a bibliographic search for studies including empirical data on the genetics of species distributed within the Chihuahuan desert in the ISI-Web of Science in December 2018, searching for papers including the word "Chihuahuan Desert" and also including any of the following words: chloroplast, genetic, mitochondria, nucleus, phylogeny. We obtained an initial list of 120 papers. This initial list was then filtered, retaining only studies including genetic structure (phylogenetic or population data) and having geographic records within the Chihuahuan desert. From the information contained in these articles, a database was made where the following data were captured: geographic coordinates, molecular markers used, taxa analyzed, main results, as well as the membership of the genetic groups. When the article did not report specific coordinates of the collections, these were inferred from the description in the text or by digitizing the maps presented in the study.

The geographic data of the genetic groups obtained in each study was contrasted against the biogeographic classification of Morafka (1977). For this, a georeferenced polygon was made. To define the geographic limits of the Chihuahuan desert, we used Morrone (2005) delimitation, which weighs the broadest geographical limits and is based on a multi taxa analysis. This polygon covers the area between the Eastern and Western Sierra Madre and in the South of the Mexican Neovolcanic Axis. However, this proposal is restricted to the political limits of Mexico, for this reason in the northern portion of the Chihuahuan desert we use the polygon of the Trans Pecos region of Morafka (1977).

Results

From the initial list of studies, the majority was discarded because they did not contain genetic or geographic data relevant for our objectives. Nine of these studies analyzed taxa with a very restricted distribution (i.e., only a small part of the Chihuahuan desert), which does not allow the detection of patterns on a regional scale or show insufficient sampling on the Chihuahuan desert (Martínez-Palacios et al. 1999; Onorato et al. 2007; De Nova et al. 2012; McGaugh 2012; Aguirre-Liguori et al. 2014; Carson et al. 2015; Magalhaes et al. 2015; Myers et al. 2018). The final list comprised the analysis of 24 studies with genetic and geographic data located in the Chihuahuan desert (Table 2.1). From this, nine involved plants (*Agave lechuguilla, A. striata, A. stricta, A. victoria-reginae, Astrophytum* spp., *Berberis trifoliata, Ephedra compacta, Larrea tridentata, Leucophyllum* spp.) and the rest were conducted in animals. The majority (17) of the studies included estimates of the dates of origin determined by molecular clock methods, while the remaining eight did not include any dating (Table 2.1).

From all the studies, only three had a geographically limited distribution between 25° and 33° latitude North (corresponding to Trans Pecos and Mapimian regions,

Table 2.1 Population genetics studies in the Chihuahuan desert

Taxa	Reference	Genetic marker	Main results
1. *Agave lechuguilla* (Asparagaceae) plant	Scheinvar et al. 2017	Chloroplast sequences	Origin of *A. lechuguilla* 4.46 MY B.P and later differentiation (Pleistocene) into four haplogroups, congruent with Morafka regions. Phylogenetic analysis shows separation in two groups (North and South) in Miocene and most recent division associated with at least five glacial refugia located south of the species' current distribution in the Pleistocene.
2. *Agave striata* (Asparagaceae) plant	Trejo et al. 2016	ISSR	Differences in genetic variation among populations between *A. striata* subsp. *striata* and *A. striata* subsp. *falcata*; correlation with differences in environmental climatic variables along their distribution. Detection of two distinct gene pools, which suggests active differentiation and perhaps incipient speciation process. No molecular dating.
3. *Agave stricta*–*Agave striata* (Asparagaceae) plant	Martínez-Ainsworth 2013	ISSR	Phylogeographical data and the historical scenarios (by ecological niche modelling) suggested that *A. stricta* presents founder effect evidence, that was probably induced by the last episodes of the Mexican Neovolcanic Axis formation as a vicariant event (Pleistocene), in the southern populations of *A. stricta* (South Chihuahuan Desert/Tehuacán Cuicatlán) that generated genetic divergence (i.e., peripatric speciation).
4. *Agave victoria–reginae* (Asparagaceae) plant	Martínez-Palacios et al. 1999	Isozymes	Principal clade endemic from Mapimian region. Differentiation east west inside of Mapimian region. No molecular dating.
5. *Berberis trifoliolata* (Berberidaceae) plant	Angulo et al. 2017	Chloroplast sequences	Origin time of principal clade during the Pleistocene, and the presence of 4 main haplogroups, one of them close to the Western Sierra Madre and three remaining in the Mapimian region.

(continued)

Table 2.1 (continued)

Taxa	Reference	Genetic marker	Main results
6. *Astrophytum* (Cacataceae) species complex plants	Vázquez-Lobo et al. 2015	Chloroplast	All recent cladogenesis events correspond to the Pleistocene. The authors propose that climatic fluctuations could be the main driver of recent differentiation.
7. *Ephedra* clade (Ephedraceae) plants	Loera et al. 2017	Chloroplast sequences	Genetic structure with six geographical groups explaining most of the variation. The best-supported phylogeographic scenario assumed population divergence and demographic expansion during the Pleistocene. Habitat stability during Pleistocene was positively associated with population genetic diversity.
8. *Leucophyllum* clade (Scrophulariaceae) plants	Gándara and Sosa 2014	Chloroplast and nuclear (ITS) sequences	The ancestral lineage was distributed in the Pacific slope (Late Miocene), from which it is dispersed to the Chih. Des. 6 MY, and then diversified during the Late Pliocene and Pleistocene.
9. *Larrea tridentata* (Zygophyllaceae) plant	Duran et al. 2005	Isozymes	Populations with major heterozygosity in the North of distribution, Trans Pecos. The differentiation events inside the principal clade occurred from South to North and distinguish Saladan, Mapimian, and Tran Pecos regions. No molecular dating.
10. *Peromyscus eremicus* (Cricetidae) rodent	Riddle et al. 2000	Mitochondrial, COIII	Monophyletic clade integrated by East and West subclades, which corresponds to south/central of the Chihuahuan desert and northern of the Chihuahuan desert respectively.
11. *Thomomys umbrinus* species complex (Geomyidae) rodents	Mathis et al. 2014	Tree mitochondrial genes	Diversification of *T. umbrinus* clade during the Pleistocene, which, in turn diversifies in two subclades, North Altiplano and South Altiplano, which diversify into 6 lineages geographically compatible with: Trans Pecos, Mapimian, Saladan, Saladan filter barrier, and Río Pánuco regions.

(continued)

Table 2.1 (continued)

Taxa	Reference	Genetic marker	Main results
12. *Dipodomys phillipsi* (Heteromyidae) rodent	Fernández et al. 2008	Mitochondrial and nucleus	The phylogeographic structure recovers a geographical structure where differentiation occurs from south to north occupying the following regions: Río Pánuco, Saladan filter barrier, and Saladan, respectively.
13. *Perognathus flavus* species complex (Heteromyidae) rodents	Neiswenter and Riddle 2010	Mitochondrial, COIII and CR	Origin time of principal clade during the Pliocene, differentiating two subclades North and South of Chih. Des., which in turn diversify into 5 lineages geographically compatible with: South of Colorado Plateau, Trans Pecos, Mapimían-Saladan-Saladan filter barrier, and Río Pánuco regions. Results consistent with models of allopatric divergence driven by Pliocene and Pleistocene geological and climatic events, particularly the late Miocene expansion of interior grasslands and Miocene–Pliocene evolution of basin and range geomorphology.
14. *Perognathus merriami* (Heteromyidae) rodent	Neiswenter and Riddle 2010	Mitochondrial, COIII and CR	Delimitation of three phylogroups distributed allopatrically, Great Plains, Northern Chihuahuan desert, and western Texas along the Río Grande, showing a differentiation pattern East-West.
15. *Cynomys ludovicianus and C. mexicanus* (Sciuridae) rodents	Castellanos-Morales et al. 2016a, b	Mitochondrial CytB and CR	Time diversification between species 2.4 MY presence of two possible refugia in the southern distribution of *C. ludovicianus*, consistent with the distribution range of *C. mexicanus*. Strong impact of Pleistocene climate changes in the distribution of both species, promoting peripatric speciation of *C. mexicanus* starting from *C. ludovicianus*.

(continued)

Table 2.1 (continued)

Taxa	Reference	Genetic marker	Main results
16. *Calothorax* spp. (Trochilidae) hummingbirds	Licona-Vera et al. 2018	Mitochondria and nuclear microsatellites	Demographic expansion, gene flow and admixture in the *C. lucifer* range, post-glacial northern expansion predicted by ecological niche modelling and Bayesian approaches, pointing towards a scenario of divergence with gene flow: a Pleistocene basal split separating *C. pulcher*; and the other two clades are derived from a second split (migratory and sedentary *C. lucifer*). Population genetic admixture was higher in localities with lower inferred stability of habitat suitability.
17. *Bufo punctatus* (Bufonidae) toad	Jaeger et al. 2005	Mitochondrial CytB	Three sister clades: Peninsular (Baja California), Western (Mojave/Sonoran Desert), and Eastern (Colorado plateau/Trans Pecos). Genetic differentiation between Central Mexican Plateau clade and Tehuacán-Cuicatlán clade, during the Pleistocene.
18. *Kinosternon flavescens* (Kinosternidae) turtle	Serb et al. 2001	Mitochondria	Greater diversity in northern populations versus southern (Chih. Des.)
19. *Crotalus molossus* species complex (Viperidae) snake	Anderson and Greenbaum 2012	Mitochondrial and nuclear genes	Origin of the principal clade in the Pliocene. Two subclades that differentiate the regions Chih. Des. from Son. Des. Diversification during the Pleistocene.
20. *Crotalus scutulatus* (Viperidae) snake	Schield et al. 2018	Mitochondrial sequences and genome-wide SNP RADseq data	Divergence events between clades registered during the Pliocene and coincident with two major regions, north integrated by Chihuahuan and Cochise/Sonoran desert, and south integrated by Central Mexican Plateau and Tehuacán Cuicatlán. Furthermore, the Chihuahuan clade shows two subclades Trans Pecos and Mapimian; while the Central Mexican Plateau clade shows Saladan/Saladan filter barrier and Río Pánuco, and Tehuacán Cuicatlán subclades. The origin of these subclades is Pleistocenic.

(continued)

Table 2.1 (continued)

Taxa	Reference	Genetic marker	Main results
21. *Campostoma ornatum* (Cyprinidae) (fish)	Schonhuth et al. 2011	Mitochondrial CytB and one nuclear gene	Northern of Chih. Des. (Nazas-Conchos), as ancestral area followed by vicariance between Nazas and Conchos portions and secondary dispersion towards Sonora and Sinaloa.
22. *Codoma* species complex (Cyprinidae) fishes	Schonhuth et al. 2015	Mitochondrial and nuclear genes	Ancestral lineage in the Conchos system extending southwards, Nazas system. Time divergence ~3 million years ago, predominantly during Pleistocene
23. *Pseudouroctonus minimus* species complex (Vaejovidae) scorpion	Bryson et al. 2013	Mitochondrial and nuclear genes	Pre-quaternary major diversification events (Late Miocene–Pliocene) and Chihuahuan desert colonization from twice independent events and regions, Mexican Highlands (SM Occidental), and Baja California. Differentiation events inside the Chih. Des, Guadalupe/Chisos Independence/Davis, registered during the Late Miocene and Pleistocene respectively.
24. *Agelenopsis aperta* (Agelenidae) spider	Ayoub and Riechert 2004	Mitochondria gene COI	Origin time of principal clade during the Pleistocene, registering the most recent differentiation between Colorado Plateau-Trans Pecos from Mapimian region

i.e., the Chihuahuan desert sensu stricto, Fig. 2.3: 14, 19, 23), four studies had wider distributions to the West and/or North, and a Southern distribution up to 25° latitude North (Fig. 2.3: 4, 15, 17, and 18), while all the others had a wider Southern distribution, extending further below 25° latitude North, representing the wider definition of the Chihuahuan desert.

From the 24 analyzed studies, we detected three main patterns (Fig. 2.2), that are not mutually exclusive: (1) An ancient differentiation of the North of the Chihuahuan desert, in the Cochise filter barrier area within the Late Miocene (~11.6–5.3 MY) supported by 3 species (Fig. 2.3, Table 2.1: 19, 21, and 23 and maybe Fig. 2.3, Table 2.1: 13, 14, 17, and 22). (2) A North/South differentiation of the Chihuahuan desert into the Altiplano Norte and Altiplano Sur, broadly congruent with the Pliocene (~5.3–2.5; Fig. 2.2b) and supported by seven species (Fig. 2.3, Table 2.1: 1, 6, 8, 14, 19, 21, and 22), and maybe by other species (see Fig. 2.3, Table 2.1: 2, 7, 9, 10, 13, and 18). (3) A set of different divergence patterns that occurred in the Pleistocene (~2.5; Fig. 2.2c), involving all of the recent events of lineage divergence, and coincident with the Morafka regionalization (Fig. 2.3, Table 2.1: 1, 2, 3, 6 to 12, 15, 16, 18, 19, 21, and 22). Within this third pattern, we could detect six organisms

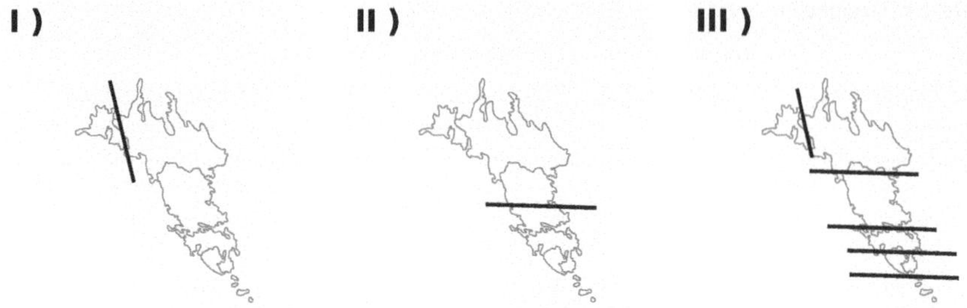

Fig. 2.2 Patterns detected in this review (black lines) and its correspondence with Morafka (1977) regionalization (gray lines in the map) (see Fig.2.1.a)

supporting the Cochise filter barrier (Pleistocene Cochise Filter Barrier), 10 the Trans Pecos region; 13 the Mapimian regiǿn, 12 the Saladan region, 5 the Saladan filter barrier, and 10 the Pánuco region (Tables 2.1 and 2.2).

Another pattern detected was the East/West differentiation within some of the Morafka regions, as can be seen in the studies of *Astrophytum* spp., *Berberis trifoliata,* and *Agave victoria-reginae* in the Mapimian region (Fig. 2.3, Table 2.1: 4–6), *Ephedra compacta* and *Leucophyllum* spp. in the Saladan (Fig. 2.3, Table 2.1: 7, 8) and *Crotalus molossus* in Trans Pecos (Fig. 2.3, Table 2.1: 19).

Discussion

Within a species and closely related species, the distribution patterns of the genetic variation and the signal left by demographic processes (growth or decrease of the populations) and by isolation or gene flow allow us to understand the historical processes that modeled these species. The comparative analyses of these phylgeographic patterns are a powerful tool to explore the generality and relevance of these historical processes in modeling the ecology and diversity of an area. The Chihuahuan desert is one of the arid deserts of North America with higher species diversity and richness of endemisms (Villaseñor 2016; Rzedowski 1993; Villarreal-Quintanilla et al. 2017; Toledo and Ordóñez 1993; Hernández and Godínez 1994; Hernández and Barcenas 1995; Ezcurra et al. 2020, this volume; Ochoterena et al. 2020, this volume) and represents an excellent system to use comparative phylogeography to help disentangle its complex geological and biological history.

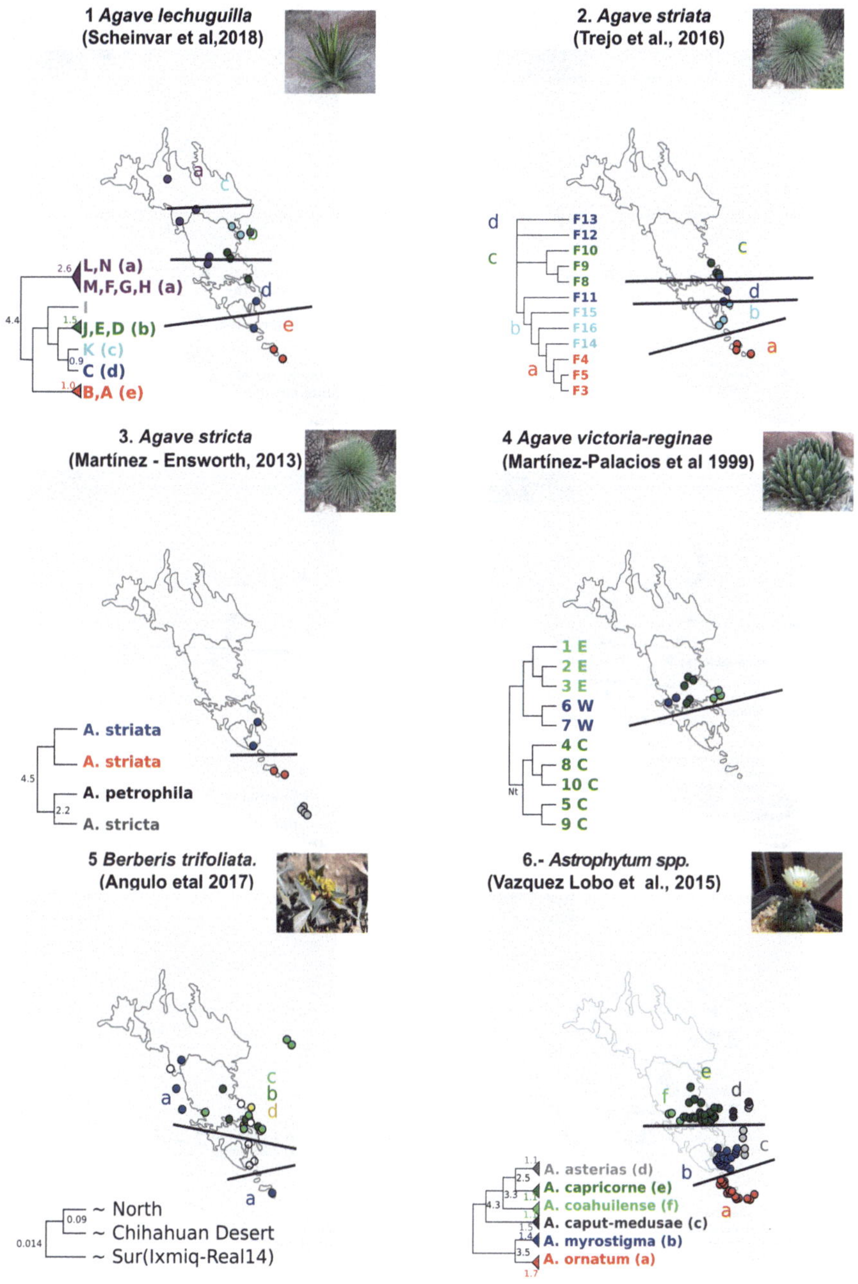

Fig. 2.3 Geographic and genetic structure (points/colors) distribution, tree structure and adjustment to the pattern proposed in this work (black lines) for each of the 24 studies analyzed. The colors in each map represent the genetic structure reported in each study. The phylogenies or distance trees (if any) were modified from the original reported in each article to simplify and make easier their interpretation. The numbers on the trees represent the estimated date of origin of the group, according to the article. The gray lines on the map indicate the regions described by Morafka (1977)

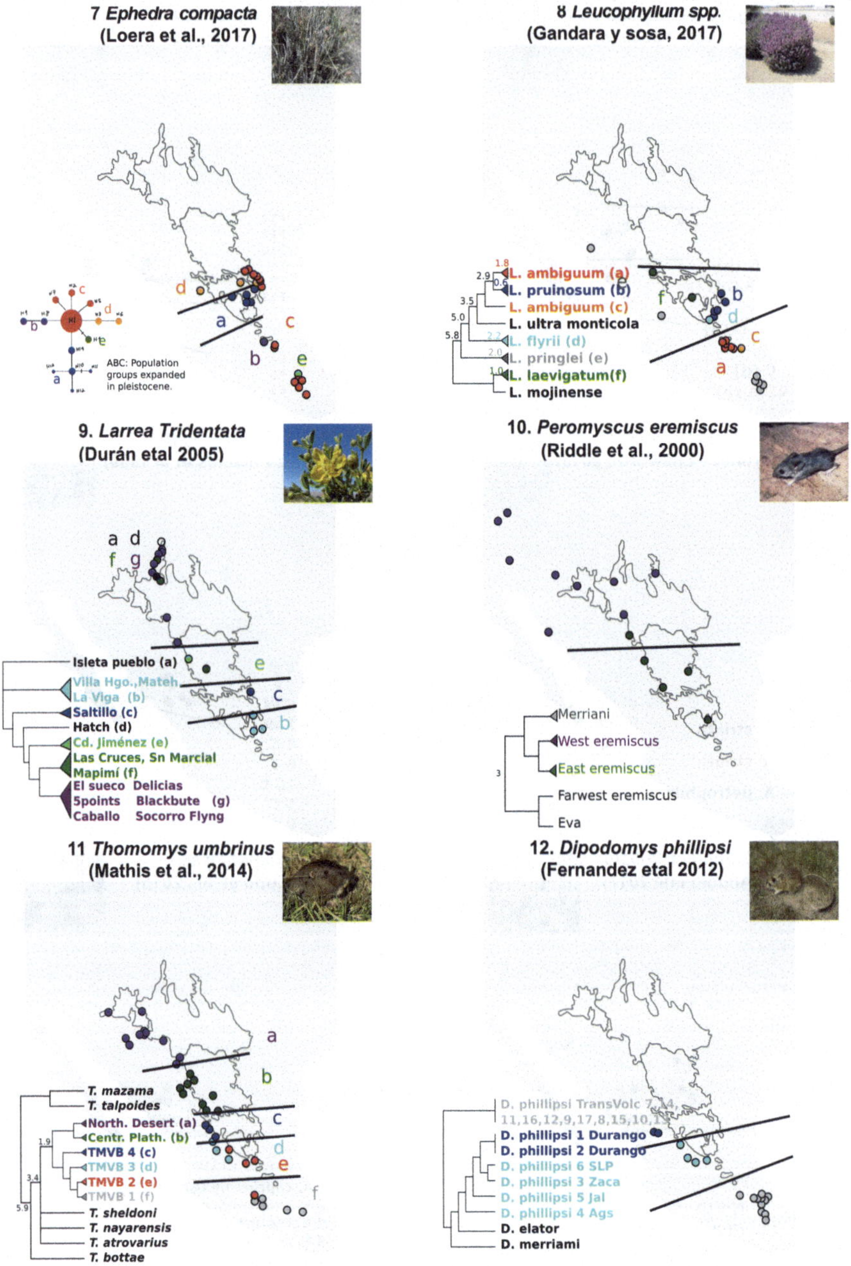

Fig. 2.3 (continued)

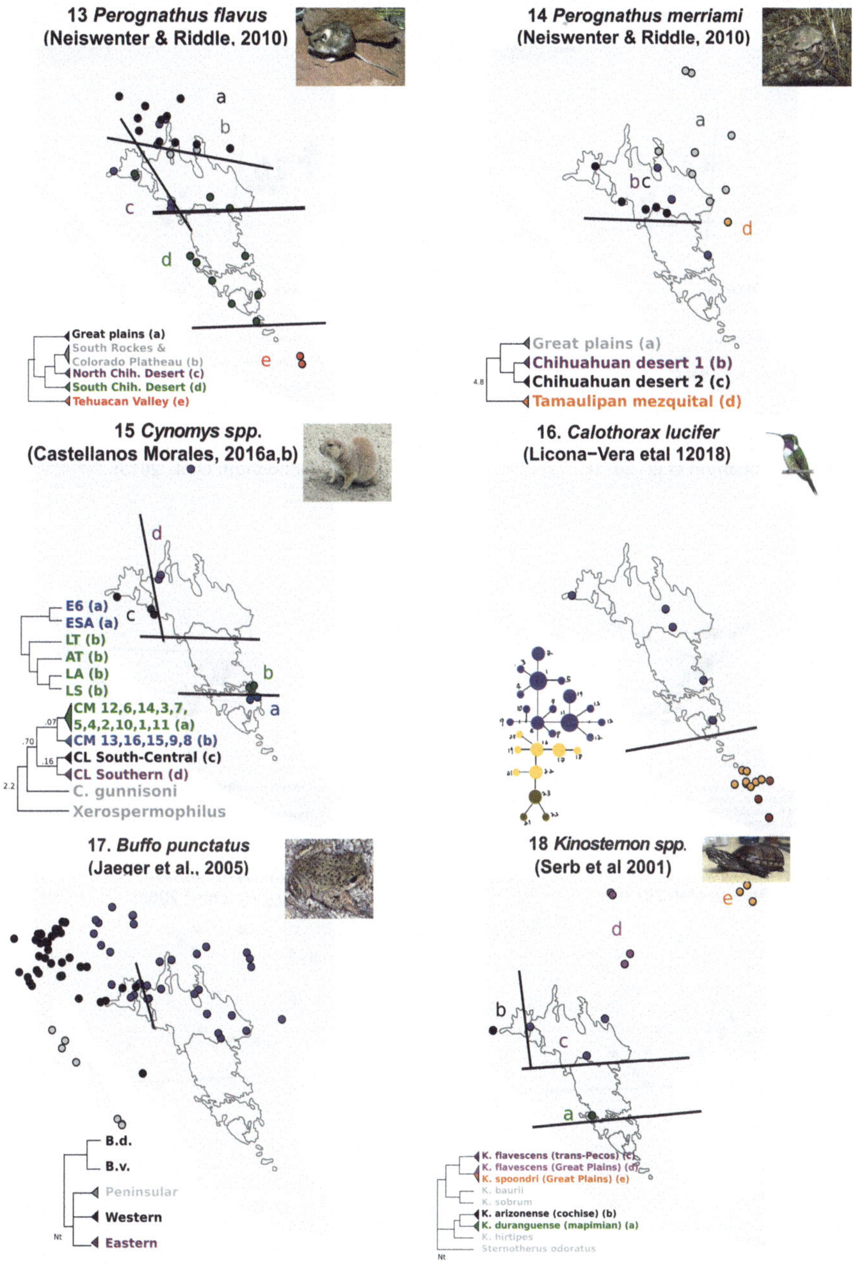

Fig. 2.3 (continued)

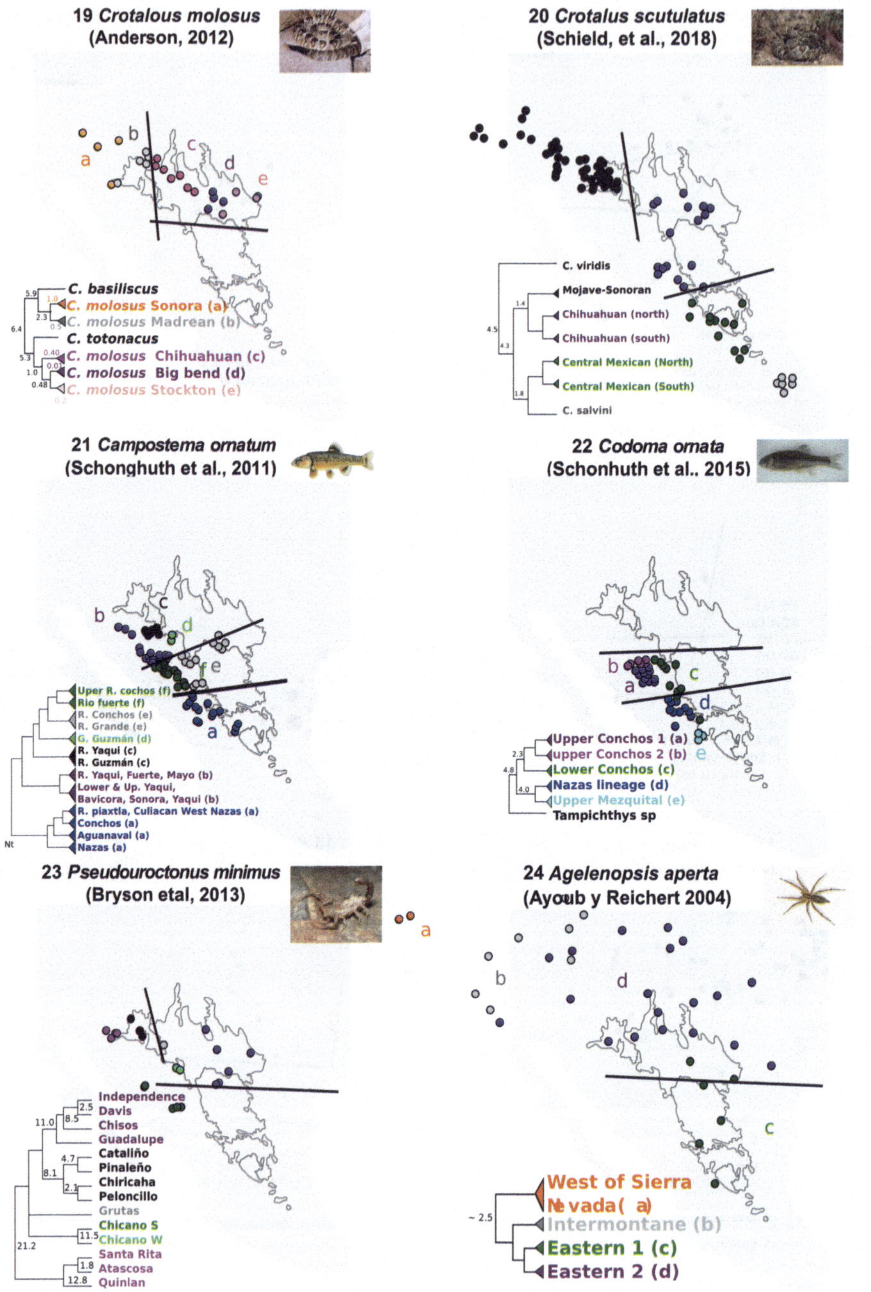

Fig. 2.3 (continued)

Table 2.2 Morafka regions detected in the analyses of genetic structure of species from the Chihuahuan desert and estimated dates of the event

Taxa/Reference	Cochise filter barrier	Trans Pecos	Mapimian	Saladan	Saladan Filter Barrier	Río Pánuco Relict Desert
1. *Agave lechuguilla* Scheinvar et al. 2017	–	Pleistocene	Pleistocene	Pleistocene	–	Pleistocene
2. *Agave striata* Trejo et al. 2016	–	–	No date	No date	–	No date
3. *Agave stricta* Martínez-Einsworth 2000	–	–	–	–	Pleistocene	Pleistocene
4. *Agave victoria-reginae* Martínez-Palacios et al. 1999	–	–	No date	–	–	–
5. *Berberis trifoliata* Angulo et al. 2017	–	–	Pleistocene	–	–	–
6. *Astrophytum* spp. Vázquez-Lobo et al. 2015	–	–	Pleistocene	Pleistocene	–	Pleistocene
7. *Ephedra compacta* Loera et al. 2017	–	–	Pleistocene	Pleistocene	Pleistocene	Pleistocene
8. *Leucophyllum* spp. Garanda & Sosa 2017	–	–	–	Pleistocene	–	Pleistocene
9. *Larrea tridentata* Duran et al. 2005	–	No date	No date	No date	–	–
10. *Peromyscus eremicus* Riddle et al. 2000	Pleistocene		Pleistocene		–	–
11. *Thomomys umbrinus* Mathis et al. 2014	Pleistocene		Pleistocene	Pleistocene	Pleistocene	
12. *Dipodomys phillipsi* Fernández et al. 2008				No date	No date	
13. *Perognathus flavus* Neiswenter and Riddle 2010	Pliocene/Pleistocene	Pliocene/Pleistocene	Pliocene/Pleistocene			Pliocene/Pleistocene

Taxa/Reference	Cochise filter barrier	Trans Pecos	Mapimian	Saladan	Saladan Filter Barrier	Río Pánuco Relict Desert
14. *Perognathus merriami* Neiswenter and Riddle 2010	Pliocene/Pleistocene		–	–	–	–
15. *Cynomys* spp. Castellanos-Morales et al. 2016a, b	Pleistocene	Pleistocene	Pleistocene	Pleistocene		–
16. *Calothorax lucifer* Licona-Vera et al. 2018	No date				–	–
17. *Bufo punctatus* Jaeger et al. 2005	Pleistocene	Pleistocene				
18. *Kinosternon* spp. Serb et al. 2001	No date	No date	No date			
19. *Crotalus molossus* Anderson and Greenbaum 2012	Pleistocene	Pleistocene				
20. *Crotalus scutulatus* Schield et al. 2018	Pleistocene	Pleistocene		Pleistocene		
21. *Campostema ornata* Schonhuth et al. 2011	No date	No date	No date	No date		
22. *Codoma* spp. Schonhuth et al. 2015			Pleistocene	Pleistocene		
23. *Pseudoerectus minimus* Bryson et al. 2013	Late Miocene	Late Miocene and Pleistocene	Pleistocene			
24. *Agelenopsis aperta* Ayoub and Riechert 2004	Pleistocene		Pleistocene			

The Chihuahuan Desert

The Chihuahuan Desert is the largest arid desert in North America (Laity 2008). Due to this extension, which ranges from 20° to 34° North latitude (14° difference) (but see below), it shows an important gradient in terms of temperature and precipitation that is accentuated with the altitudinal effect generating climatically differentiated regions (Fig. 2.1; Zavala-Hurtado and Jiménez 2020, this volume). Understanding the processes that gestated and shaped genetic diversity in these areas will allow us to propose appropriate strategies for their conservation in a planet in which climate change and increased aridity are a reality to which we must act. Thus, it is relevant to mention that the delimitation of the Chihuahuan desert is complicated in part because the geological, orographical, climatic and biological components have been constantly changing over time. Besides the changes in geology—that can affect the conformation of an area—a given locality can experience drastic environmental changes in time. These changes will affect the population dynamics by modifying the composition of species that can be found in it, as well as population sizes, dispersal and gene flow patterns within each species.

In 1977, Morafka made a detailed study of the distribution patterns and endemisms of reptiles and amphibians from Central Mexico and delimited the area of the Chihuahuan desert as a herpetofaunal province conformed by three subprovinces: Trans Pecos, Mapimian, and Saladan and three transition zones: Cochise filter barrier, Saladan filter barrier, and Rio Panuco Relict Desert (Fig. 2.1a). This entire province is located in the central area of Mexico between the Eastern and Western Sierra Madre to the East and West, the Trans-Mexican Volcanic Belt to the South and to the North, South of Texas and New Mexico.

In 1979, Schmidt used an aridity index and defined the climatic limits of the Chihuahuan desert in an *strict sense* as the area between the Eastern and Western Sierra Madre and above 25° North latitude in the states of Coahuila, Norwest Durango, Chihuahua, Texas, and New Mexico, excluding the Saladan, Saladan filter barrier, and the Panuco River Relict Desert of Morafka, Schmidt called it the "Real Chihuahuan Desert." For instance, from the species distribution analyzed in this chapter, only three studies (Fig. 2.3: 4, 19, 23) showed a limited distribution above 25° and below 33° North latitude according to the climatic limits established by Schmidt (1979) (Martínez-Palacios et al. 1999; Anderson and Greenbaum 2012; Bryson et al. 2013). The rest of the species analyzed have distributions that go further South or in some cases North, supporting the wider limits proposed by Morafka (1977) and by ourselves.

Comparative Phylogeography of the Chihuahuan Desert

We found six groups of species that supported an ancient integration of the northern Chihuahuan desert (Cochise filter barrier) with the Sonoran Desert (Riddle et al. 2000; Jaeger et al. 2005; Anderson and Greenbaum 2012; Bryson et al. 2013;

Schield et al. 2018) and nine studies that supported a North-South division in the Pliocene (Riddle et al. 2000; Neiswenter and Riddle 2010; Mathis et al. 2014; Schonhuth et al. 2015; Vázquez-Lobo et al. 2015; Castellanos-Morales et al. 2016a, b; Scheinvar et al. 2017; Schield et al. 2018). Of the 24 analyzed studies, practically all distinguished at least one of the subprovinces of Morafka: 9 supported the differentiation of the Cochise filter barrier, 10 the Trans Pecos, 14 that of the Mapimian, 11 the Saladan, 3 that of the Saladan filter barrier, and 8 that of the Río Pánuco Relict desert.

When analyzing the temporal relationship of the emergence of these areas, we found that although six studies did not present temporal data, more than half of the papers included molecular dating (17), reporting or inferring the formation of any of these areas in the last 2.5 MY. When environmental conditions were adequate, we expected local adaptation processes to increase population sizes, generating larger effective sizes with populations connected through genetic flow. When environmental conditions changed and were not suitable for local adaptation processes, the population sizes reduced, leaving few isolated survivors in relict populations in which the environmental conditions were less unfavorable and could survive. During these periods when populations were reduced, genetic drift acted, increasing the processes of differentiation and divergence among populations. If population isolation did not last long enough to give rise to a new species, subsequent climate changes again generated favorable conditions, in which the genotypes improved their fitness and thus populations could grow and could enter into secondary contact, and incorporate new variants generated in other populations during isolation.

Although in the present chapter we cannot yet establish a causal relationship of the observed patterns, we know that the last 2.5 MY were times of intense environmental climatic changes. During this period, 11 climatic cycles of growth and decrease of the Polar ice-cap occurred in North America affecting firstly the global climatic conditions and secondly the composition of species associated with that environment. During these periods, species populations decreased their sizes, being reduced to the so-called Pleistocene refuges in which they survived these difficult periods of climate change. Each one of the regions described by Morafka (1977) has very different environmental conditions (Fig. 2.1), so each one could host different refuges for a single species, in which drift and local adaptation could have acted, generating adaptative and non-adaptative divergence among populations. When the environmental conditions changed again, the populations located in each of the different refuges could expand and enter into secondary contact, increasing intra-population variation and generating complex patterns of genetical variation structure. These refuges represent areas with high genetic diversity, like Cuatro Ciénegas and the Mapimí Valley, that have been identified in several occasions as refuges of the biota of North America.

Additionally, the increase in elevation of the mountain ranges to the North and West of what is now the Chihuahuan desert, generated a rainfall orographic shadow effect that increased the aridity, stopping the moisture sources from the north and promoting the formation of new arid zones in North America (Axelrod 1985; Wilson and Pitts 2010; Bryson and Riddle 2012; Loera et al. 2012). These geological changes increased the fragmentation of the biota between the cold and warm deserts (Gámez et al. 2017), promoting a climatic affinity between the deserts of Sonora and Chihuahua, which explains the Cochise filter barrier's pattern as a transition zone between the Chihuahuan and the Sonoran deserts.

Comparing Patterns in the Chihuahuan Desert

We detected two general patterns of genetic structuring (Fig. 2.1: I and II) from ancient events in the Miocene according to molecular clock analyses. It is known that important geological events occurred that could have affected species distribution. In the late Miocene the North American Cordillera rose, doubling in size over the last 15 MY (Raymo and Ruddiman 1992), the late uplift of the Sierra Madre Occidental to the North occurred (Ruddiman et al. 1989; Wilson and Pitts 2010) and initiated a series of early magmatic events that formed the Trans-Mexican volcanic belt to the South (Ferrari 2011). Gámez et al. (2017) performed a cladistic biogeographic analysis of 13 areas of North America, relating this with previously published phylogenetic analyzes of five lineages of arthropods, 24 lineages of vertebrates, and 13 lineages of plants, finding that the distribution patterns and the associated biota of the Chihuahuan desert have changed over time. In that study, it was found that over the Late Miocene (~11.6–5.3 MY)/Pliocene (~5.3–2.6 MY), the co-distribution groups of the species show a North/South pattern in which species of the present Northern section of the Chihuahuan desert (Cochise filter barrier, Trans Pecos, and Mapimian) share distribution patterns with the species currently found in the deserts of Sonora, California, and Baja California, while the distribution patterns of the species located in the Southern section of the Chihuahuan desert (Saladan and Rio Pánuco) show association with species distributed in the Sierra Madre Oriental (Gámez et al. 2017). We detected this same old signal pattern with the old differentiation of the Cochise filter barrier (Fig. 2.2: I) and also with the later dividing signal of the North Altiplano from the Southern Altiplano (Fig. 2.2: II).

Arriaga et al. (1997) reported a consensus biogeographic classification resulting from the superposition of morphotectonic features, floristic, herpetofaunal, and mastofaunal provinces. This classification results in the division of the Chihuahuan desert area into two biogeographical provinces: Altiplano Norte (Chihuahuense) and Altiplano Sur (Zacatecano-Potosino). Morrone (2009) recommended treating them as two districts of the province of the Mexican Altiplano. In this review, we also detected the association pattern described by Arriaga et al. (1997) (Fig. 2.1: II), but when we related it to the molecular dating, we noticed that this classification detects diversification events that happened in the Miocene, while the Morafka (1977) classification reflects more recent events. As mentioned above, although in the present review, we cannot establish a causal relationship of the observed patterns, we currently know that the last 2.5 MY were times of intense environmental climatic changes, affecting in the first place the global climatic conditions and secondly the composition of associated species.

While the fossil record in the southern region of the Chihuahuan desert is scarce, there is evidence of a change in the composition of plant communities in the northern part of the Chihuahuan desert. During the Late Wisconsin (27–11 Kyrs B.P) the Big Bend and New Mexico/Arizona borderline were dominated by woodland of paper-shell pinyon and juniper, with few Chihuahuan scrub elements as sotol (*Dasylirion* spp.), lechuguilla (*Agave lechuguilla*), and prickly pears (*Opuntia* spp.) (Betancourt et al. 1990). During the same period, the Mapimian region assemblages were dominated by coniferous/juniper forest, while scrub Chihuahuan desert elements had a scarce to medium abundance. This is an indicator that during the Pleistocene there was a differentiated pattern in the composition of plant communities and that the establishment of Chihuahuan desert scrub as a dominant element was recorded later from 8 Kyr B.P. (Betancourt et al. 1990; Holmgren et al. 2003).

Eight of the analyzed studies support the relatively close association of the Mexican highland area populations with populations of the Tehuacán-Cuicatlán desert area, on the other side of the Trans-Mexican Volcanic Belt; in consequence, more robust biogeographical and genetic inferences are obtained when considering this area outside the Chihuahuan desert *sensu lato* definition (Fernández et al. 2008; Neiswenter and Riddle 2010; Martínez-Ainsworth 2013; Gándara and Sosa 2014; Mathis et al. 2014; Loera et al. 2017; Licona-Vera et al. 2018; Schield et al. 2018).

Conclusions and Perspectives

The present work is an effort to understand the genetic structure of populations of species located in the Chihuahuan desert. The number of population genetics studies that have been made in the Chihuahuan desert is still insufficient, maybe because of the extension and complexity of the area and we still lack data for many families and groups of plants and animals characteristic of this desert. However, different patterns begin to emerge, and we can start to understand the processes that were forged over time.

The patterns of genetic differentiation detected in this review can in general be explained in terms of Pleistocene climate dynamics, along altitudinal and latitudinal gradients of locally adapted populations that underwent cyclic processes of contraction, isolation, and divergence followed by expansion and secondary contact processes and can serve as a starting point to test local adaptation with new genomic tools.

Current sequencing strategies along with ABC type coalescent analysis and modern niche modeling tools will allow us to explore the historical and adaptive structure of populations in terms of the environmental change and to test specific models of historical population dynamics.

References

Ackermann M, Doebeli M (2004) Evolution of niche width and adaptive diversification. Evolution 58(12):2599–2612

Aguirre-Liguori GBS, Jaramillo Correa JP, Tenaillon MI, Montes Hernández S, García Oliva F, Hearne SJ, Eguiarte LE (2019) Divergence with gene flow is driven by local adaptation to temperature and soil phosphorus concentration in teosinte subspecies (*Zea mays parviglumis* and *Zea mays mexicana*). Mol Ecol 28:2814–2830

Aguirre-Liguori J, Scheinvar E, Eguiarte LE (2014) Gypsum soil restriction drives genetic differentiation in *Fouquieria shrevei* (Fouquieriaceae). Am J Bot 101(4):730–736

Anderson CG, Greenbaum E (2012) Phylogeography of northern populations of the black-tailed rattlesnake (*Crotalus molossus* Baird and Girard, 1853), with the revalidation of *C. ornatus* Hallowell, 1854. Herpetol Monogr 26(1):19–57

Angulo DF, Amarilla LD, Anton AM, Sosa V (2017) Colonization in North American arid lands: the journey of agarito (*Berberis trifoliolata*) revealed by multilocus molecular data and packrat midden fossil remains. PLoS One 12(2):e0128559

Arriaga L, Aguilar C, Espinosa-Organista D, Jiménez R (1997) Regionalización ecológica y biogeográfica de México. Taller de la Comisión Nacional para el Conocimiento y Uso de la Biodiversidad (Conabio), México, DF, Mexico

Axelrod DI (1985) Rise of the grassland biome, central North America. Bot Rev 51(2):163–201

Ayoub N, Riechert SE (2004) Molecular evidence for Pleistocene glacial cycles driving diversification of a North American desert spider, *Agelenopsis aperta*. Mol Ecol 13(11):3453–3465. https://doi.org/10.1111/j.1365-294X.2004.02335.x

Betancourt JL, Van Devender TR, Martin PS, Paul S (1990) Packrat middens: the last 40,000 years of biotic change. University of Arizona Press, Tucson, AZ, USA

Brooks DR, McLennan DA (2012) The nature of diversity: an evolutionary voyage of discovery. University of Chicago Press, Chicago, IL, USA

Bryson RW, Riddle BR (2012) Tracing the origins of widespread highland species: a case of Neogene diversification across the Mexican sierras in an endemic lizard. Biol J Linn Soc 105(2):382–394

Bryson RW, Savary WE, Prendini L (2013) Biogeography of scorpions in the *Pseudouroctonus minimus* complex (Vaejovidae) from south-western North America: implications of ecological specialization for pre-Quaternary diversification. J Biogeogr 40(10):1850–1860

Carson EW, Souza V, Espinosa-Pérez H, Turne TF (2015) Mitochondrial DNA diversity and phylogeography of *Lucania interioris* inform biodiversity conservation in the Cuatro Cienegas Basin, Mexico. Western North American Naturalist 75(2):200–208

Castellanos-Morales G, Gámez N, Castillo-Gámez RA, Eguiarte LE (2016a) Peripatric speciation of an endemic species driven by Pleistocene climate change: the case of the Mexican prairie dog (*Cynomys mexicanus*). Mol Phylogenet Evol 94:171–181

Castellanos-Morales G, Gutiérrez-Guerrero YT, Gámez N, Eguiarte LE (2016b) Use of molecular and environmental analyses for integrated in situ and ex situ conservation: the case of the Mexican prairie dog. Biol Conserv 204:284–295

Castoe TA, Spencer CL, Parkinson CL (2007) Phylogeographic structure and historical demography of the western diamondback rattlesnake (*Crotalus atrox*): a perspective on North American desert biogeography. Mol Phylogenet Evol 42(1):193–212

De Nova JA, Medina R, Montero JC, Weeks A, Rosell JA, Olson ME, Eguiarte LE, Magallón S (2012) Insights into the historical construction of species-rich Mesoamerican seasonally dry tropical forests: the diversification of *Bursera* (Burseraceae, Sapindales). New Phytol 193:276–287

Duran KL, Lowrey TK, Parmenter RR, Lewis PO (2005) Genetic diversity in Chihuahuan Desert populations of creosotebush (Zygophyllaceae: *Larrea tridentata*) . Am J Botany 92(4):722–729

Ezcurra E, Martínez-Berdeja A, Villanueva-Almanza L (2020) The evolution of North American deserts and the uniqueness of Cuatro Ciénegas. In: Mandujano MC, Pisanty I, Eguiarte LE (eds) Plant diversity and Ecology in the Chihuahuan desert. Springer International, Cham, Switzerland

Fernández JA, Cervantes FA, Hafner MS (2008) Molecular systematics and biogeography of the Mexican endemic kangaroo rat, *Dipodomys phillipsii* (Rodentia: Heteromyidae). J Mammal 93(2):560–571

Ferrari L (2011) Avances en el conocimiento de la Faja Volcánica Transmexicana durante la última década. Bol Soc Geol Mex 53:84–92

Ferrusquía-Villafranca I (1993) Geology of Mexico: a synopsis. In: Ramamoorthy TP, Bye R, Lot A, Fa J (eds) Biological diversity of Mexico: origins and distribution. Oxford University Press, New York, NY, USA, pp 3–108

Gámez N, Nihei SS, Scheinvar E, Morrone JJ (2017) A temporally dynamic approach for cladistic biogeography and the processes underlying the biogeographic patterns of North American deserts. J Zool Syst Evol Res. 55(1):11-18 https://doi.org/10.1111/jzs.12142

Gándara E, Sosa V (2014) Spatio-temporal evolution of *Leucophyllum pringlei* and allies (Scrophulariaceae): a group endemic to North American xeric regions. Mol Phylogenet Evol 76:93–101

Gentry HS (2004) Agaves of continental North America. University of Arizona Press, Tucson, AZ, USA

Halffter G (1987) Biogeography of the montane entomofauna of Mexico and Central America. Annu Rev Entomol 32:95–114

Hernández HM, Godínez HA (1994) Contribución al conocimiento de las cactaceas mexicanas amenzadas. Acta Botánica Mexicana 26:33–52

Hernández HM, Barcenas RT (1995) Endangered cacti in the Chihuahuan Desert: I. distribution pattern. Conserv Biol 9(5):1176–1188

Hernández HM, Gómez-Hinostrosa C, Bárcenas RT (2001) Diversity, spatial arrangement, and endemism of Cactaceae in the Huizache area, a hot-spot in the Chihuahuan Desert. Biodivers Conserv 10:1097–1112

Hijmans RJ, Cameron SE, Parra JL, Jones PG, Jarvis A (2005) Very high resolution interpolated climate surfaces for global land areas. Int J Climatol 25:1965–1978

Holmgren CA, Peñalba MC, Rylander KA, Betancourt JL (2003) A 16,000 14C yr B.P packrat midden series from the USA-Mexico borderlands. Quat Res 60:319–329

Holt RD (1990) The microevolutionary consequences of climate change. Trends Ecol Evol 5(9):311–315

Jaeger JR, Riddle BR, Bradford DF (2005) Cryptic Neogene vicariance and Quaternary dispersal of the red-spotted toad (*Bufo punctatus*): insights on the evolution of North American warm desert biotas. Mol Ecol 14(10):3033–3048

Laity J (2008) Deserts and desert environments. Wiley-Blackwell, Chichester, UK

Lanner RM, Van Devender TR (1981) Late Pleistocene Piñon Pines in the Chihuahuan Desert. Quat Res 15:278–290

Licona-Vera Y, Ornelas JF, Wethington S, Bryan KB (2018) Pleistocene range expansions promote divergence with gene flow between migratory and sedentary populations of *Calothorax* hummingbirds. Biol J Linn Soc 124(4):645–667

Loera I, Ickert-Bond SM, Sosa V (2017) Pleistocene refugia in the Chihuahuan Desert: the phylogeographic and demographic history of the gymnosperm *Ephedra compacta*. J Biogeogr 44(12):2706–2716

Loera I, Sosa V, Ickert-Bond SM (2012) Diversification in north American arid lands: niche conservatism, divergence and expansion of habitat explain speciation in the genus *Ephedra*. Mol Phylogenet Evol 65(2):437–450

Magalhaes IS, Ornelas-García CP, Leal-Cardin M, Ramírez T, Barluenga M (2015) Untangling the evolutionary history of a highly polymorphic species: introgressive hybridization and high genetic structure in the desert cichlid fish *Herichtys minckleyi*. Mol Ecol 24(17):4505–4520

Martínez-Palacios A, Eguiarte LE, Furnier GR (1999) Genetic diversity of the endangered endemic *Agave victoria-reginae* (Agavaceae) in the Chihuahuan Desert. Am J Bot 86(8):1093–1098

Martínez-Ainsworth N (2013) Genética de poblaciones de *Agave stricta* Salm-Dyck, especie endémica al Valle de Tehuacán-Cuicatlán. México, Tesis de Licenciatura, Facultad de Ciencias, UNAM, México, DF, Mexico

Mathis VL, Hafner MS, Hafner DJ (2014) Evolution and phylogeography of the *Thomomys umbrinus* species complex (Rodentia: Geomyidae). J Mammal 95(4):754–771

McGaugh SE (2012) Comparative population genetics of aquatic turtles in the desert. Conserv Genet 13:1561–1576

Morafka DJ (1977) A biogeographical analysis of the Chihuahuan Desert through its herpetofauna, Biogoegraphica, vol 9. Springer Science & Business Media

Morrone JJ (2001a) Biogeografía de América Latina y el Caribe. Manuales y Tesis SEA 3, Zaragoza, España

Morrone JJ (2001b) Toward a cladistic model for the Caribbean subregion: delimitation of areas of endemism. Caldasia 23:43–76

Morrone JJ (2004) Homología biogeográfica: las coordenadas espaciales de la vida. Instituto de Biología, UNAM, México, DF, Mexico

Morrone JJ (2005) Hacia una síntesis biogeográfica de México. Revista Mexicana de Biodiversidad 76(2):207–252

Morrone JJ (2009) Evolutionary biogeography: an integrative approach. Columbia University Press, New York, NY, USA

Myers EA, Bryson RW Jr, Robert H, Aardema ML, Lazcano D, Burbrink FT (2018) Exploring Chihuahuan Desert diversification in the gray-banded kingsnake, *Lampropeltis alterna* (Serpentes: Colubridae). Mol Phylogenet Evol 131:211–218

Neiswenter SA, Riddle BR (2010) Diversification of the *Perognathus flavus* species group in emerging arid grasslands of western North America. J Mammal 91(2):348–362

Ochoterena H, Flores-Olvera H, Gómez-Hinostrosa C, Moore, MJ (2020) Gypsum and plant species: a marvel of Cuatro Ciénegas and the Chihuahuan Desert. In: Mandujano MC, Pisanty I, Eguiarte LE (eds) Plant diversity and Ecology in the Chihuahuan desert. Springer International, Cham, Switzerland

Olson DM, Dinerstein E (2002) The Global 200: priority ecoregions for global conservation. Ann Mo Bot Gard 89(2):199–224

Onorato DP, Hellgren EC, Van Den Bussche RA, Doan-Crider DL, Skiles JR (2007) Genetic structure of American black bears in the desert southwest of North America: conservation implications for recolonization. Conserv Genet 8(3):565–576

Peterson AT, Soberón J, Pearson RG, Anderson RP, Martínez-Meyer E, Nakamura M, Araújo MB (2011) Ecological niches and geographic distributions. Princeton University Press, Princeton, NJ, USA

Raymo M, Ruddiman F (1992) Tectonic forcing of late Cenozoic climate. Nature 359:117–122

Riddle BR, Hafner DJ, Alexander LF (2000) Phylogeography and systematics of the *Peromyscus eremicus* species group and the historical biogeography of North American warm regional deserts. Mol Phylogenet Evol 17(2):145–160. https://doi.org/10.1006/mpev.2000.0841

Ruddiman, WF, Prell WL, y Raymo ME. (1989). Late Cenozoic uplift in Southern Asia and the American West' rationale for general circulation modeling experiments. J Geophys Res, 94, 379-391

Rzedowski J (1978) Vegetación de México. Limusa, México, DF, Mexico

Rzedowski J (1993) Diversity and origins of the phanerogamic flora of Mexico. In: Ramamoorthy TP, Bye R, Lot A, Fa J (eds) Biological diversity of Mexico: origins and distribution. Oxford University Press, New York, NY. USA, pp 129–144

Scheinvar E, Gámez N, Castellanos-Morales G, Aguirre-Planter E, Eguiarte LE (2017) Neogene and Pleistocene history of *Agave lechuguilla* in the Chihuahuan Desert. J Biogeogr 44(2):322–334
Schield DR, Adams RH, Card DC, Corbin AB, Jezkova T, Hales NR, Castoe TA (2018) Cryptic genetic diversity, population structure, and gene flow in the Mojave rattlesnake (*Crotalus scutulatus*). Mol Phylogenet Evol 127:669–681
Schmidt RH Jr (1979) A climatic delineation of the "real" Chihuahuan Desert. Journal of Arid Environment 2:243–250
Schonhuth S, Blum MJ, Lozano-Vilano L, Neely DA, Varela-Romero A, Espinosa H, Mayden RL (2011) Inter-basin exchange and repeated headwater capture across the Sierra Madre Occidental inferred from the phylogeography of Mexican stonerollers. J Biogeogr 38(7):1406–1421
Schonhuth S, Lozano-Vilano L, Perdices A, Espinosa H, Mayden RL (2015) Phylogeny, genetic diversity and phylogeography of the genus *Codoma* (Teleostei, Cyprinidae). Zool Scr 44(1):11–28
Serb JM, Phillips CA, Iverson JB (2001) Molecular phylogeny and biogeography of *Kinosternon flavescens* based on complete mitochondrial control region sequences. Mol Phylogenet Evol 18(1):149–162
Toledo VM, Ordóñez MJ (1993) The biodiversity scenario of Mexico: a review of terrestrial habitats. In: Ramamoorthy TP, Bye RA, Lot A, Fa J (eds) Biological diversity of Mexico: origins and distribution. Oxford University Press, New York, NY, USA, pp 757–777
Trejo L, Alvarado-Cardenas LO, Scheinvar E, Eguiarte LE (2016) Population genetic analysis and bioclimatic modeling in *Agave striata* in the Chihuahuan Desert indicate higher genetic variation and lower differentiation in drier and more variable environments. Am J Bot 103(6):1020–1029
Van Devender TR (1977) Holocene woodlands in the Southwestern deserts. Science 198(4313):189–192
Van Devender TR, Burguess TL (1985) Late Pleistocene woodlands in the Bolson de Mapimi: a refugium for the Chihuahuan desert biota? Quat Res 24:346–353
Van Devender TR, Spaulding WG (1979) Development of vegetation and climate in the Southwestern United States. Science 204(4394):701–710
Vázquez-Lobo A, Morales GA, Arias S, Golubov J, Hernández-Hernández T, Mandujano MC (2015) Phylogeny and biogeographic history of *Astrophytum* (Cactaceae). Syst Bot 40(4):1022–1030
Villaseñor JL (2016) Checklist of the native vascular plants of Mexico. Revista Mexicana de Biodiversidad 87(3):559–902
Villarreal-Quintanilla JA, Bartolomé-Hernández JA, Estrada-Castillón E, Ramírez-Rodríguez H, Martínez-Amador SJ (2017) El elemento endémico de la flora vascular del Desierto Chihuahuense. Acta Botánica Mexicana 118:65–96
Williford D, DeYoung RW, Honeycutt RL, Brennan LA, Hernández F, Heffelfinger JR, Harveson LA (2014) Phylogeography of the Gambel's Quail (*Callipepla gambel*ii) of western North America. Wilson Journal of Ornithology 126(2):218–235
Wilson JS, Pitts JP (2010) Illuminating the lack of consensus among descriptions of earth history data in the North American deserts: a resource for biologists. Prog Phys Geogr 34(4):419–441
Zavala-Hurtado JA, Jiménez M (2020) Diversity and uniqueness at its best: vegetation of the Chihuahuan Desert. In: Mandujano MC, Pisanty I, Eguiarte LE (eds) Plant diversity and Ecology in the Chihuahuan desert. Springer International, Cham, Switzerland
Zink RM, Blackwell-Rago RC, Ronquist F (2000) The shifting roles of dispersal and vicariance in biogeography. Proceedings of the Royal Society of London. Series B: Biological Sciences 267(1442):497–503

Chapter 3
The Evolution of North American Deserts and the Uniqueness of Cuatro Ciénegas

Exequiel Ezcurra, Alejandra Martínez-Berdeja, and Lorena Villanueva-Almanza

Abstract With an area of less than 600 km^2, the Cuatro Ciénegas Basin harbors one of the most diverse desert landscapes in Mexico. On the general aridity pattern of the Chihuahuan Desert, limited by a very short season of scant summer rain, a unique geology of gypsum soils is superimposed, dotted by montane sky islands where relicts of temperate vegetation that occupied the Mexican Plateau still survive. Plants here show remarkable adaptations to this unique desert environment, including tolerance to gypsum and salinity, seed retention (serotiny) during the dry season, thick, isolateral leaves, and a panoply of different life-forms, including desert annuals, stem- and leaf-succulents, microphyllous shrubs, desert perennial with photosynthetic stems, and drought-deciduous trees, and phreatophytes, tapping the deep with powerful pivot roots. This chapter analyzes the evolution of this diversity of life-forms in the heart of the Chihuahuan Deserts, and discusses the challenges and opportunities it offers for conservation and sustainable resource use.

Keywords Gypsophily · Sky islands · Relictual vegetation · Desert annuals · Serotiny · Leaf isolaterality · Nurse plants

---

E. Ezcurra (✉) · L. Villanueva-Almanza
Department of Botany and Plant Sciences, University of California, Riverside, Riverside, CA, USA
e-mail: exequiel.ezcurra@ucr.edu; lorena.villanuevaalmanza@email.ucr.edu

A. Martínez-Berdeja
Evolution and Ecology Department, University of California, Davis, Davis, CA, USA
e-mail: amberdeja@ucdavis.edu

M. C. Mandujano et al. (eds.), *Plant Diversity and Ecology in the Chihuahuan Desert*, Cuatro Ciénegas Basin: An Endangered Hyperdiverse Oasis,
https://doi.org/10.1007/978-3-030-44963-6_3

Introduction

With an area of less than 600 km^2, the Cuatro Ciénegas Basin harbors one of the most diverse landscapes in Mexico, including (a) aquatic habitats maintained by the basin's lowland *pozas* (artesian pools); (b) desert plains; (c) gypsum dunes and flats; (d) basin grasslands; (e) montane chaparral; (f) oak and oak-pine woodlands, and (g) montane conifer forests. Plants here have to deal with a panoply of environmental factors limiting their growth, including aridity in the desert plains, seasonal flooding around the *pozas*, salinity in the lower playas, shifting substrates in the dunes, gypsum soils, and freezing conditions in the mountains. This heterogeneity has been the driving factor of the region's remarkable biodiversity: With a flora of 860 vascular plants, the valley contains 49 species whose type specimens were found in the valley, and 23 species that are fully endemic to the basin. Endemism is particularly high in the lower parts of the basin, where extreme conditions have forced desert species to develop unique adaptations (Pinkava 1984).

Thus, the taxonomic diversity of the area is complemented by an extraordinary array of morphologies and life histories. Psammophily, the adaptation to dune substrates, is combined here with gypsophily, the ability to grow on gypsum soils saturated by calcium sulfate. Desert plants show a remarkable array of drought-evading strategies such as deep taproots, microphylls with exceptionally small leaves, water-storing succulents, and short-lived ephemerals that retain seeds in their dead tissues until the next rains arrive, among many others. The Cuatro Ciénegas Basin (henceforward, CCB) provides a deep insight into the evolution of desert plants.

Geologic Origin

During the middle of the Cretaceous Period, some 100 million years ago, North America was a set of two major land masses called by geologists Laramidia (the western section) and Appalachia (the eastern one). The central part of what is now North America was part of a gigantic depression where the Arctic Ocean and the Gulf of Mexico met to form a large seaway—the Western Interior Seaway—800 m deep, 1000 km wide, and over 3000 km long. Different sediments accumulated in the floor of this ancient sea—in some parts, organic debris and mud, in others, sand or the remains of the calcareous exoskeletons and shells of algae, diatoms, crinoids, foraminifera, corals, bryozoans, and mollusks (Cochran et al. 2003).

The tectonic forces pushing the North American Plate produced the uplift of the Cretaceous Seaway and folded the coast of Mexico into the Eastern Sierra Madre, around 65 million years ago at the end of the Cretaceous. The old ocean floors were raised and the ancient marine sediments were hoisted into layers of shale, sandstone, and limestone that can be seen today from the Great Plains along

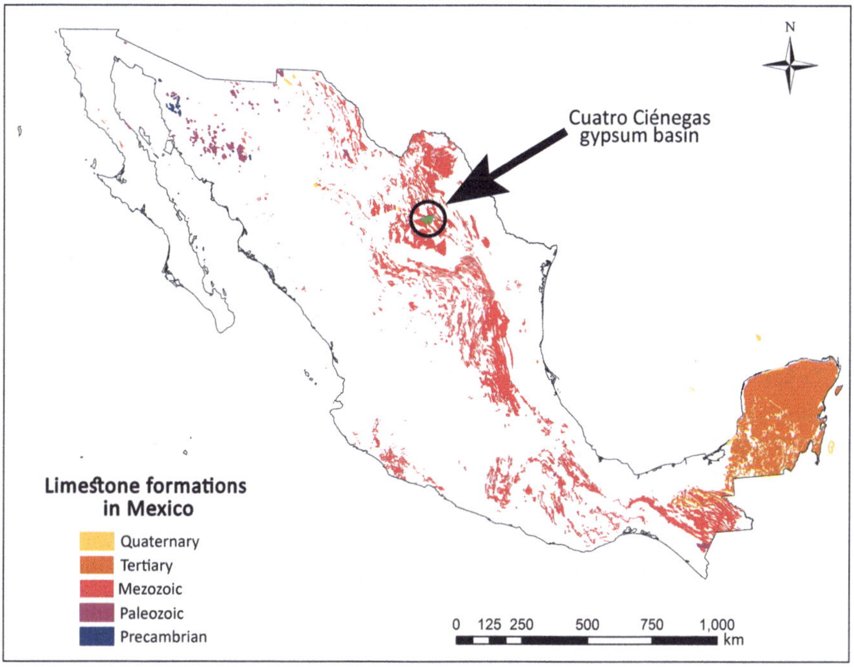

Fig. 3.1 Distribution of limestone substrates in Mexico (modified from INEGI's Geologic chart of Mexico, available at https://www.inegi.org.mx/temas/mapas/geologia/). Note the large corridor of Mezozoic rocks (66–251 My BP) of marine origin that form the Eastern Sierra Madre, and the location of the Cuatro Ciénegas gypsum basin in a closed watershed, or "bolsón," in the arid north of the calcareous ranges

Mexico's Sierra Madre all the way to the Guatemalan border (Cochran et al. 2003). As a consequence, the system of mountains and ranges that form Mexico's Eastern Sierra Madre are rich in calcareous sediments transformed into limestone rocks (Fig. 3.1).

In some closed valleys within these massive calcareous sierras gypsum deposits have been formed over the years by solubilization of calcium salts and their re-deposition combined with sulfate solutions from hydrothermal veins. The closed basins allow waters high in calcium and sulfate content to slowly evaporate and be regularly replenished with new sources of water. The result is the accumulation of large beds of sedimentary gypsum, and Cuatro Ciénegas Basin (CCB) is one of the most spectacular examples of this process in the world. Furthermore, because gypsum dissolves over time in water, gypsum sand dunes are only found in extremely dry environments such as Cuatro Ciénegas in México (Ochoterena et al. 2020, this volume) and White Sands in the US State of New Mexico, both within the northern Chihuahuan Desert.

The Causes of Aridity

Two large deserts occupy the northern states of Mexico: (a) the Sonoran Desert, which runs along the coastal states of Mexico's Pacific Northwest (Sinaloa, Sonora, Baja California, and Baja California Sur), and (b) the Chihuahuan Desert, which occupies the Mexican northern *Altiplano*, i.e., the highland plateau that stretches between Mexico's two large range systems, the Eastern and the Western Sierra Madre. Together with the Mojave Desert in the USA, they form the large subtropical deserts of North America, lying at 25–35° latitude and some 2000 km away from the tropical rainforests of southern Mexico and Central America (Ezcurra et al. 2006; Ezcurra and Mellink 2013).

Deserts in North America occur in these precise latitudes because of the general thermodynamics of our planet: Solar radiation hits the Earth with highest intensity near the equator. Because the Earth's axis is tilted 23.5° with respect to the plane of its orbit, during a part of the year, the zone of maximum solar interception shifts northwards, towards the Tropic of Cancer, and during the other part it moves southwards, towards the Tropic of Capricorn. Thus, the warm tropics form a belt around the equator from latitude 23°N to latitude 23°S, where the tropical heat generates rising, unstable air. As the rising air cools in its ascent, it condenses the moisture evaporated from the warm tropical seas and forests producing the heavy downpours that characterize the wet tropics. Having lost its moisture, the air moves in the upper atmosphere away from the tropical belt, and starts to descend around 25–30° latitude. This stable, dry air forms the mid-latitude arid fringes that run north and south of the tropical belt, forming corridors of stable atmosphere—known as the "horse" latitudes—where calm air dominates. The closed circulation of air, ascending in the tropics into the upper atmosphere to descend in the subtropical latitudes and moving again towards the tropics, is known as the "Hadley Cells" (in honor of the British climatologist George Hadley). Hadley Cells, the low-latitude overturning circulations that have air rising at the equator and sinking at 30° latitude, are responsible for the trade winds in the Tropics and control low-latitude weather patterns. Because of the stable atmosphere in the polar-ward side of the cells not only winds are slack but also rainstorms seldom develop; and this is the reason why most of the world's large deserts occur at these latitudes both in the northern and southern hemisphere (Goudie and Wilkinson 1977; McGinnies et al. 1977).

Mexico's topographic heterogeneity also contributes to the formation of drylands, especially within the country's tropical belt. When the moisture-laden tropical trade winds reach continental mountain ranges they cool as they ascend, condensing as fog and drizzle that feed montane cloud forests. Once the winds pass the mountain divide, they start compressing and warming-up again in their descent, but, having left behind their original moisture, they become hot and dry. Thus, while the windward slopes of most tropical mountain ranges are covered by cloud forests, the leeward part, also known as the "rain shadow" of the mountains, is covered by arid scrubs. The rain shadow effect is largely responsible for the intriguing occurrence of tropical drylands in areas where one would expect tropical forests, such as the Tehuacán Valley desert in the states of Puebla and Oaxaca, Mexico, a hotspot for cactus biodiversity. Because the Chihuahuan Desert lies between the two Sierra Madre ranges, potential sources of precipitation are also trapped in the rain shadow of these mountains.

Desert Sky Islands and Climates of the Past

At a very large scale, two very distinct types of desert geomorphologies can be recognized: large, mostly flat "shield" deserts, and highly folded, topographically heterogeneous "basin-and-mountain" deserts (Cooke et al. 1993). Shield deserts have developed on ancient crystalline bedrocks and are mainly composed of very old land systems unearthed and leveled by erosion forming relatively flat landscapes with continuous desert vegetation, such as the Sahara or the Arabian deserts. Basin-and-mountain deserts are formed by recent and present-day tectonic forces and are largely composed of tall mountain ranges or *sierras* that emerge from alluvial plains. The deserts in North America are mostly of this type. In basin-and-mountain deserts the desert sierras form islands of moister, colder ecosystems with temperate vegetation, and conifer forests surrounded by continuous plains of alluvial sediments covered by true desert vegetation. Alluvial fans are common at the point where sediment-laden streams leave the mountain front and spread out over the plain or *bajada* zone (Flores-Vázquez et al. 2020, this volume). Many of these desert basins do not drain onto larger rivers but have a closed drainage system, forming a geomorphologic landscape known as a *bolsón*, with salty playas at their bottom. CCB is one of the most outstanding of these bolsones in the Chihuahuan Desert.

This landscape of temperate mountains, or "sky islands," surrounded by hot subtropical desert vegetation is an evolutionary witness of past climate changes. During the last 2 million years (the Pleistocene period) the Earth underwent a series of alternating cycles of cooling and warming, induced by variations in the planet's orbit and in the inclination of its axis. During the colder periods—known as the "Ice Ages"—most of the high-latitude regions of the world became covered by massive glaciers and temperate ecosystems such as cold grasslands and conifer forests moved southwards. During these glacial periods the tropical belt narrowed and the mid-latitude hot deserts shrunk, replaced by grasslands, semiarid scrubs, and open woodlands. The desert biota found refuge in dry subtropical valleys where arid conditions persisted under the rain shadow of large mountain ranges. The last glaciation ended around 15,000 years ago, and we are still living in the warm interglacial period that followed—the Holocene (Betancourt et al. 1990).

When, 20,000 years ago, the ice sheets started to retreat, most of the temperate flora and fauna started to move back into higher latitudes. A subset of those species, however, managed to survive in the mid-latitude regions by climbing up the rugged and cool mountain ranges that emerge like islands from the desert plains. Establishing higher-up with each passing generation as the climate warmed, the ice-age organisms were able to persist in the cool mountain environments where they found a climate similar to the one they had enjoyed in the lower plains during the ice ages. As they ascended into the isolated desert mountains, the communities of the desert "sky islands" became separated from other mountains by a sea of harsh desert plains. Like antediluvian castaways, the ice-age species now survive high-up in the cool refuges of the sky islands. The temperate forests and scrubs that prosper in the mountains that dapple the Chihuahuan Desert in North America are a magnificent example of these relictual ecosystems; a biological memory of bygone evolutionary history surviving high-up in the mountains like a ghost of climates past (Dimmit 2000), covering the slopes with pinyon, juniper, and oaks, and the wetter and colder canyons with majestic conifers such as *Pinus strobiformis* and *Abies coahuilensis* (Pinkava 1984).

Cycles of Abundance and Scarcity

Three climatic signals bring precipitation to Mexican deserts: (a) Convective winds coming across the Pacific from the northwest bring winter rains to Baja California and the northern Sonoran Desert. (b) Thunderstorms driven by low-pressure centers that develop in summer when the continent becomes hot bring monsoon-type rainfall to the southern Sonoran and the Chihuahuan deserts, including CCB. (c) Lastly, hurricanes and tropical storms that develop in fall (September–November) over the tropical Pacific Ocean may bring autumnal downpours to southern Sonora and Baja California's Cape Region. While the Sonoran Desert receives mostly winter rains in its northwestern reaches, near the Mojave, and is fed by the Mexican summer monsoon in its tropical southern reaches (Robichaux 1999), the Chihuahuan Desert, in contrast, is almost entirely driven by summer rains (Caso et al. 2007).

The intensity of the atmospheric and oceanic signals that drive these patterns may vary significantly from year to year, driven by oscillations in the intensity of the Hadley cells and semi-cyclical changes in oceanic water temperature. When the temperature of the California current increases, winter precipitation in Mexican deserts increases in what we call "El Niño" seasons. Similarly, when water temperatures decrease in the tropical Pacific Ocean, the Mexican monsoon increases in strength and the likelihood of strong downpours in CCB increases (Caso et al. 2007).

These pulses of abundance and scarcity of resources are a major force in the ecological organization of deserts. During pulses of bounty, the fragile seedlings of desert plants can germinate, establish, and prepare for long droughts burying their roots deep into the desert soils. Ephemerals can replenish their seed banks, desert toads can reproduce in extraordinary numbers before entering again into their waterless torpor, and granivorous rodents, such as the kangaroo rats (*Dipodomys* spp.) and the pocket mice (*Chaetodipus* spp.), can stock up their underground caches. The desert becomes renewed, and ready to face again years, or even decades, of extreme hardship.

Annual plants—often called ephemerals for their short and opportunistic life cycle—survive the long drought periods in the form of seeds, bulbs, or tubers, and

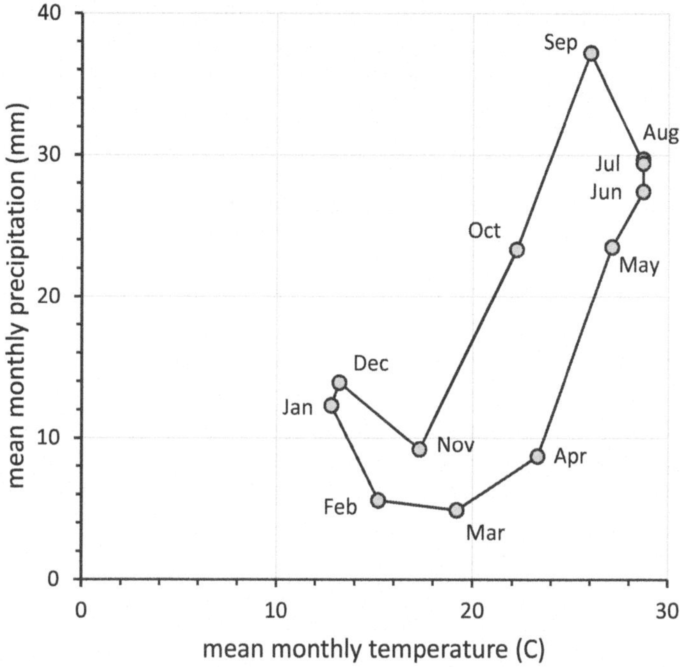

Fig. 3.2 Climatic diagram for Cuatro Ciénegas (data from the Servicio Meteorológico Nacional online database, available at http://smn.cna.gob.mx/es/climatologia/informacion-climatologica/normales-climatologicas-por-estado). Note the monsoon-type weather pattern driven by summer rainfall, revealed by the positive correlation between monthly mean precipitation and temperature

quickly sprout during the narrow window of opportunity that the desert rains provide. Perennial plants often show extensive networks of shallow roots, as many plants compete to extract water from the soil immediately after the rain has fallen. In monsoon deserts, like Cuatro Ciénegas, rainfall pulses coincide with adequate temperatures for plant growth (Fig. 3.2). Plants with succulent, water-storing tissues, like cacti and agaves, are well adapted to accumulate water and are common in CCB.

Life-Forms and Adaptations

In its origins, life on Earth evolved in water, and water is the most crucial element for the survival of all organisms (Souza et al. 2018). Thus, it is no surprise that some of the most remarkable adaptations for survival are found in deserts, i.e. in the environments where water is most scarce (Louw and Seely 1982). The short pulses of abundance that contrast sharply with the background condition of aridity and scarcity are the major force that has driven evolution, natural selection, and adaptation in desert biota. Plants and animals are adapted to these seasonal strokes. Natural selection and evolution have molded in very precise ways the life-forms of desert organisms to their harsh and unpredictable environment. Furthermore, because most deserts of the world have evolved recently and in relative isolation from each other, many of their constituent species have evolved from different ancestors (Morton 1979). Thus, deserts are prime ecosystems to study and understand the phenomenon of convergent evolution—the development of similar growth forms and adaptations derived from different ancestors.

Adaptations of Plants to Aridity

Most desert species have found remarkable ways to survive by evading drought. Desert succulents, such as cacti or agaves (century plants), can evade dry spells by accumulating moisture in their fleshy tissues. They have an extensive system of shallow roots that allows them to capture soil water only a few hours after it has rained. Their photosynthesis is modified to exchange gases and fix carbon dioxide (CO_2) during the night, when evaporative demand is low, and to accumulate the fixed carbon in the form of malic acid, which is later used by the plant as the building blocks of more complex organic molecules (this photosynthetic pathway is called "Crassulacean Acid Metabolism" or CAM). Additionally, cacti are leafless and commonly have vertically erect, green trunks that maximize light interception during the early and late hours of the day, but avoid the midday sun, when excessive heat may damage, or even kill, the plant tissues. Vertically-oriented photosynthetic tissues are noticeable, for example, in the columnar *Grusonia bradtiana* (Plascencia-López et al. 2020, this volume) in barrel cacti such as *Echinocereus freudenbergeri*, or in flat or cylindrical-stemmed prickly pears and chollas such as *Opuntia anteojoensis* (=*Cylindropuntia anteojoensis*). A similar behavior is observed in the central leaves of agave rosettes, such as *Agave lecheguilla*, which have a predominantly vertical orientation.

Woody desert trees, such as acacias, cannot store much water in their trunks, but many of them evade drought by shedding their leaves as the dry season sets in, entering into a sort of drought-induced latency. Many of these desert species, such as the mesquite *Prosopis glandulosa*, also have deep pivotal roots that allow them to tap into the desert's aquifer, the underground water layer accumulated deep beneath the soil. Some of these deep-tapping, deciduous trees also possess green stems, such as the Texas paloverde *Parkinsonia texana*. Green stems allow these trees to maintain a low level of photosynthesis during long periods of drought, recycling the CO_2 produced by respiration and maintaining the plant alive without the need to risk water loss through the exchange of gases (Ávila-Lovera and Ezcurra 2016).

Other group of trees, called pachycauls, have convergently evolved a mixture of these strategies: they can store water in gigantic trunks and have a smooth bark that can do some cactus-like photosynthesis during dry periods. When it rains, however, they produce abundant green leaves and shift their metabolism towards that of normal-leaved plants. This group is formed by trees with famously "bizarre" trunks, such as the Baja-Californian cirio or Boojum-tree (*Fouquieria columnaris*) and elephant-trees (*Bursera* and *Pachycormus*). Pachycauls, however, dominate in hot tropical deserts and are not present in CCB, but some taxonomically-related species can be found, such as the highly-ramified Shreve's ocotillo *Fouquieria shrevei*, with deciduous leaves and photosynthetic stems that can burst with new leaves only hours after heavy rains, and it is endemic to CCB (Nova et al. 2020, this volume).

A third group of plants, the "true xerophytes" or true drought-tolerant plants, have simply adapted their morphology and their metabolism to survive extremely long droughts. These species have remarkably low osmotic potentials in their tissue, which means that they can still extract moisture from the soil when most other plants cannot do so. True xerophytes, such as the creosote bush (*Larrea tridentata*) (Fig. 3.3), are mostly shrubs with small, leathery leaves that are protected from excessive evaporation by a dense cover of hairs or a thick varnish of epidermal resin (Ezcurra et al. 1991). Their adaptive advantage lies in their capacity to extract a fraction of soil water that is not available to other life-forms. However, because their leaves are so small and protected from transpiration by a thick layer of resin, their gas-exchange metabolism is comparatively inefficient after rain pulses when moisture is abundant. In consequence, these species are extremely slow growers but very hardy plants that can bear always a green foliage.

A particular group of desert plants, linked to the true xerophytes, is found in some mosses, ferns, and fern allies, which have the ability to completely desiccate and lose all their moisture without facing the death of the cell, and to recover their cellular activity and re-sprout within a few hours of receiving moisture. The phenomenon, known as *revivescence*, is typically found in desert mosses, in some ferns, and in the "resurrection plants" of the genus *Selaginella* (a fern-ally within the *Licopodium* group), such as *Selaginella lepidophylla*, frequent in the rocky desert scrubs of CCB that looks dry and dead during times of drought but can open their rosette-arranged fronds, and turn bright green and photosynthetically active in a few hours after a good rain.

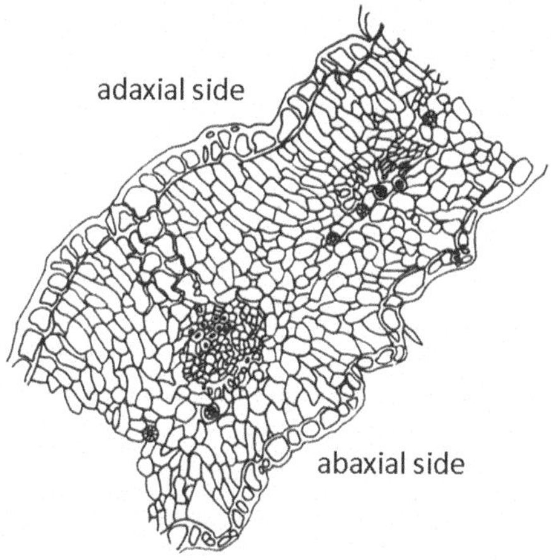

Fig. 3.3 Leaf xeromorphism in *Larrea*. A leaf transversal section of creosote bush or gobernadora (*Larrea tridentata*) leaves shows palisade tissue on both sides of the leaf (isolaterality) as well as the presence of stomata on both the adaxial and the abaxial sides (amphistomaty). Redrawn from Pyykkö (1966)

Finally, one of the most effective drought-survival adaptations for many species is the evolution of an ephemeral life cycle. Selection for a short life and for the capacity to leave behind resistant forms of propagation is perhaps one of the most important evolutionary drivers in most deserts, found not only in plants but also in many invertebrates. Desert ephemerals are extraordinarily rapid growers capable of reproducing at a remarkably high rate during good seasons, leaving behind myriad resistance forms that persist during adverse periods. In CCB, short-lived desert plants in the genus *Gilia* (Polemoniaceae), and various genera in the family Boraginaceae (e.g., *Nama, Phacelia, Tiquilia*) embellish the desert plains with their showy flowers after a good rainy season. Their population numbers simply track environmental bonanzas: Their way to evade critical periods is to die-off, leaving behind immense numbers of propagules (seeds or bulbs in the case of plants, eggs in the case of insects) that will restart the life cycle when conditions ameliorate. These opportunistic species play an immensely important role in the ecological web of deserts: A myriad organisms, like ants, rodents, and birds, survive the dry spells by harvesting and consuming the seeds left behind by the short-lived ephemeral and other perennial plants. Granivory (the consumption of seeds) and not herbivory (the consumption of leaves) is at the base of the food chain in most deserts, as those few plants that maintain leaves during dry spells usually endow them with toxic compounds or protect them with spines. The onset of rainy periods brings to the desert a reproduction frenzy of desert annuals, and a subsequent seed-pulse that drives the entire food web for years (Brown 1979).

Coping with Unpredictability: The Evolution of Serotiny in Deserts

In most deserts rainfall is not only scarce but also highly unpredictable (Noy-Meir 1973). Desert annuals have developed precise evolutionary responses to cope with environmental unpredictability by avoiding taking excessive risk when a single rain falls. These ephemeral plants spend most of their life cycle as seeds, and germinate and grow only when there is available moisture. But a false signal, like a single rain not followed by additional moisture, can drive all the seeds that germinate following that false queue to die before reaching reproductive age. In order to avoid this risk, most desert annuals have evolved fractioned germination, where not all the seeds of a cohort will germinate following a water queue. Some will germinate easily with the first rain, and, if more rains continue, they will have a competitive edge of having taken early advantage of available water. Others will not germinate with a single rain, but may require successive rains to accumulate enough moisture in the environment. These more prudent germinators will not have the adaptive advantage of early sprouters, but will face none of the risks of a false rain cue. For a single plant, producing highly variable seeds, ranging from quick sprouters to extremely cautious germinators will increase its chances of survival in extremely unpredictable deserts (Gutterman 1993; Mulroy and Rundel 1977).

Timing of seed dispersal also allows desert annuals to cope with environmental variability. Many desert plants retain the seeds within the maternal tissue in capsules or dry fruits, and release them gradually into the risky desert environment only when enough moisture softens the seed-retaining structure. Serotiny—the ability to retain seeds in the mother plant—allows plants to reduce risk by retaining seed safely within protected maternal structures and releasing them gradually into the environment as rains arrive (Martínez-Berdeja et al. 2015).

Finally, while plants in more predictable environments have well-defined life history phases (seedling, sapling, vegetative, and reproductive phases), many desert plants start producing flowers shortly after germination and keep on flowering as they grow, until the resumption of dry conditions ends their ephemeral life. If the rainy season is short, they will have produced at least a few fruits, and if the rainy season continues they will opportunistically keep on producing seeds to replenish their seed banks and reinitiate their ephemeral life cycle when future rains arrive.

In Cuatro Ciénegas, plants with woody, gradually-opening capsules, such as agaves and yuccas, a number of species in the Onagraceae (*Calylophus*, *Gaura*, *Oenothera*), or the jimsonweed (*Datura wrightii*, Solanaceae) typically retain seeds in the dry flowering scape for more than one season, and shed them gradually into the environment as the suture of the carpels in the dry capsule become weathered (Fig. 3.4). Similarly, desert families endowed with many-carpelled schizocarps—or

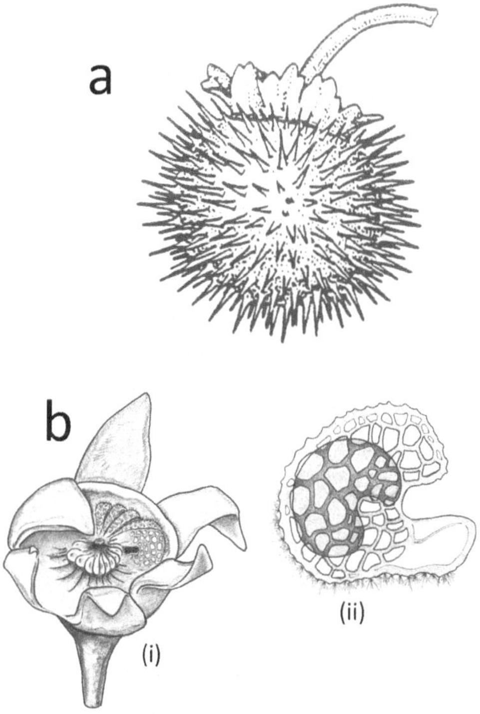

Fig. 3.4 Delayed dispersal (serotiny) in desert plants. (**a**) The lignified capsules of the jimson weed or toloache (*Datura wrightii*) stay attached to the dry mother plant for months, or even years, gradually opening along their carpel sutures and slowly releasing their seeds into the environment. (**b**) The schizocarps (i) of the desert mallows (*Spheralcea* sp.) can gradually release some of their carpels while retaining other carpels attached to the mother plant. Each detached carpel, called a mericarp (ii), contains a single seed inside and functions as the basic dispersal unit. Redrawn from Martínez-Berdeja et al. (2015)

"divided fruits," as in the Malvaceae (*Spheralcea*) and Zygophyllaceae (*Larrea*)—also have fruits that become lignified and gradually detach in parts from the mother plant, shedding seeds continuously for many months or even years. The mechanism of serotinous fruit retention also occurs in some cacti, such as *Grusonia bradtiana* (Rosas Barrera et al. 2020, this volume) and the sunken fruits of the nipple-cacti *Mammillaria* (Peters et al. 2011).

Leaf Xeromorphism: Shifting Photosynthesis to Favorable Hours

Most angiosperms possess leaves with a bifacial or dorsiventral structure. The upper (adaxial) side normally harbors a layer of palisade tissue, formed by a chloroplast-rich parenchyma of tightly-packed columnar cells under the upper epidermis. Between the palisade tissue and the lower (abaxial) epidermis there is a spongy mesophyll, with cells widely separated from each other so that the circulation of CO_2 entering the leaf through the abaxial stomata and diffusing on to the palisade tissue above, is enhanced. In short, most angiosperms show some level of functional specialization in their leaf sides, the upper surface being specialized in the capture of light, and the lower one being specialized in the exchange of gases with the surrounding atmosphere (Smith et al. 1998).

In arid environments, however, it is common to observe plants that have lost the dorsiventral specialization showing instead isolateral leaves with palisade tissue on both sides. Isolaterality in dryland plants is often accompanied by amphistomaty (the presence of stomata in roughly equal density in both sides of the leaf), as well as by a vertical orientation of the leaf laminae and increased leaf thickness (Smith et al. 1998). Vertically-oriented leaves allow desert plants to shift photosynthesis to the early hours of the morning and the late hours of the afternoon, when temperatures are lower and water vapor pressure is higher, allowing the plants to photosynthesize with less water loss and, hence, with a higher water-use efficiency.

Leaf xeromorphism (joint amphistomaty and isolaterality, coupled with thick, resinous or waxy leaves) is very strikingly visible in many dominant desert plants such as desert legumes in the genera *Senna*, *Parkinsonia*, *Hoffmanseggia*, and *Dalea*, or the creosote bush dominant in all North and South American deserts (Pyykkö 1966; Gibson 1998). Indeed, many studies suggest that the xeromorphic leaf anatomy (isolateral, amphistomatic leaves, often thick and mostly vertically-oriented) might be dominant in most drylands (Wood 1932; Mott et al. 1982; Arambarri et al. 2011).

Species Interactions

The harsh conditions of desert ecosystems have promoted the evolution of a complex set of relations among desert organisms, a surprising number of which are positive interactions. Desert shrubs in general and woody legumes in particular, create microhabitats that are critical for the survival of other species. Small animals seek the shade of desert trees and shrubs, birds find refuge and nesting sites in their canopies and many small plants recruit their juveniles under the nitrogen-rich canopy of desert legumes such as acacias, carobs, and mesquites. Because of their CAM metabolism, desert succulents such as agaves and cacti are poor thermoregulators as young seedlings, and cannot survive the harsh ground-level midday temperatures (Martorell and Portilla-Alonso 2020, this volume). For this reason, they can germinate and establish only under the protective shade of shrubby "nurse plants" that act as true cornerstone species in desert conservation (Franco and Nobel 1989). If the desert trees and shrubs are cut, all the accompanying biota soon disappears (Callaway 1995).

Additionally, many desert plants have very specific requirements in terms of their pollinators and seed dispersers. Although some desert ephemerals are truly unspecific in their requirements and produce thousands of seeds, the slow-growing desert perennials are frequently highly specialized in their reproductive habits, and depend strictly on co-evolved animals to help them out in their sexual and reproductive processes. Many cacti and agaves produce sugar-rich nocturnal flowers that engage the pollinating services of nectar-eating bats. Red tubular flowers attract hummingbirds and giant sphinx-moths. The sweet pulp of prickly pears (*Opuntia* spp.) lures birds to disperse their seeds miles away (Mandujano et al. 2010).

Deserts and Agriculture

Because desert ephemerals grow so fast and produce so much seeds in just a few weeks, it comes as no surprise that the earliest archeological records of agriculture come from dryland regions and that the first domesticated crops evolved from desert annuals. Indeed, the first records of cultivated wheat and barley (two dryland ephemerals) come from the Fertile Crescent of the Middle-East some 7000–9000 years ago. In the American Continent, the first agricultural records come from the Tehuacán Valley in southern Mexico, a hot tropical dryland where corn and squash (two annual, drought-tolerant fast growers) were first domesticated. To a large extent deserts have been the cradle of agriculture, and humans have been using desert environments for thousands of years (Cloudsley-Thompson 1979, 1996).

Concluding Remarks

For the untrained eye, deserts look scrubby and poor in biological richness, especially during dry periods. However, because of their evolution in relative geographic isolation, most deserts of the world are rich in rare and endemic species, and are hence highly vulnerable to biological extinction and environmental degradation. In spite of their remarkable convergence in adaptation, all deserts are different in their origin and their evolutionary history (Pipes 1998; Ricciuti 1996).

The incredible variation of the world's deserts in rainfall patterns, continentality, temperature regime, and evolutionary history have all contributed not only to their biological uniqueness, but also to their wondrous wealth of life-forms and adaptations. This adaptive diversity—what Darwin, strongly influenced by deserts himself, called "forms most beautiful and most wonderful"—is what makes deserts so unique. In the hot deserts, we may find giant cacti and trees with mammoth fleshy stems coexisting with some of the toughest hardwoods; ground-creeping succulents side by side with fog-harvesting rosettes, incredibly fast-growing annuals together with some of the hardiest drought-resistant perennials ever known; aromatic shrubs of enticing odors with some of the nastiest, spiniest plants ever. Very few places on Earth contain a richer collection of natural adaptations and such a unique array of evolutionary histories that deserts possess (Davis 1998), and, of these, few are as diverse and rich as Cuatro Ciénegas.

The fragmented evolutionary history of the deserts of the world has been the driving force of their biological rarity, of adaptation to local conditions, of specialization to isolated environments. After millions of years in isolation, the forces of evolution and fragmentation have yielded unique life-forms in each desert, strangely shaped desert plants and extraordinary animals. The world's deserts have been indeed almost biological and cultural islands, lands of fantasy and adventure, habitats of surprising, often bizarre growth forms, and territories of immense natural beauty.

References

Arambarri AM, Novoa MC, Bayón ND et al (2011) Anatomía foliar de arbustos y árboles medicinales de la región chaqueña semiárida de la Argentina. Dominguezia 27:5–24

Ávila-Lovera E, Ezcurra E (2016) Stem-succulent trees from the old and new world tropics. In: Goldstein G, Santiago LS (eds) Tropical tree physiology. Adaptations and responses in a changing environment. Springer, New York, NY, USA, pp 45–65

Betancourt JL, Van Devender TR, Martin PS (eds) (1990) Packrat middens: the last 40,000 years of biotic change. University of Arizona Press, Tucson, AZ, USA, 467 pp

Brown JH (1979) Granivory in desert ecosystems. Annu Rev Ecol Syst 10:201–227

Callaway RM (1995) Positive interactions among plants. Bot Rev 61:306–349

Caso M, González-Abraham C, Ezcurra E (2007) Divergent ecological effects of oceanographic anomalies on terrestrial ecosystems of the Mexican Pacific coast. Proc Natl Acad Sci U S A 104:10530–10535

Cloudsley-Thompson JL (1979) Man and the biology of arid zones. Edward Arnold, London, UK

Cloudsley-Thompson JL (1996) Biotic interactions in arid lands. Springer, Berlin, Heidelberg, Germany

Cochran JK, Landman NH, Turekian KK, Michard A, Schrag DP (2003) Paleoceanography of the late cretaceous (Maastrichtian) Western interior seaway of North America: evidence from Sr and O isotopes. Palaeogeogr Palaeoclimatol Palaeoecol 191:45–64

Cooke R, Warren A, Goudie A (1993) Desert geomorphology. UCL Press, University College London, London, UK

Davis W (1998) Shadows in the sun. Travels to landscapes of spirit and desire. Island Press, Washington, DC, USA

Dimmit MA (2000) Biomes & communities of the Sonoran Desert region. In: Phillips SJ, Comus PW (eds) A natural history of the Sonoran Desert. Arizona-Sonora Desert Museum and University of California Press, Tucson AZ, USA, pp 3–18

De-Nova JA, Aguirre-Liguori JA, Eguiarte LE (2020) How did *Fouquieria* come to the Chihuahuan Desert? Phylogenetic and phylogeographic studies of *Fouquieria shrevei* and *F. splendens*, and the role of vicariance, selection and genetic drift. In: Mandujano MC, Pisanty I, Eguiarte LE (eds.) Plant diversity and Ecology in the Chihuahuan desert. Springer International, Cham, Switzerland

Ezcurra E, Mellink E (2013) Desert ecosystems. In: Levin SA (ed) Encyclopedia of biodiversity, vol 2, 2nd edn. Academic Press, Waltham, MA, USA, pp 457–478

Ezcurra E, Montaña C, Arizaga S (1991) Architecture, light interception, and distribution of *Larrea* species in the Monte Desert, Argentina. Ecology 72:23–34

Ezcurra E, Mellink E, Wehncke E et al (2006) Natural history and evolution of the world's deserts. In: Ezcurra E (ed) Global deserts outlook. United Nations Environment Programme (UNEP), Nairobi, Kenya, pp 1–26

Flores-Vázquez JC, Rosa Barrera MD, Golubov J, Sánchez-Gallen I, Mandujano MC (2020) Ecological importance of Bajadas in the Chihuahuan Desert. In: Mandujano MC, Pisanty I, Eguiarte LE (eds.) Plant diversity and Ecology in the Chihuahuan desert. Springer International, Cham, Switzerland

Franco AC, Nobel PS (1989) Effect of nurse plants on the microhabitat and growth of cacti. J Ecol 77:870–886

Gibson AC (1998) Photosynthetic organs of desert plants. Bioscience 48:911–920

Goudie A, Wilkinson J (1977) The warm desert environment. Cambridge University Press, Cambridge, UK

Gutterman T (1993) Seed germination in desert plants. Adaptations of desert organisms. Springer, Berlin, Germany

Louw GN, Seely MK (1982) Ecology of desert organisms. Longman, London, UK

Mandujano MC, Carrillo-Ángeles I, Martínez-Peralta C, Golubov J (2010) Reproductive biology of Cactaceae. In: Ramawat KG (ed) Desert plants—biology and biotechnology, chap 10. Springer, Berlin, Germany, pp 197–230

Martínez-Berdeja A, Ezcurra E, Sanders AC (2015) Delayed seed dispersal in California deserts. Madrono 62(1):21–32

Martorell C, Portilla-Alonso RM (2020). The ages of life: The changing forms and history of Coryphantha werdermanii throughout its development. In: Mandujano MC, Pisanty I, Eguiarte L (eds.) Plant diversity and Ecology in the Chihuahuan desert. Springer International, Cham, Switzerland

McGinnies WG, Goldman BJ, Paylore P (eds) (1977) Deserts of the world. University of Arizona Press, Tucson, AZ, USA

Morton RR (1979) Diversity of desert-dwelling mammals: a comparison of Australia and North America. J Mammal 60:253–264

Mott KA, Gibson AC, O'Leary JW (1982) The adaptive significance of amphistomatic leaves. Plant Cell Environ 5:455–460

Mulroy WT, Rundel PW (1977) Annual plants: adaptations to desert environments. Bioscience 27:109–114

Noy-Meir I (1973) Desert ecosystems: environment and producers. Annu Rev Ecol Syst 4:25–41

Ochoterena H, Flores-Olvera H, Gómez-Hinostrosa C, Moore MJ (2020) Gypsum and plant species: a marvel of Cuatro Ciénegas and the Chihuahuan Desert. In: Mandujano MC, Pisanty I, Eguiarte LE (eds) Plant diversity and ecology in the Chihuahuan Desert. Springer International, Cham, Switzerland

Peters EM, Martorell C, Ezcurra E (2011) The effects of serotiny and rainfall-cued dispersal on fitness: bet-hedging in the threatened cactus *Mammillaria pectinifera*. Popul Ecol 53:383–392

Pinkava DJ (1984) Vegetation and flora of the bolson of Cuatro Cienegas region, Coahuila, Mexico: IV. Summary, endemism and corrected catalogue. J Ariz Nev Acad Sci 19(1):23–47

Pipes R (1998) Hot deserts (world habitats). Raintree, New York, NY, USA

Plascencia-López L, Rojas-Aréchiga M, Mansujano, MC (2020) Reproductive biology of *Grusonia bradtiana* (Cactaceae): a dominant species and endemic clonal Cactus from Cuatro Ciénegas Basin and contiguous areas in the Chihuahuan Desert. In: Eguiarte L, Mandujano MC, Pisanty I (eds) Plant diversity and ecology in the Chihuahuan Desert. Springer International, New York

Pyykkö M (1966) The leaf anatomy of East Patagonian xeromophic plants. Ann Bot Fenn 3:453–622

Ricciuti ER (1996) Desert (biomes of the world). Benchmark Books, New York, NY, USA

Robichaux RH (ed) (1999) Ecology of Sonoran Desert plants and plant communities. University of Arizona Press, Tucson, AZ, USA

Rosas Barrera MD, Golubov J, Pisanty I, Mandujano MC (2020) Effect of reproductive modes on the population dynamics of an endemic cactus from Cuatro Ciénegas In: Mandujano MC, Pisanty I, Eguiarte LE (eds.) Plant diversity and Ecology in the Chihuahuan desert. Springer International, Cham, Switzerland

Smith WK, Bell DT, Shepherd KA (1998) Associations between leaf structure, orientation, and sunlight exposure in five western Australian communities. Am J Bot 85:56–63

Souza V, Olmedo-Álvarez G, Eguiarte LE (eds) (2018) Cuatro Ciénegas ecology, natural history and microbiology. Vol. I of series Cuatro Ciénegas Basin: an endangered hyperdiverse oasis. Springer, Cham, Switzerland, 134 pp

Wood JG (1932) The physiology of xerophytism in Australian plants. Aust J Exp Biol Med Sci 10:89–95

Chapter 4
Diversity and Distribution of Cacti Species in the Cuatro Ciénegas Basin

José Guadalupe Martínez-Ávalos, Rodolfo Martínez-Gallegos, Antonio Guerra-Pérez, Jorge Ariel Torres Castillo, Christian S. Venegas-Barrera, Ariadna Maía Álvarez-González, and Fortunato Garza Ocañas

Abstract Diversity and geographic distribution of cacti in the Cuatro Ciénegas Basin (CCB), in the state of Coahuila, Northeast Mexico, were assessed using the complementary index (CI) and the geographic expansion index (GEI) in order to define priority zones for conservation from January 2017 to December 2018. The study area was divided into 12 plots, each of an area of approximately 192.93 km^2. For each site, two rectangular plots were defined, each 1 km long and 40 m wide (i.e., a total 8 hectares of sample area per site), where cactus diversity, vegetation and soil types, and climate were recorded. Results showed the presence of 21 genera and 65 taxa including 42% and 12% of the taxa mentioned in the literature for the state of Coahuila, highlighting the richness of this area in the state. *Cylindropuntia leptocaulis* had the widest geographic distribution in CCB with GEI = 1.00, followed by eight taxa, whereas *Echinomastus mariposensis, Echinomastus warnockii,* and *Epithelantha micromeris* subsp. *bokei* had a narrower distribution with GEI = 0.1667 in CCB. Plot 3 (El Churince 1) was considered as a first priority site with a CI = 85% (55 taxa), followed by plot 2 (La Jara) with 8 taxa (CI = 8), and by site 7 San Juan with two additional taxa (CI = 3). Regarding priority sites for conservation, plot 2 (La Jara) had the highest number of threatened taxa (19) and a CI = 76%; this was followed by plot 3 (Churince 1) with 5 taxa endangered,

J. G. Martínez-Ávalos (✉) · A. Guerra-Pérez · J. A. Torres Castillo · A. M. Álvarez-González
Instituto de Ecología Aplicada, Universidad Autónoma de Tamaulipas, Cd. Victoria, Mexico
e-mail: jmartin@uat.edu.mx

R. Martínez-Gallegos
Centro de Investigación en Geografía y Geomática "Ing. Jorge L. Tamayo" (CentroGeo), Mexico City, Mexico

C. S. Venegas-Barrera
Tecnológico Nacional de México/Instituto Tecnológico de Cd. Victoria, Cd. Victoria, Mexico
e-mail: crystian_venegas@itvictoria.edu.mx

F. Garza Ocañas
Facultad de Ciencias Forestales, Universidad Autónoma de Nuevo León, Linares, Mexico

M. C. Mandujano et al. (eds.), *Plant Diversity and Ecology in the Chihuahuan Desert*, Cuatro Ciénegas Basin: An Endangered Hyperdiverse Oasis,
https://doi.org/10.1007/978-3-030-44963-6_4

(CI = 20%) and plot 7 was the last with only 1 taxa (CI = 4%). These plots are located in two vegetation types, the rosetophyllic desertic thornscrub and the microphyllic desertic thornscrub, with a semi-arid desertic climate (BWhw); soils include regosol, calcisol, and leptosol. Due to these environmental characteristics, CCB should be considered a hotspot zone regarding cacti species diversity in northeastern Mexico, and it is important to establish biodiversity conservation actions in the plots in CCB with the highest concentration of cacti species.

Keywords Taxa · Cactaceae · Complementary index · Geographic · Expansion index · Conservation Cuatro Ciénegas · Coahuila · Diversity

Introduction

Cuatro Ciénegas Basin (CCB) in the state of Coahuila, Mexico, is a region characterized by its high diversity and plant and animal endemism (Contreras-Balderas 1984; Villarreal-Quintanilla and Encina-Domínguez 2005), with a notable presence of many aquatic and microbial species (Peimbert et al. 2012; Souza et al. 2004, 2008, 2012). Therefore CCB is considered as very important area within the Chihuahuan Desert Eco region (Medellín 1982; Dinerstein et al. 1999).

The Cactaceae family is well represented in northeastern Mexico (Pinkava 1984; Hernández and Gómez-Hinostrosa 2011, 2015), where cactus grow in arid and semi-arid conditions (Hernández et al. 2004, 2008). A recent study regarding the cacti flora from the state of Coahuila showed that this family is represented by a high diversity that includes 144 species and subspecies (Flores-Valdés 2016). After the pioneer study carried out by Pinkava (1984), there are not many studies analyzing species richness, their geographic affinities, and their distribution patterns of CCB cacti. Nevertheless, CCB stands out as an area with a high diversity of cacti species, and recent explorations are discovering new records of species, increasing the number of known species from this family (Hernández and Bárcenas 1995, 1996; Hernández and Gómez-Hinostrosa 2011, 2015).

Even if the climatic cycles from the Pleistocene might have had an influence on the diversity and geographic distribution of species around the world, there are still unanswered questions regarding how these climatic fluctuations affected the habitat of organisms living in the semi-arid environment (Scheinvar et al. 2020, this volume).

Some researchers argue that during the glacial period, organisms adapted to arid environments were forced to stay in refuge areas from the Sonora and Chihuahua deserts (Van Devender and Spaulding 1979; Thompson and Anderson 2000). Refuge places, such as the CCB, are areas with a stable climate and thus they should be considered as high priority areas for conservation of genetic diversity of the species (Wilson and Pitts 2012; Médail and Diadema 2009; Scheinvar et al. 2020, this volume).

In this chapter, we analyzed the distribution area and floristic elements of the Cactaceae family at CCB to determine the geographic range of the species. Possible relationships of richness, endemism, and the change in the composition of species were studied. A complementarity analysis was made to describe the total number of sampling sites in the CCB area and suggest priority zones for the conservation of cacti in this region was used.

Methods

Floristic Inventory

In order to determine the number of species in the Flora and Fauna Protection Area of Cuatro Ciénegas, we divided into 12 plots of 7 min by 30 s of lat/long (.192.93 km^2), establishing in each one two transects of 1000 m × 40 m (Fig. 4.1). With the number of species recorded along the transects, a binary matrix (presence/absence) of *n* species in each plot (24 in total) was built, and with this data the floristic contingent was analyzed. The nomenclature used for vegetation types proposed by Villarreal-Quintanilla and Encina-Domínguez (2005) was followed, whereas for the Cactaceae family we used the one proposed by Guzmán et al. (2003) and Anderson (2001).

Cacti Diversity

The Simpson Diversity Index was used to calculate cacti diversity per plot in the CCB, as this index, based on the abundance of species per site, (Magurran 2004), is frequently used for ecological studies focused on species diversity from a geographic location. The index is calculated with the following formula:

$$D = \frac{1 - \Sigma n(n-1)}{N(N-1)}$$

where *n* is the number of individuals of one species
N = the total number of all individuals

Geographic Expansion Index (GEI)

The GEI was used in order to make the quantitative estimate of cacti species from the Flora and Fauna Protection Area of Cuatro Ciénegas. This methodology has been reported in many studies but particularly in those by Hernández and Bárcenas (1995, 1996), Gómez-Hinostrosa and Hernández (2000), Martínez-Ávalos and Jurado (2005), Santa Anna del Conde Juárez et al. (2009), and Hernández-Magaña

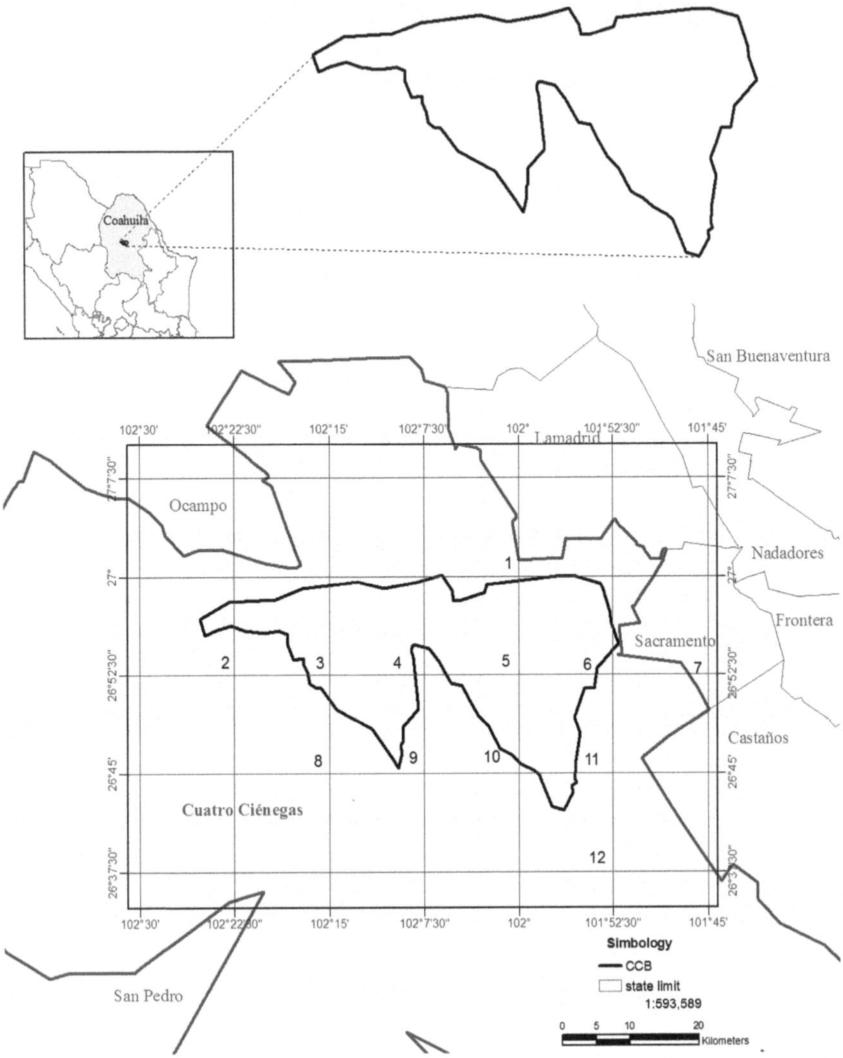

Fig. 4.1 Geographic location of the 12 plots used for the floristic inventory of cacti in the CCB, Coahuila, Mexico

et al. (2012). For this analysis, we used the same number of sample plots (12), in each, we traced two sampling sites giving a total of 24 sites studied in an area of eight hectares (80,000 m^2).

All records obtained from this study were integrated in the database of cacti at the Instituto de Ecología Aplicada, Universidad Autónoma de Tamaulipas. In order to complete the presence of other species at the study zone, an exhaustive review of the databases from other national and international herbaria (i.e., Texas University in Austin (UT), New York Botanical Garden (NYBG), and the Missouri Botanical Garden (MO)) was carried out.

Presence or absence of species in all the quadrats was confirmed. The GEI was measured following the formula proposed by Gómez-Hinostrosa and Hernández (2000).

$$\text{GEI} = \text{Ss} / \text{Sm}$$

where Ss = plot where each taxa was found

Sm = plot with the higher distribution range of the most common taxa.

Cylindropuntia leptocaulis proved to be the most frequent species, since it was registered in the 12 plots; this was used as reference to obtain the GEI of the other taxa. The total number of plots where each taxa was recorded was divided by the total number of plots where the most frequent taxa was found (i.e., *Cylindropuntia leptocaulis*).

According to these criteria, some taxa with a restricted distribution, such as *Echinomastus mariposensis* were recorded with a GEI of 0.500. *Cylindropuntia leptocaulis* showed a GEI = 1.00 as it was found in all the plots (12 plots).

Complementary Index (CI)

The complementary index has been used in other studies to determine priority sites for conservation due to their high diversity of species of plants and animals. Thus, it has been used for the conservation of flora and fauna of specific areas, as it allows the recognition of priorities to optimize the conservation of biodiversity (Humphires et al. 1991; Pressey et al. 1993). It has been used to identify areas with a higher number of cacti species (Goettsch et al. 2015; Hernández and Bárcenas 1995, 1996; Gómez-Hinostrosa and Hernández 2000; Martínez-Ávalos and Jurado 2005; Santa Anna del Conde Juárez et al. 2009; Hernández-Magaña et al. 2012). Based on the same analysis, other areas are considered as second and third priority due to the fact that they have a low biodiversity.

This methodology was used in the 12 plots of the CCB area and the values obtained from the complementary index were calculated according to the following formula:

$$\text{CI} = \text{CR} \times 100 / \text{CO}$$

where CI is the complementary index, corresponding to the total number of taxa for the plots.

CR is called complement residue and corresponds to the number of taxa not found in the first priority plots. CO is considered as the complement, i.e., the percentage of taxa corresponding to the number of unique taxa found in each priority quadrant.

This index is very similar to the Simpson diversity index since it considers the species richness in a given site (Magurran 2004).

For instance, plot 3 (El Churince 1) was considered as the first priority at the CCB, for it had the highest number of taxa (55), and the higher CI values (IC = 85%), followed by plot 2 (La Jara), considered the second priority, with eight additional taxa and with an IC value of 12%. The next eight additional taxa were then added to the first priority plot, giving as a result a new larger, composite plot, with the highest number of additional taxa. This is repeated until no more taxa are added. For this study, it was only possible to group all cacti species (threatened and no-threatened) into three plots of conservation priority.

Results and Discussion

Diversity and Richness

A total of 21 genera and 65 cacti taxa (including species or subspecies) were determined for the CCB region. *Escobaria* and *Mammillaria* were the genera with the higher richness, with eight taxa each, and *Coryphantha* and *Echinocereus* followed them with 6 taxa each. There were eight genera with low numbers of taxa, and *Ansistrocactus* and *Ariocarpus* had only one species (Fig. 4.2). Twenty-five taxa were reported as endangered according to the Mexican national federal law norm NOM 059 SEMARNAT (2010), representing 38% of the cacti flora in the CCB. From the total of taxa considered in danger of becoming extinct, nine of them are endemic to the study area.

The rosetophyllic desertic thornscrub (Rdt) is the vegetation type with the higher number of cacti taxa (63), followed by the microphyll desert thornscrub, with 53 taxa. Some studies have shown that in arid and semi-arid ecosystems of northern Mexico, the variety of soil types and the climatological diversity of a site promote the establishment of different types of vegetation, as well as an increase in plant diversity (Maya et al. 2002; Arriaga et al. 2000). This could explain the high diversity of cacti found in these sites with different types of climate, vegetation, and soil in the CCB. Table 4.1 shows that from the 12 plots studied, plot 3 (El Churince 1) had the higher diversity values ($D = 0.9818$), followed by plot 2 (La Jara) and plot 7 (San Juan) with $D = 0.9808$ and $D = 09808$, respectively.

Geographic Distribution Pattern

According to the results of this study, the CCB is considered the most important region in diversity and richness of cacti in the state of Coahuila because here we can find 42% (46) of the genera and 46% (65) of the cacti taxa reported for the state

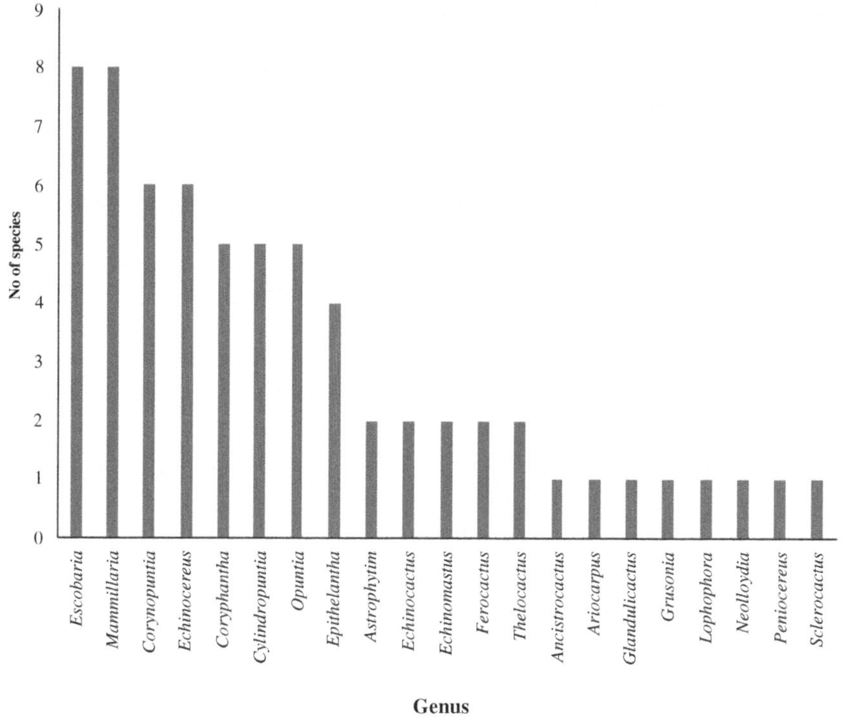

Fig. 4.2 Number of cacti species per genus in the CCB, Coahuila, Mexico

Table 4.1 Simpson Diversity Index (*D*) which shows that El Churince 1 (in bold) is the site with the greatest diversity of cacti in the 12 plots sampled at the CCB. *N* = the total number of all individuals

Plots	Name	*N*	*D*
1	Las Antenas	38	0.9737
2	La Jara	52	0.9808
3	**El Churince 1**	55	**0.9818**
4	El Carrizal	34	0.9706
5	Cuatro Ciénegas	21	0.9524
6	Tecla Santa	40	0.975
7	San Juan	52	0.9808
8	Sacramento	38	0.9737
9	El Chiquero	48	0.9792
10	La Becerra	29	0.9655
11	Altamira	32	0.9688
12	La Colorada	30	0.9667

(Fig. 4.2). From the 12 study plots, plot 3 (El Churince 1) is ranked as the most important and this plot contains the highest number of taxa (51), followed by plots 2 (La Jara) and 7 (San Juan) with 50 and 48 taxa, respectively, following the previously described diversity estimates.

According to other studies (Contreras-Balderas 1984; Pinkava 1984; Hernández and Gómez-Hinostrosa 2011, 2015; Villarreal-Quintanilla and Encina-Domínguez 2005), the CCB is considered a region with a high endemism of flora and fauna species, due to its climatic conditions. This could explain the high diversity of species presented in the first two plots, as they have a semi-arid climate with summer rains (BWhw), considered one of the driest in the study area (INEGI 2013a). In addition, these sites present different soil types including calcisol, leptosol, and regosol, as well as different vegetation types, including rosethophilous desert scrub (Rdt), microphyllous desert thornscrub (Mdt), halophyte vegetation (Hv), and crassicaule scrub (Ct), among others (INEGI 2013b) (Table 4.2).

Geographic Expansion Index (GEI)

This index is an important tool used for conservation of species in their distribution area (Arita et al. 1997; Hernández et al. 2010). This index uses the median value ($M = 0.5833$) to group the species in two different categories: those with a wide distribution above the median and those with a restricted distribution below the median (Gómez-Hinostrosa and Hernández 2000; Martínez-Ávalos and Jurado 2005). From the 65 taxa recorded in the CCB, about 51% have a restricted distribution, while the 49% had a wide distribution. Among some of the cacti with a very restricted distribution in CCB we have: *Echinomastus mariposensis*, *E. warnockii,* and *Epithelantha micromeris* subsp. *bokei*, all of them with a GEI = 0.1667. In contrast, *Cylindropuntia leptocaulis* had the widest distribution range, with a GEI = 1.000 followed by *Corynopuntia schottii,* seven taxa with a GEI = 0.9167, and 14 taxa with GEI around 0.8, highlighting the old man cactus, *Grusonia bradtiana*, a dominant species of *bajadas* (Fig. 4.3; Flores Vázquez et al. 2020, this volume; Plasencia López et al. 2020, this volume; Rosas Barrera et al. 2020, this volume).

Complementary Index

The complementary index (CI) for the 12 plots shows that plot 3 (El Churince 1) is the first priority site, with an IC = 85% and a residual value of 55 taxa. This means that 85% of the taxa are located in this site of the CCB. Plots 2 and 7 are the second and third priority, with eight and two taxa, and with residual value of CI = 12% and CI = 3%, respectively. Regarding the endangered taxa, plot 2 (La Jara) is the first priority (CI = 76% and a residual value of 19 taxa),

Table 4.2 Richness of cacti species and climate types, soils, and vegetation present at 12 analyzed at the CCB, Coahuila, Mexico (see text for details)

Plots	Name	Richness (number of species)	Climate	Soils	Vegetation
1	Las Antenas	36	BSOkw	Leptosol	Tc
2	La Jara	50	BWhw	Regosol, Leptosol	Tc, Mdt
3	El Churince 1	51	BWhw	Arenosol, Regosol, Leptosol, Gleysol, Calcisol, Cambisol, Solonchak	Mdr, Rdt, Hv, Mc
4	El Carrizal	32	BWhw	Solonchak, Leptosol, Arenosol	Tc, Mdt, Hv, Gv
5	Cuatro Ciénegas	21	BWhw, BWhw′, BW(h′)hw	Solonchak, Gypsisol	M, Mdt, Hv, Hg
6	Tecla Santa	38	BWhw′, BW(h′)hw	Solonchak, Gypsisol, Calcisol	M, Gv
7	San Juan	48	BWhw′	Calcisol	M
8	Sacramento	36	BWhx	Regosol, Gleysol	Mdr
9	El Chiquero	44	BWhw	Regosol, Leptosol, Gleysol, Calcisol, Cambisol, Solonchak, Gypsisol	Tc, Mdt, Rdt, Ig, M, Hv, Gv
10	La Becerra	27	BSOkw′, BW(h′)hw	Regosol, Luvisol, Calcisol, Solonchak, Gypsisol	M Mdt, Hv, Hg
11	Altamira	31	BWhw′, BW(h′)hw	Calcisol, Solonchak, Gypsisol	Rdt
12	La Colorada	27	BWhw′, BW(h′)hw	Calcisol	Mdt, Hg, M, Hv

Simbology: Climate: **BSOkw** (Dry steparian climate with rains in the summer); **BWhw** (Semi-arid desert climate with rains in summer); **BWhw′** (Calid desert climate with scarce rains all year around), **BW(h′)hw** (Desert semi-arid climate with summer rains). Vegetation: Crasscaule thorn-scrub (**Ct**); Microphyllous desert thornscrub (**Mdt**); Haolphytic vegetation (**Hv**); Rosetophyllic desert thornscrub (**Rdt**); Mezquital (**M**); Halophyllic grassland (**Hg**); Gypsophillic Vegetation (**Gv**); Induced grassland (**Ig**)

followed by the second and third priority plot 3 (El Churince 1) and plot 7 (San Juan) with a CI = 20% and 4%, respectively (Table 4.3). These three plots were also the ones with the highest diversity and richness of taxa, as indicated above (Table 4.3).

The establishment of high priority plots for conservation could be fundamental for the conservation of species from a determined area (Arita et al. 1997; Koleff 2009; Razola et al. 2006; Suarez-Mota and Téllez-Valdés 2014). Likewise, those taxa with a restricted distribution should be integrated to the list of threatened species from Mexico. Williams et al. (1991) suggest that genetic characteristics of the species should also be considered for the conservation of priority sites, as they could

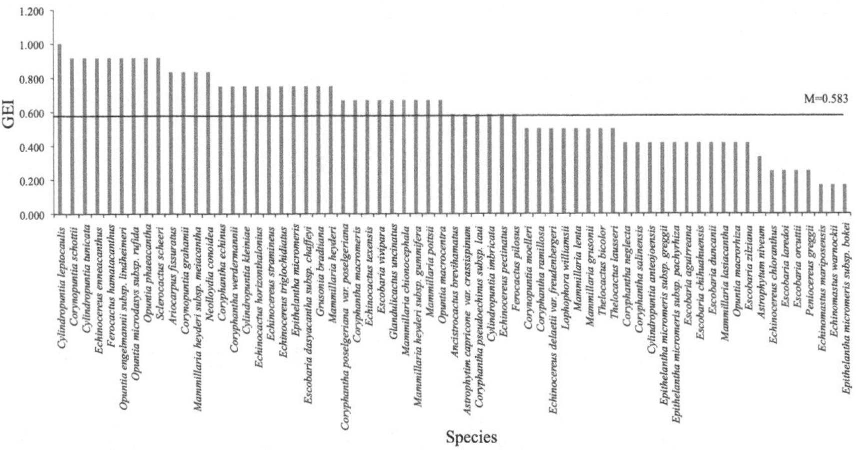

Fig. 4.3 Geographic Expansion Index calculated for cacti species from the CCB Coahuila, Mexico

Table 4.3 Priority sites for conservation of cacti estimated with the Complementary Index (see text for details), in the CCB, Coahuila, Mexico

Priority	Name	Plot	No. of unique species	Complementarity residual value	Complementarity value (%)
First	El Churince 1	3	55	55	85
Second	La Jara	2	8	8	12
Third	San Juan	7	2	2	3
Species in danger					
First	La Jara	2	19	19	76
Second	El Churince 1	3	5	5	20
Third	San Juan	7	1	1	4

help to understand hierarchical knowledge through cladistics and other phylogenetic based analyses for conservation. Until recently, knowledge of cactus phylogenetic relationships was limited. Today, different phylogenetic studies for cacti are available (Arias et al. 2003; Nyffeler 2002; Vázquez-Sánchez et al. 2013; Hernández-Hernández et al. 2011) and they might, together with all other studies, contribute to the conservation of the species. The results of this study can generate information focused on the conservation of already established protected natural areas, such as the Natural Protected Area for the Protection of Flora and Fauna in Cuatro Ciénegas in the state of Coahuila, Mexico.

Conclusions

In Coahuila, CCB can be considered as the most important region for cacti, as it houses 42% of the genera (21) and 43% of the total of taxa of cacti (65 taxa) from the state. *Escobaria* and *Coryphanta* are the more diverse taxa in CCB, with eight species each. Plots 3 (El Churince 1), 2 (La Jara), and 7 (San Juan) stand out as the most important priority sites for conservation of biodiversity of cacti within the area, with more than 90% of their taxa either common or endangered in the CCB. Also, the microphyll and rosetophyllic desert thornscrubs are the most important vegetation types for the diversity of cacti in CCB and they should be specially considered for the conservation of this family. Diversity of climatic and edaphic conditions seem to be partially responsible for the existence of the high diversity of cacti in this area of Mexico.

Acknowledgements To the Cactus and Succulent Society of America (CSSA) for the financial support to the project: Cacti diversity and conservation strategies for Northeast Mexico. To the National Commission for Natural Protected Areas "Comisión Nacional de Áreas Naturales Protegidas" (CONAMP) and Presidencia Municipal de Cuatro Ciénegas, for all their support to carry out this study.

References

Anderson, EF (2001) The Cactus Family. Timber Press, Portland, OR, USA,776 pp.
Arias S, Terrazas T, Cameron K (2003) Phylogenetic analysis of *Pachycereus* (Cactaceae, Pachycereeae) based on chloroplast and nuclear DNA sequences. Syst Bot 28(3):547–557
Arita H, Figueroa F, Frisch A, Rodríguez P, Santos-Del Prado K (1997) Geographical range size and the conservation of Mexican mammals. Conserv Biol 11:92–100
Arriaga L, Espinoza JM, Aguilar C, et al (coord) (2000) Regiones terrestres prioritarias de México. Comisión Nacional para el Conocimiento y uso de la Biodiversidad. México, DF, Mexico, 810 pp.
Contreras-Balderas S (1984) Environmental impacts in Cuatro Cienegas, Coahuila, Mexico: A Commentary. J Ariz Nev Acad Sci 19:85–88
Dinerstein E, Olson D, Atchley J, Loucks C, Contreras-Balderas S, Abell R, Iñigo E, Enkerlin E, Williams C E, Castilleja G (eds) (1999) Eco region-based conservation in the Chihuahuan Desert: A biological assessment and biodiversity vision. World Wild Fund, Comisión Nacional para el Conocimiento y uso de la Biodiversidad, Pronatura A.C. e Instituto Tecnológico y de Estudios Superiores de Monterrey. Washington, DC, USA, 489 pp
Flores-Valdés A (2016) Nueva guía de Cactáceas del Estado de Coahuila. Secretaría de Medio Ambiente de Coahuila, Gobierno del Estado de Coahuila. Edit. Quintanilla Ediciones Primera Edición. Saltillo, Coahuila, Mexico, 222 pp
Flores-Vázquez JC, Rosas Barrera MD, Golubov J, Sánchez-Gallen I, Mandujano MC (2020) Ecological importance of Bajadas in the Chihuahuan Desert. In: Mandujano MC, Pisanty I, Eguiarte LE (eds) Plant diversity and Ecology in the Chihuahuan desert. Springer International, Cham, Switzerland
Goettsch B, Hilton C, Cruz G, et al (2015) High proportion of cactus species threatened with extinction. Nat Plants 142:1–7

Gómez-Hinostrosa C, Hernández HM (2000) Diversity, geographical distribution, and conservation of Cactaceae in the Mier y Noriega region, Mexico. Biodivers Conserv 9:403–418
Guzmán U, Arias S, Dávila P (2003) Catálogo de cactáceas mexicanas. Universidad Nacional Autónoma de México y Comisión Nacional para el Conocimiento y Uso de la Biodiversidad, México, DF, Mexico
Hernández HM, Bárcenas RB (1995) Endangered cacti in the Chihuahuan Desert: I. Distribution patterns. Conserv Biol 9(5):1176–1188
Hernández HM, Bárcenas RT (1996) Endangered cacti in the Chihuahuan Desert: II. Biogeography and conservation. Conserv Biol 10(4):1200–1209
Hernández HM, Gómez-Hinostrosa C (2011) Areas of endemism of Cactaceae and the effectiveness of the protected area network in the Chihuahuan Desert. Oryx 45(2):191–200
Hernández HM, Gómez-Hinostrosa C (2015) Mapping the Cacti of Mexico. Part II *Mammillaria*. CONABIO, DHBooks. Milborne Port, UK, 189 pp
Hernández HM, Gómez-Hinostrosa C, Goettsch B (2004) Check list of Chihuahuan Desert Cactaceae. Harv Pap Bot 9:51–68
Hernández HM, Goettsch B, Gómez-Hinostrosa C, Arita HT (2008) Cactus species turnover and diversity along a latitudinal transect in the Chihuahuan Desert region. Biodivers Conserv 17:703–720
Hernández H, Gómez-Hinostrosa C, Hoffmann G (2010) Is geographical rarity frequent among the cacti of the Chihuahuan Desert? Revista Mexicana de Biodiversidad 81:163–175
Hernández-Hernández T, Hernández HM, De-Nova JA, Puente R, Eguiarte LE, Magallón S (2011) Phylogenetic relationships and evolution of growth form in cactaceae (Caryophyllales, Eudicotyledoneae). Am J Bot 98(1):44–61
Hernández-Magaña R, Hernández Oria JG, Chávez R (2012) Datos para la conservación florística en función de la amplitud geográfica de las especies en el semidesierto Queretano, México. Acta Bot Mex 99:105–140
Humphires C, Vane-Wright D, Willams P (1991) Biodiversity reserves: Setting new priorities for the conservation of wildlife. Parks 2:34–38
INEGI (2013a) Cartografía digital de clima. Escala 1:250 000. Instituto Nacional de Estadística y Geografía. [En línea] [Consultado: martes 29 abril 2019]. Disponible en: http://gaia.inegi.org.mx/mdm5/viewer.html
INEGI (2013b) Cartografía digital de vegetación. Escala 1:250 000. Instituto Nacional de Estadística y Geografía. [En línea] [Consultado: martes 29 abril 2019]. Disponible en: http://gaia.inegi.org.mx/mdm5/viewer.html
Koleff P (2009) Identificación de prioridades y análisis de vacíos y omisiones en la conservación de la biodiversidad de México. In: Drizo R, González R, March IJ (eds) Capital Natural de México. CONABIO, México, DF, Mexico, pp 651–718
Magurran AE (2004) Measuring biological diversity. Blackwell Publishing, Sheffield, UK.
Martínez-Ávalos JG, Jurado E (2005) Geographic distribution and conservation of Cactaceae from Tamaulipas Mexico. Biodivers Conserv 14:2483–2506
Maya Y, López-Cortés A, Soeldner A (2002) Cyanobacterial microbiotic crusts in eroded soils of a tropical dry forest in the Baja California Peninsula, Mexico. Geomicrobiol J 19:515–518
Médail F, Diadema K (2009) Glacial refugia influence plant diversity patterns in the Mediterranean Basin. J Biogeogr 36:1333–1345
Medellín F (1982) The Chihuahuan Desert. In: Bender GL (ed) Reference handbook on the deserts of North America. Greenwood Press, Westport, CT, USA, pp 321–381
Nyffeler R (2002) Phylogenetic relationships in the cactus family (Cactaceae) based on evidence from trnK / matK and trnLtrnF sequences. Am J Bot 89:312–326
Peimbert M, Alcaraz LD, Bonilla-Rosso G, Olmedo-Álvarez G, García-Oliva F, Segovia L, Eguiarte LE, Souza V (2012) Comparative Metagenomics of two microbial Mats at Cuatro Ciénegas Basin I: Ancient lessons on how to cope with an environment under severe nutrient stress. Astrobiology 12(7):648–658
Pinkava DJ (1984) Vegetation and Flora of the Bolsón of Cuatro Ciénegas region, Coahuila, Mexico: IV. Summary, endemism and corrected catalogue. J Ariz Nev Acad Sci 19(1):23–47

Plasencia-López L, Rojas-Aréchiga M, Mandujano MC (2020) Reproductive Biology of *Grusonia bradtiana* (Cactaceae): a dominant species and endemic clonal Cactus from Cuatro Ciénegas Basin and contiguous areas in the Chihuahuan Desert. In: Mandujano MC, Pisanty I, Eguiarte LE (eds) Plant diversity and Ecology in the Chihuahuan desert. Springer International, Cham, Switzerland

Pressey R, Humphries C, Margules C, Van-Wright R, Williams P (1993) Beyond opportunism: key principles for systematic reserve selection. Trends Ecol Evol 8:124–128

Razola I, Rey-Benayas JM, de la Montaña E, Cayuela L (2006) Selección de áreas relevantes para la conservación de la biodiversidad. Ecosistemas 15(2):34–41

Rosas Barrera MD, Golubov J, Pisanty I, Mandujano MC (2020) Effect of reproductive modes on the population dynamics of an endemic cactus from Cuatro Ciénegas In: Mandujano MC, Pisanty I, Eguiarte LE (eds) Plant diversity and Ecology in the Chihuahuan desert. Springer International, Cham, Switzerland

Santa Anna del Conde Juárez H, Contreras-Medina R, Luna-Vega I (2009) Biogeographic analysis of endemic cacti of the Sierra Madre Oriental, Mexico. Biol J Linn Soc 97:373–389

Scheinvar E, Gámez N, Moreno-Letelier A, Aguire-Planter E, Eguiarte LE (2020). Phylogeography of the Chihuahuan Desert: diversification and evolution over the Pleistocene. In: Mandujano MC, Pisanty I, Eguiarte LE (eds) Plant diversity and Ecology in the Chihuahuan desert. Springer International, Cham, Switzerland

SEMARNAT (Secretaría del Medio Ambiente y Recursos Naturales) (2010) Norma Oficial Mexicana NOM-059-SEMARNAT-2010, Protección ambiental – Especies nativas de México de flora y fauna silvestres – Categorías de riesgo y especificaciones para su inclusión, exclusión o cambio – Lista de especies en riesgo. Diario Oficial de la Federación. 30 de diciembre de 2010, Segunda Sección, México, DF, Mexico

Souza V, Escalante A, Espinoza L, Valera A, Cruz A, Eguiarte LE, García PF, Elser J (2004) El Valle de Cuatro Ciénagas, un Laboratorio natural de astrobiología. Ciencias 75:4–12

Souza V, Eguiarte LE, Siefert J, James J, Elser J (2008) Microbial endemism: Does phosphorus limitation enhance speciation? Nat Rev Microbiol 6(7):559–564

Souza V, Siefert JL, Escalante AE, Elser JJ, Eguiarte LE (2012) The Cuatro Ciénegas Basin in Coahuila, Mexico: An Astrobiological Precambrian Park. Astrobiology 12(7):641–647

Suárez-Mota ME, Téllez-Valdés O (2014) Red de áreas prioritarias para la conservación de la biodiversidad del eje volcánico transmexicano analizando su riqueza florística y variabilidad climática. Polibotánica 38:67–93

Thompson JN, Anderson KH (2000) Biomes of western North America at 18 000, 6000 and 0 14C yr BP reconstructed from pollen and packrat midden data. J Biogeogr 27:555–584

Van Devender TR, Spaulding WG (1979) Development of vegetation and climate in the Southwestern United States. Science 204:701–710

Vázquez-Sánchez M, Terrazas T, Arias S, Ochoterena H (2013) Molecular phylogeny, origin and taxonomic implications of the tribe Cacteae (Cactaceae). Syst Biodivers 11(1):103–116

Villarreal-Quintanilla JA, Encina-Domínguez JA (2005) Plantas vasculares endémicas de Coahuila y algunas áreas adyacentes, México. Acta Botánica Mexicana 70:1–46

Williams PH, Humphries CJ, Vane-Wright RI (1991) Measuring biodiversity: taxonomic relatedness for conservation priorities. Aust Syst Bot 4:665–667

Wilson JS, Pitts J (2012) Identifying Pleistocene refugia in North American cold deserts using phylogeographic analyses and ecological niche modelling. Divers Distrib 18(11):1139–1152

Chapter 5
Reproductive Biology of *Grusonia bradtiana* (Cactaceae): A Dominant Species and Endemic Clonal Cactus from Cuatro Ciénegas Basin and Contiguous Areas in the Chihuahuan Desert

Lucía Plasencia-López, Mariana Rojas-Aréchiga, and María C. Mandujano

Abstract *Grusonia bradtiana* "viejito" (old man cactus) is an endemic species from Cuatro Ciénegas Basin and nearby areas. *Grusonia* includes 17 clonal species distributed along North American deserts which grow in dense cushion or shrubs. *Grusonia bradtiana* reproduces sexually by flowering and fruiting, forming seeds with new genetic combinations (new genets). The species clones by fragmentation of stems of different sizes that root independently, producing genetically identical offspring (ramets). Clonal species develop complex reproductive interactions as pollination output depends on pollen transfer between genetically different genets or identical ramets. The hypothesis is that clonality negatively affects sexual reproduction as floral traits are adaptations to promote cross-pollination (among genets) and have evolved to reduce negative effects of inbreeding. We studied the reproductive biology of *Grusonia bradtiana* and assessed the effect of clonality upon its reproductive success with controlled pollination. We also determined the frequency and taxonomic identity of floral visitors, to assess the pollination syndrome. Flowering occurs once a year during spring. Flowers are diurnal with a life span of 8 h; they are yellow, with radial symmetry, yellow-white lobulated stigma, and produce nectar. Flowers have thigmonastic stamens with red filaments supporting anthers that contain high amount of viable pollen. The flower is perfect, there is no separation of sexual functions in time (dichogamy), but there is in space (herkogamy), attributes that allow selfing and may reduce sexual interference, respectively. The fruit is dry, possibly a trait unique to *Grusonia*. Pollinators are solitary

L. Plasencia-López
Comisión Nacional para el Conocimiento y Uso de la Biodiversidad, CONABIO, Mexico City, Mexico

M. Rojas-Aréchiga · M. C. Mandujano (✉)
Departamento de Ecología de la Biodiversidad, Instituto de Ecología, Universidad Nacional Autónoma de México, UNAM, Mexico City, Mexico
e-mail: mcmandujano@iecologia.unam.mx

M. C. Mandujano et al. (eds.), *Plant Diversity and Ecology in the Chihuahuan Desert*, Cuatro Ciénegas Basin: An Endangered Hyperdiverse Oasis,
https://doi.org/10.1007/978-3-030-44963-6_5

bees (*Diadasia* and *Melissodes*), a melittophily pollination syndrome. The species require the pollinator services to set fruit and seeds, and it suffers inbreeding depression if self-pollinated, so pollination among ramets decreases seed set, representing a cost of clonality. The selfing rate is high when plants are big or if clonality is frequent, as pollinators tend to visit nearby flowers increasing geitonogamy. The species has a mixed mating system with an outcrossing tendency, where delayed selfing (autogamy and geitonogamy) is a mechanism that ensure reproduction when outcrossing fails.

Keywords Autogamy · Cactaceae · Clonality · Geitonogamy · Pollination

Reproduction in Plants

Several species that inhabit harsh environments display facultative reproduction whereby new offspring are recruited either by sexual and or by clonal means (Harper 1977; Abrahamson 1980). In clonal species, sexual reproduction promotes the production of genetically distinct seeds that are usually capable of long distance dispersal, which potentially will establish genetically new individuals (i.e., genets; Harper 1977; Harder and Barrett 1995), while clonality produces offspring that are an exact genetic copy of the parent plant (i.e., ramets; Harper 1977), usually with limited dispersal (García-Morales et al. 2018), which can be susceptible to diseases or changes in environmental conditions (Harper 1977).

Among angiosperms, a reproductive or sexual system is constituted by the breeding, mating, and pollination systems (Barrett 2013); the first refers to the arrangement in space and time of reproductive structures, while the mating system refers to how and with whom plants mate (Holsinger 2000). Reproductive success of most flowering plants depends on pollination mediated by animals or by abiotic vectors, including bees, mammals, birds, water, and wind. In species pollinated by animals, the frequency of selfing (i.e., autogamy—self-pollination of an hermaphrodite flower and geitonogamy—pollination among flowers of the same plant, de Jong et al. 1993), outcrossing, and final sexual reproductive output are usually dependent on flower phenology (Goulson 2000; Ishii and Sakai 2002), floral displays (Harder and Barrett 1995), floral rewards (Golubov et al. 1999), and resource limitation (Piña et al. 2007) that affect pollinator foraging behavior and in this way determine the amount and destiny of transported pollen (e.g., Snow et al. 1996; Eckert 2000). Hence, pollinator mediated selection of reproductive traits is often considered the main factor that drives evolution of sexual systems (Barrett 2013). Furthermore, diversification in lineages and life history traits in some angiosperm's families as Agavaceae, Orchidaceae, and Cactaceae are attributed to their reproductive versatility (Mandujano et al. 2010; Barrett 2013).

Plants tend to produce much more flowers than fruits (Stephenson 1981), but some species in Cactaceae have high fruit set - frequency of a hermaphrodite or pistillate flowers to become fruit (ca. 50–98%; Mandujano et al. 1996, 2010) in comparison with other angiosperms (ca. >0–35%; Sutherland 1987), which has been related to the constancy and behavior of pollinators (Mandujano et al. 1996; Pimienta-Barrios and Del Castillo 2002), but may also be due to a large store and availability of resources in these species. For example, small to medium sized solitary bees are usually the main pollinators of several species of Opuntioideae subfamily (Cactaceae) (Grant and Grant 1979; Grant et al. 1979; Ordway 1987; Del Castillo and González-Espinosa 1988; McFarland et al. 1989; Mandujano et al. 1996), and of these bees, the genus *Diadasia* is one of the most important because of its behavior, size, and abundance (Mandujano et al. 2010). These solitary bees can move from 50 m up to 200 m from their nesting areas (Schlising 1972; Neff et al. 1982; Piña et al. 2007) which is assumed as the distance of pollen transfer (Piña et al. 2007).

The flowers of Opuntioideae species are diurnal and floral life span is from one to 3 days; they open early in the morning and close in the afternoon (Trujillo-Argueta and González-Espinosa 1991; Mandujano et al. 1996; Pimienta-Barrios and Del Castillo 2002). Floral presence is usually restricted to spring, before the rainy season (Grant and Grant 1979; Grant et al. 1979; McFarland et al. 1989; Mandujano et al. 1996). Most Opuntioideae species are hermaphrodite with perfect functional flowers, and dichogamy is absent (Reyes-Agüero et al. 2006). Even when pollen is released before the stigma becomes receptive, several species seem to be self-compatible with a mixed mating system where fruiting relays on pollinators (Mandujano et al. 2010).

All Opuntioideae species produce clonal offspring through some mechanism: agamospermy (seeds), stems (joints—cylindrical stems or cladodes—flattened stems), roots, or plantlets (Mandujano et al. 1996; Palleiro et al. 2006; Carrillo-Ángeles et al. 2011), and clonality is a common strategy to establish crops, for example, several prickly pear species with economic importance are cultivated by planting artificially selected cladodes (Reyes-Agüero et al. 2006); and in the wild, there are extreme cases in which the species sustain the population by clonality as sexual reproduction no longer exist (Grant and Grant 1980), but is more common that clonal species have populations with intermingled individuals of sexual and clonal origin that generate complex clonal architectures at variable scales (Carrillo-Ángeles et al. 2011; García-Morales et al. 2018).

Clonality affects population size, effective population size, density, clonal architecture, and sexual reproduction (Eckert 2000; Charpentier 2002; Carrillo-Ángeles et al. 2011; García-Morales et al. 2018). Furthermore, it is proposed that geitonogamy will increase and reduce plant fitness (Handel 1985; Eckert 2000; Charpentier 2002). Sexual reproduction is affected by clonality as transfer of pollen occurs at different scales: within a flower (intra-flower self-pollination), between flowers of the same plant (geitonogamy), between flowers of spatially independent plants that share the same genotype— ramets (inter-ramet geitonogamy) or among flowers of genetically different plants (cross pollen, outcrossing between genets) (Handel 1985; Mandujano et al. 1996; Eckert 2000). It is expected that any combination of self-pollination will negatively affect the progeny in comparison with outcrossing, that is, it will cause inbreeding depression (Charlesworth and Charlesworth 1987).

In this chapter we describe the reproductive biology and the floral visitors of the clonal cactus *Grusonia bradtiana* (Opuntioideae, Cactaceae), to determine the effect of clonality upon its sexual reproductive success. We determine the implications that both floral biology and breeding systems can have in the life cycle of this endemic cactus, which in turn is very useful to establish management strategies for this species.

Grusonia bradtiana

Grusonia is a genus that comprises 17 species which are distributed along North American deserts, and Mexico is the center of diversity with 13 species (Guzmán et al. 2003). The boundaries of this genus are not yet clear, but seed studies, and morphological, cytological, and molecular analyses suggest that they are a separate group (Anderson 2001; Pinkava 2002; Guzmán et al. 2003). Bárcenas et al. (2011) confirm that *Grusonia bradtiana* is not within the *Opuntia* clade, but it belongs to a polyphyletic unresolved group with other genera of tribe Cylindropuntieae (*Cylindropuntia* -chollas' group, *Grusonia*, *Corynopuntia*, and *Pereskiopsis*). All species of *Grusonia* form dense cushion plants or shrubs and as other species that inhabit unpredictable environments, they exhibit sexual and clonal reproduction.

Grusonia bradtiana (J. M. Coult.) Britton & Rose [= *Opuntia bradtiana* (J. M. Coult.) Brandegee] (Guzmán et al. 2003), locally known as "viejito," and in English as "old man cactus," are short-branched cacti that can form dense shrubs (Fig. 5.1). This cactus reproduces sexually by flowering and fruiting, each fruit contains around 20 seeds (Rosas Barrera et al. 2020, this volume), and a sexual offspring has a new genetic combination (genet, Harper 1977). The species clones by fragmentation of stems (Rosas Barrera et al. 2020, this volume), this generate newborns of different sizes that root independently, and clonal offspring (ramet) has identical genetic composition as parent plant (Harper 1977). Stems are light green, of about 1 m height, from 4 to 7 cm in diameter, with 8–10 short, longitudinal, and tuberculated ribs. Areoles are 3–5 mm in diameter with white wool when young. Leaves, which soon decay, are succulent and green. Yellowish brown spines are found when young, turning white with age. Flowers are yellow, typical of Opuntioideae, 3–4 cm long (Bravo-Hollis 1978) (Fig. 5.2a,b).

Grusonia bradtiana and *Grusonia moelleri* are endemic species of CCB and contiguous areas of Mapimi in the Chihuahuan desert; locally abundant in calcareous soils (Bravo-Hollis 1978; Pinkava 1984; Mandujano and Golubov 2000; Guzmán et al. 2003; Martínez-Ávalos et al. 2020, this volume). The study site in CCB, 8 km south from La Becerra at San Marcos y Pinos (26° 45′ 47.8″ N and at 102° 9′ 10.3″ W), has a mean annual precipitation of 200 mm and a temperature that ranges from 0 °C in winter to 44 °C in summer (Marsh 1984; Montiel González et al. 2018).

Grusonia bradtiana establishes in slopes between the alkaline floor of the basin characterized by pastures and gypsum dunes, and oak-pine forests in higher

Fig. 5.1 *Grusonia bradtiana* growing at Cuatro Ciénegas Basin, Coahuila, Mexico. There are flowers and fruits on the tip of the branches. Photograph by Erick García Morales

a b

Fig. 5.2 Flower of *Grusonia bradtiana*. Segments of perianth are bright yellow (i.e., petals). It has numerous stamens with red filaments and yellow anthers containing abundant pollen, the yellow lobulated stigma is located at flower's center (**a**). Photograph by María C. Mandujano. Longitudinal section of a flower fixed with FAA (10 : 50 : 5 : 35 formalin, 95% ethanol,acetic acid, distilled water) (**b**), Letters indicate A: stigma, B: anthers and C: ovary chamber with numerous ovules. Photograph by Lucía Plasencia-López

elevations with arid shrubs dominated by *Larrea tridentata*, *Prosopis glandulosa* var. *torreyana, Acacia greggii*, *Fouquieria splendens*, *Suaeda mexicana*, and *S. suffruticosa* (Pinkava 1984; Mandujano and Golubov 2000; Flores-Vázquez et al. 2020, this volume), and other succulents, like *Opuntia rufida, Dasylirion* sp., *Echinocereus engelmannii, Agave lechuguilla, Epithelantha micromeris, Mammillaria pottsii, Euphorbia antisyphilitica, Cylindropuntia imbricata, C. leptocaulis,* and *Jatropha dioica* (Pinkava 1984; Flores-Vázquez et al. 2020, this volume; Martínez-Ávalos et al. 2020, this volume).

Floral Behavior and Flower Production of *Grusonia bradtiana* in CCB

The flowering period of *G. bradtiana* was determined in six visits to the study site (March-April 2000, October 2000, June and October 2001, April 2017, and October 2017) and from data of herbaria specimens (MEXU and ENCB-IPN). During June 2001, we measured the diameter of the perianth (corolla) in one flower from each of 40 sampled plants, at six different times from 8:30 h, before the flowers become active, to 20:30 h, when flowers were already closed.

Grusonia bradtiana flowers from May to June, before the main rainy season, and mature fruits appeared in September and October (after the rainy season, Fig. 5.3). This flowering phenology follows the typical pattern of other Opuntioideae and most cacti, showing seasonality and displaying a flowering peak in a specific time of the year (Grant et al. 1979; Mandujano et al. 2010); although in some cactus genera flowering may extend throughout the year (e.g., *Ferocactus histrix*; Del Castillo 1988; *F. robustus*, Piña 2000).

Flowers of *G. bradtiana* were produced on the tip of the stem segments, each producing 1–3 flowers which showed a diurnal flowering cycle: anthesis started at 10:00–10:30 h (1.6 cm ± 0.1 S.E., corolla-perianth opening) and flowers closed at 18:00–18:30 h (1.1 cm ± 0.1 corolla opening), with maximum corolla aperture

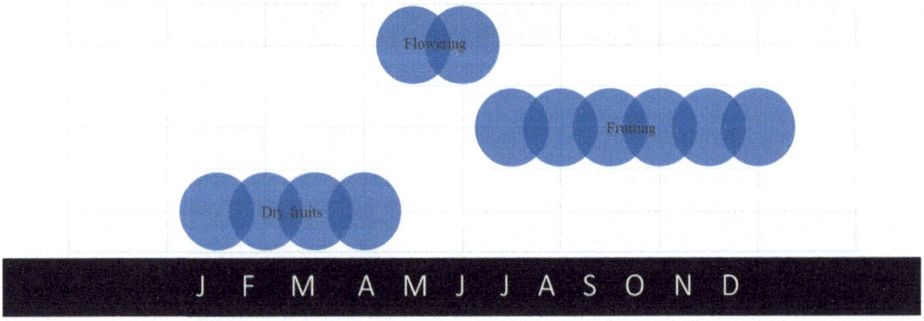

Fig. 5.3 Yearly reproductive phenology of *Grusonia bradtiana* at Cuatro Ciénegas Basin, Coahuila, Mexico. Dry fruits can remain attached to the stems up to three reproductive seasons

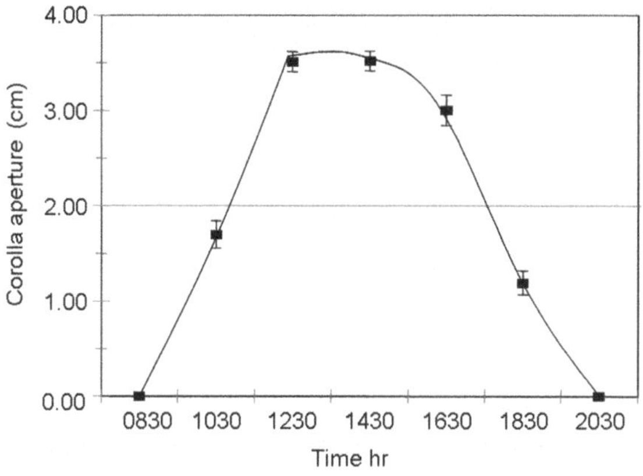

Fig. 5.4 Floral behavior of *Grusonia bradtiana* during June 2001. Flower opening represents the corolla diameter (cm) over time ($N = 40$ flowers, mean ± SE)

occurring between 12:30 and 14:30 h (Fig. 5.4). The flowers were active ca. 8 h for a single day. Similar reproductive patterns have been described in related Opuntioideae, such as *Cylindropuntia imbricata, Opuntia basilaris, O. lindheimeri, O. rastrera, O. robusta,* and *O. spinosissima* (Bravo-Hollis 1978; Grant et al. 1979; McFarland et al. 1989; Mandujano et al. 1996; Negrón-Ortiz 1998).

Flowers of *G. bradtiana* produce yellow and abundant pollen displaying thigmonastic stamens (i.e., the stamens always move inwards and toward the center (style) of the flower when are touched, Braam 2005; Cota-Sánchez et al. 2013) that could promote cross-pollination (Ren and Tang 2012) or self-pollination (Mandujano et al. 1996; Nagy et al. 1999), a phenomenon also reported in related Opuntioideae, including *O. rastrera, O. brunneogemmia, O. viridirubra,* and *O. polyacantha* (Mandujano et al. 1996; Schlindwein and Wittman 1997; Cota-Sánchez et al. 2013), but absent in *O. cochenillifera* and *Brasiliopuntia brasiliensis* (Cota-Sánchez et al. 2013). The stamens movement in *G. bradtiana* is triggered by activity of floral visitors or mechanical stimuli, in other species the movement is provoked by abiotic environmental stimuli like temperature or light as well as visitors; the stamens movement regulates the rate of pollen dispensation in species of Loasaceae, and thigmonastic stamens optimize pollen transfer (Henning and Weigend 2013).

A linear regression between the number of branches per plant against reproductive structures (fruits, flowers, and buds) showed that flower production in *G. bradtiana* increased with the plant size ($N = 40$; $r^2 = 0.66$; $P < 0.001$). Small plants can have one to 20 stems, up to 4000 stems the largest. Plants start producing flowers when they have around 50 stems, but they can clone at any size (Rosas Barrera et al. 2020, this volume).

Fruits are formed rapidly after pollination in July, but they are ripe (i.e., dry and with mature seeds) by the end of September (Fig. 5.3). Fruits of *G. bradtiana* are dry or semidry and indehiscent, which contrast with many other Opuntioideae species with fleshy and sweet cactus pears (Reyes-Agüero et al. 2006). It is possible that other *Grusonia* species have dry fruits but most species descriptions lack of details of reproductive traits.

Breeding System and Pollen Viability in *Grusonia bradtiana* in CCB

The breeding system and the outcrossing index (OCI) were determined following Cruden (1977). The outcrossing index consists on an estimation of pollen:ovule (P:O) ratios and three characteristics of floral morphology and floral behavior. OCI is the sum of the assigned values of (1) the diameter of the flower (assigned to one of three classes; corolla up to 1 mm wide = 0, 1–6 mm wide = 2, more than 6 mm wide = 3), (2) temporal separation between sexual functions (dichogamy), where homogamy and protogyny received a value of 0 and protandry a value of 1, and (3) spatial relationship of stigma and anthers (herkogamy), if there was contact between stigma and anthers the value was 0, and if they were spatially separated and contact seemed unlikely the value was 1 (Cruden 1977). The spatial or temporal separation between anther dehiscence and stigma receptivity (herkogamy or dichogamy, respectively) was evaluated in one flower per plant in a sample of 40 plants in the spring 2001. The following measurements were made at maximum corolla aperture (i.e., from 12:30 h to 14:30 h): stamen and style length (cm), and corolla, as well as pericarpel size (cm) (Table 5.1, Fig. 5.2b). Each flower was observed every 2 h from 8:30 to 20:30 h to visually evaluate both stigma receptivity (humidity and stickiness of the stigma's surface) and anther dehiscence (Mandujano et al. 1996). Finally, herkogamy was analyzed through a *t*-test of paired values of stamen and style lengths.

We did find herkogamy (spatial separation between sexual organs) but no dichogamy (temporal separation between sexual functions) in *G. bradtiana*, although some flowers present inverse herkogamy, that is, longer stamens than stigma ($t = 4.81$; df = 37; $P < 0.0001$; Barrett et al. 2000). Herkogamy is uncommon within Cactaceae but has been reported in *Hylocereus undatus, Nopalea* spp. and all species of *Ariocarpus* (Pimienta-Barrios and Del Castillo 2002; Martínez-Peralta et al. 2014). Herkogamy is usually interpreted as a floral adaptation that reduces pollen-stigma interference and as a mechanism to avoid self-pollination (Barrett et al. 2000) and to export pollen in self-incompatible species (Martínez-Peralta et al. 2014).

The stigma of some flowers was receptive 30 min after floral opening, and 60% of the flowers showed an apparently receptive stigma at 10:30 h. Anthers started

Table 5.1 Flower measurements (N = 38 flowers, mean ± SE) of *Grusonia bradtiana* in Cuatro Ciénegas, Coahuila, Mexico.

Floral trait	Size (cm)
Pericarpel width	2.4 ± 0.06
Pericarpel length	2.6 ± 0.07
Corolla width	3.5 ± 0.10
Corolla length	2.02 ± 0.8
Stamen length	1.18 ± 0.09
Style and stigma length	1.61 ± 0.05

releasing pollen at the same time (45%, 10:30 h) and after 2 h, all flowers had both receptive stigma and dehiscent anthers, thus showing absence of dichogamy.

Most cacti flowers are perfect (i.e., bisexual), even though some species show functional dioecism (i.e., flower that produce mainly pollen (male) and flowers that produce mainly seeds (female); Anderson 2001). The ancestral condition of the Opuntioideae and probably for all the Cactaceae is the presence of hermaphroditic flowers with mixed mating systems (Pimienta-Barrios and Del Castillo 2002).

Grusonia bradtiana flowers are homogamous, male and female organs mature at the same time inside the flower; like results were found in several Opuntioideae (Bravo-Hollis 1978; Grant et al. 1979; Grant and Grant 1979; Mandujano et al. 1996; Negrón-Ortíz 1998) and other cacti like *Ferocactus robustus* (Piña 2000). Homogamous flowers allow self-fertilization (Wyatt 1983), and particularly for *G. bradtiana,* as will be shown below, some individuals have autonomous pollination (Nagy et al. 1999).

Pollen-ovule ratio was estimated from a random sample of one flower from 32 different plants in spring of 2001 and 2002. The flower was longitudinal-sectioned, and the number of ovules counted from one half. The number of anthers per flower was counted from ¼ flower. We quantified the number of pollen grains from these 32 flowers by taking a closed anther from each flower. Each anther was placed in a 1.5 ml microtube and 1 ml of alcohol was added. Samples were homogenized by stirring and an aliquot of 10 μl of the solution was placed in a Neubauer chamber for quantification under a stereoscopic microscope. The relation of the number of pollen grains in 10 μl and in 1000 μl was calculated, and the number of anthers per flower was multiplied by the resulting value. The final P:O was the overall average of all sampled flowers (Kearns and Inouye 1993). Anthers produced large numbers of pollen grains (mean = 287, 023 ± SE = 48,760) per ovule (mean = 156 ± SE =12) in each flower giving a P/O = 2490 ± 601.

The final value of OCI for *G. bradtiana* is 4, with the following characteristics: flower diameter above 6 mm, spatial separation between sexes, as well as absence of protandry and protogyny. The P/O and the OCI indicated that the breeding system of *G. bradtiana* is facultative xenogamous. Species of Cactaceae tend to display high P/O ratios which is associated with pollen waste during transfer and with the amount of pollen that is consumed by vectors (Cruden 1977), such as bats in columnar cacti and solitary bees (use to provision their offspring) in Opuntioideae and Cactoideae subfamilies (Fleming et al. 1994; Mandujano et al. 2010; Camacho-Velázquez et al. 2016).

Pollen viability was estimated from a random sample of 15 plants from which one anther of a single flower was collected and placed in 1.5 ml microtubes. The percentage of viable pollen grains from the sampled pollen was obtained as follows: a quarter anther was sectioned and placed in a Petri dish with one drop of water, plus one of lactophenol-aniline blue stain (Kearns and Inouye 1993). After a minute, the pollen samples were observed in a stereoscopic microscope, viable (stained dark blue) and sterile pollen grains (stained light blue) were counted. In general, all sampled anthers showed a large amount of viable pollen grains (85% ± SE 13%). High pollen viability (stainability) correlates with successful set of fruit and seeds in controlled pollination (Dafni and Firmage 2001). Fruit set in control flowers was close to 100% and number of seeds was the highest (Fig. 5.5).

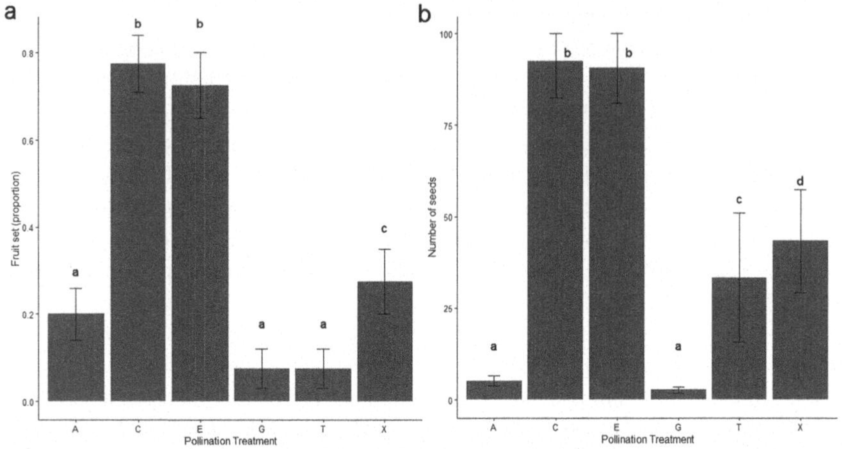

Fig. 5.5 Mean (± s.d.) fruit set (**a**) and number of seeds per fruit (**b**) from manual pollination treatments on *Grusonia bradtiana* in Cuatro Ciénegas Basin, Coahuila, Mexico. 40 different plants were used in a randomized block design (N = 40 flowers per treatment). Pollination treatments were A = forced self-pollination, C = control, E = supplement of pollen, G = automatic self-pollination, T = geitonogamy, and X = outcrossing. Different letters above bars indicate groups that had a significant difference between them and groups with the same letter had not detectable difference ($P < 0.005$ see Table 5.2)

Mating System and Pollen Limitation in *Grusonia bradtiana* in CCB

Mating system of *G. bradtiana* was determined in a manual pollination experiment using a random sample of 40 plants (blocks), during reproductive season of spring 2001. Seven flowers in each plant were randomly assigned to one of the following treatments (Table 5.2 for explanation): apomixis (P), automatic autogamy- autonomous pollination (G), forced autogamy (A), geitonogamy (T), pollination of flowers with pollen of another flower, but from the same ramet, within genet, outcrossing (X) (between genets), supplemental cross pollen (E), and an untreated control (C, Table 5.2). Treatments T, G, and A involved self-pollination at different spatial scales.

Once a flower was treated, it was labeled with its corresponding treatment and attached with a string to the stem to avoid fruit removal by herbivores. Ripe fruits were collected 4 months later, from which we determined fruit set and count of seeds per fruit. Generalized linear models were used to compare seed set (with Poisson distribution) and fruit set percentages (binomial distribution) using the statistical software GLIM (Generalized Linear Models; Crawley 1993). Differences among treatments were calculated with contrast X^2 evaluating each model without effect of each treatment (Crawley 1993). Inbreeding depression (δ) was estimated following Charlesworth and Charlesworth (1987) as $\delta = 1 - ws/wo$, the inverse of the ratio of value of selfed fruit or seed (progeny, *ws*) to value for outcrossed

Table 5.2 Pollination treatments applied to flowers of *Grusonia bradtiana* in Cuatro Ciénegas Basin, Coahuila, Mexico

Symbol	Treatment	Method
T	Geitonogamy	Flower emasculation and addition of pollen from another flower from the same ramet. The flower was covered and labeled.
G	Automatic autogamy	The flower bud was covered and labeled to evaluate selfing in absence of pollen vectors. The treatment allows to determine autonomous pollination.
A	Forced autogamy	The flower bud was covered and as soon as the flower opened, a load of its own pollen was applied on its receptive stigma* with a small brush. The flower was covered and labeled.
X	Outcrossing (inter-genet)	A pollen mixture was prepared from pollen from at least ten different flowers of different putative genets (10 m apart from the focal plant). When the flower opened it was emasculated and we applied ethanol at 50% with a small brush in the internal segments of the perianth to sterilize the adhered pollen, with a second brush the pollen mixture was added to receptive stigma. The flower was covered and labeled.
P	Apomixis (pollen exclusion)	While the flower was opening a straw was placed around the pistil to completely cover the stigma and was emasculated. The flower was covered and labeled.
C	Control	A flower was left with no treatment to receive visitors. The flower was covered and labeled after anthesis.
E	Supplemental cross pollen (pollen limitation)	At the time of maximum perianth aperture a supplement load of pollen mixture of ten genets as donors was added to the receptive stigma and the flower was left uncovered to receive visitors. The flower was covered and labeled after anthesis.

Flowers or floral buds were covered with mesh bags (10 × 10 cm) to exclude flower visitors, and bags were attached to the stem to prevent the loss of samples. Pollen deposition on hand pollination treatments was made with a small brush that was cleaned before each pollination with ethanol at 90%

[a]Receptive stigma is humid, and pollen easily adheres to it when receptive

progeny (*wo*); values towards 1 indicate high inbreeding depression and towards 0 that inbreeding depression is low.

We found significant differences in fruit and seed production among treatments (Fig. 5.5 a,b; $X^2 = 92.58$; df = 5; $P < 0.001$; $X^2 = 1736$; df = 5; $P < 0.0001$, respectively). Fruit and seed production were higher in cross-pollination treatments (X, C, and E) than in self-pollination treatments (A, G, and T, Fig. 5.5a) and fruit set in apomixis (P) was nil. The geitonogamy treatment (T), that is, pollination among flowers of the same ramet showed a significant higher seed production than the other self-pollination treatments, but lower than the cross-pollination treatments (Fig. 5.5a). Fruits that were produced via self-pollination produced less seeds than the fruits of outcrossing-pollination (Fig. 5.5b). Moreover, they were significantly different between contrasts involving the self-pollination group and outcrossing group treatments (Fig. 5.5b, $X^2 = 150$; df = 1; $P < 0.0001$). Manual crossing experiments allowed to conclude that the species is self-compatible, as both fruits and seeds are produced by forced self-pollination (A) and by automatic autogamy (G) treatments, and do not need pollinators to develop some seeds, but more seed and fruit are always produced by outcrossing, involving pollinators. This is a pollination insurance strategy, where autonomous pollination allows the species to produce progeny under unpredictable conditions, being scarce of resources, partners, or pollinators (Nagy et al. 1999). However, there is a marked reduction on fruit set and seed loss by selfing compared with outcrossing as the species has high inbreeding depression as the estimated values are close to one (δ fruit set = 0.74, δ seed set = 0.85; Charlesworth and Charlesworth 1987). However, manual pollinations indicated that *G. bradtiana* is not limited by pollen and pollen transfer mediated by pollinators is efficient as control treatment showed the highest fruit and seed set. Even though *G. bradtiana* is self-compatible, as it produces some fruits and seeds by autogamy, self-pollination is clearly less successful than outcrossing (3 and 18 times less fruit and seeds, respectively). The production of significantly more seeds in the control treatment than in supplemental pollen treatment indicates no pollen limitation. Accordingly, Larson and Barrett (2000) reported that pollen limitation for 224 species they analyzed is less intense in self-compatible, autogamous, and nectariferous species than in self-incompatible, non-autogamous, and nectarless species.

Floral Visitors and Nectar Availability *Grusonia bradtiana* in CCB

Bee species that visited the flowers of *G. bradtiana* in the study area in CCB were collected, registering date and hour of collection during 2001. The specimens were identified according to Michener et al. (1994) and kept at the Museum of Zoology "Alfonso L. Herrera" at the Facultad de Ciencias (School of Sciences), Universidad Nacional Autónoma de México, Mexico City.

To determine the frequency and type of flower visitors of *G. bradtiana* we took a random sample of flowers within four areas with patches of reproductive plants. From each area we randomly selected ten flowers, in each of which we recorded the number of visits and the species of floral visitors. These observations were made during a single day from 10:30 h to 18:30 h in two-hour intervals lasting 20 min each, until we completed six and a half hours of observations. We calculated the proportions of visits by means of a contingency table (χ^2) and adjusted residuals to determine if significant differences existed among visitors and hours of visits (Everitt 1977).

In addition, we approximated pollinator efficiency in seed production within the population using the same 40 followed flowers. At the end of the day we covered each flower with a fine mesh, and 4 months later we collected the fruits to determine seed production with respect to the number of visits. Data were analyzed using a non-parametric regression (Zar 1996) with JMP (version 3.2.1, SAS Institute 1995).

Flowers of *G. bradtiana* were visited as soon as they began to open (around 10:00 h, see Figs. 5.4 and 5.6). The flowers are visited by species of Hymenoptera,

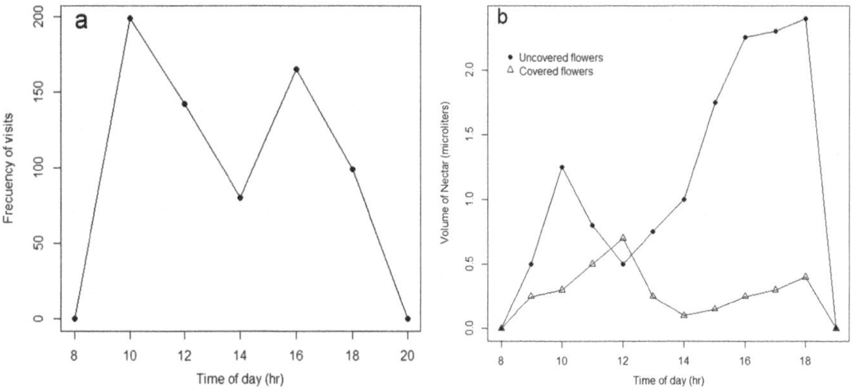

Fig. 5.6 (**a**) Frequency of bee visits to flowers of *Grusonia bradtiana* over time ($N = 40$ flowers). (**b**) Nectar production (in microliters ± SE) in two different treatments, covered flowers ($N = 35$) and uncovered flowers ($N = 35$)

Table 5.3 Floral visitors of *Grusonia bradtiana* and their activity

Order	Family	Species	Sample by gender	Activity
Hymenoptera	Anthophoridae	*Diadasia rinconis*	3♂ y 4♀	Pollinator
Hymenoptera	Anthophoridae	*Melissodes* sp.1	1♂	Pollinator
Hymenoptera	Anthophoridae	*Melissodes tristis*	1♂ y 2♀	Pollinator
Hymenoptera	Megachilidae	*Dianthidium* sp.1	1 ♀	Pollinator
Hymenoptera	Andrenidae	*Perdita* spp.	1 ♂ y 1♀	Self-pollinator
Hymenoptera	Megachilidae	*Ashmeadiella* sp.1	1 ♀	Pollinator
Coleoptera	Nitidulidae	Species1	-	Thief
Orthoptera	Grillidae	Species 1	-	Pollen thief

Coleoptera, and Orthoptera and especially by solitary bees from the genera *Diadasia* and *Melissodes* (Table 5.3). *Diadasia* sp. and *Melissodes* spp. (two species) were the most frequent visitors (552 visits), followed by *Ashmeadiella* sp. (92 visits) and *Perdita* sp. (25 visits). *Diadasia, Melissodes,* and *Dianthidium* showed typical pollinator behavior, landing on stigma before searching the base of the flower for nectar, causing their body to touch the anthers, nectar, and pollen assuring the uptake and deposition of pollen grains in the stigma of another flower, and due to their size, they necessarily made contact with the anthers and stigma of the flowers, visiting the flowers in a radius of approximately 10 m. *Ashmeadiella* and *Perdita* bees did not show clear contact with the stigmatic surface, but probably stimulate self-pollination by the thigmonastic sensitivity of the stamens (Table 5.3).

In *Opuntia* species, as well as in *Echinocereus viridiflorus* and *Ferocactus histrix*, a similar behavior has been observed and other large pollinators such as *Lithurge, Megachile, Xylocopa,* and *Bombus* have been reported to be effective pollinators for *Opuntia* species (Grant and Grant 1979; Grant et al. 1979; Leuck and

Miller 1982; Simpson and Neff 1983; Del Castillo and González-Espinosa 1988; McFarland et al. 1989; Del Castillo 1994; Mandujano et al. 1996; Mandujano et al. 2010).

Beetles (Coleoptera) were also found within the flowers throughout the floral life span, but mostly at 16:00 h (Table 5.3). Sap beetles, family Nitidulidae, were abundant but do not promote self-pollination; visitors from this group stayed in the lower part of the perianth and it seems that they do not produce stamen movements. Sap beetles and grasshoppers were classified as nectar and/or pollen thieves and florivores (Table 5.3). Sap beetle behavior as nectar thieves has also been described for *O. basilaris*, *O. lindheimeri*, *O. robusta,* and *F. histrix* (Grant and Grant 1979; Grant et al. 1979; Rowley 1980; Del Castillo and González-Espinosa 1988; Del Castillo 1994) and can be considered parasites of *Opuntia* flowers (as it has been found for *O. compressa* and *O. imbricata;* McFarland et al. 1989). Grasshoppers were florivores because during the evening they eat flower anthers, a phenomenon also observed in *Echinocactus platyacanthus* and *Ariocarpus trigonus* (Mandujano et al. 2010; Cárdenas-Ramos and Mandujano 2018).

In *G. bradtiana* there were differences among visitor frequencies ($X^2 = 39.8$, df = 8, $P < 0.001$), and visits are more frequent during the morning (195 visits at 10:30 h), when nectar production was the higher between 12:30 and 14:30 h (78 visits) and increases again after 16:30 h (Fig. 5.6a, b; 161 visits). High frequency of visits during the morning is clearly linked to nectar production. Medium and small-sized visitors of the genera *Perdita* and *Ashmeadiella* stimulate thigmonastic movements all day, but preferentially in the evening which could probably favor self-pollination in *G. bradtiana,* as has been seen in *O. littoralis* and *O. rastrera* (Grant and Grant 1979; Mandujano et al. 1996). In contrast with what happens in *G. bradtiana*, small bees like *Ashmeadiella* and *Dialictus* are considered nectar thieves in *O. robusta* (Del Castillo and González-Espinosa 1988).

Nectar is one of the most important rewards for pollinators (Nicolson 2007), so we quantified accumulated nectar in 70 randomly selected flowers. Nectar was measured using 5 μl capillary tubes, which were inserted into the nectar chamber. To assess the nectar's consumption of the flower visitors, half of the flowers were covered with a fine mesh to keep visitors away from flowers and the other half ($n = 35$ flowers) remained uncovered. To avoid non-independence among samples, each flower was only used once. The quantity of nectar for each flower was quantified from seven flowers of each group from 10:30 to 20:30 h in two-hour intervals. Floral visitors consume nectar soon after flower opening. However, we found available nectar during all the floral cycle in uncovered flowers, suggesting that the rate of nectar production exceeds consumption. In covered flowers we found a decrease in nectar production at 12:30 h (Fig. 5.4b) followed by an increase between 16:30 and 18:30 h, which could be explained either by nectar evaporation or nectar reabsorption. *Grusonia bradtiana* is pollinated by bees and there is an important effect of frequency of visits, the more visits a flower experiments the more seeds it sets (Spearman Rho = 0.3528, $P < 0.05$). Bees are searching for rewards such as pollen and nectar, which are produced in abundance by all flowers.

General Discussion, Conclusions, and Perspectives

We found that the endemic cactus of the CCB and nearby desert areas *G. bradtiana* has a complex sexual system combining selfing and outcrossing. It is also able to self-fertilize without pollinators, but the production of seeds decreases considerably, so the species greatly benefits from animal pollination mediated by solitary bees. In addition, control pollination proves that sexual reproduction of the species is highly successful and selfing (autonomous pollination, forced selfing, and geitonogamy) decreases sexual output. Therefore, clonality may limit the production of fruits and seeds as is the case of other clonal species (Liao and Harder 2014) if pollen transfer occurs among genetic similar or identical partners. Our results also indicate that *G. bradtiana* has a mixed mating system with a tendency towards cross-fertilization, that is, a facultative xenogamous system (P/O = 2490.3,) according to P:O ratio indicated by Cruden (1977). Pollination experiments reinforced the premise that outcrossing is the dominant form of crossing in *G. bradtiana*, where it is favored by both the fact that pollinators are moving pollen among genets and adaptations to select outcross pollen.

Most cacti, including *G. bradtiana,* have several adaptations favoring outcrossing, meaning that most of the seeds are produced by pollen that comes from different individuals for fertilization, although self-pollination occurs in some groups (Ross 1981; Clark-Tapia and Molina-Freaner 2004; Mandujano et al. 2010; Camacho-Velázquez et al. 2016). Nevertheless, relatively few species have been studied, considering the large number of species in the family (less than 2% of 2000 species, Mandujano et al. 2010; Camacho-Velázquez et al. 2016) and results show a variety of breeding systems within the cactus family. For example, the genus *Frailea* and some species of *Melocactus* are cleistogamous, flowers do not open and self-fertilize (Anderson 2001) while, as we discussed above, dichogamy occurs in *Hylocereus* spp. cultivars (Mandujano et al. 2010; Camacho-Velázquez et al. 2016). Herkogamy has been reported for *Ariocarpus* genus (Martínez-Peralta et al. 2014). Trioecy where individuals in the populations have staminate-male, pistillate-female, and hermaphrodite flowers occurs in *Pachycereus pringlei* (Fleming et al. 1994; Cervantes 2001); functional dioecy, species that show incomplete morphological differentiation as appeared to be hermaphrodite flowers, yet they have complete functional differentiation between male and female flowers, is seen in *Mammillaria dioica* and *Echinocereus coccineus* (Bravo-Hollis and Sánchez-Mejorada 1991; Hoffman 1992); gynodioecy, where hermaphrodite and female plants coexist within a population, has been documented in *M. blossfeldiana* var. *shurliana* (Rebman 2001; Camacho-Velázquez et al. 2016) and dioecy in *Opuntia stenopetala* and *Pereskia zinniflora* (Anderson 2001; Reyes-Agüero et al. 2006).

Grusonia bradtiana is pollinated by bees, mainly by large bees of the genera *Diadasia*, *Melissodes,* and *Dianthidium,* and there is an important effect of frequency of visits, as more visited flowers produce more seeds. Bees visit the flowers by searching for rewards such as pollen and nectar that are produced in abundance by all *Grusonia bradtiana* flowers.

Bee pollination is very common within the Cactaceae (46% of the genera) and is considered a primitive feature of the family (Anderson 2001; Pimienta-Barrios and Del Castillo 2002; Wallace and Gibson 2002). Bee pollination is typical for the Pereskioideae and Opuntioideae subfamilies, but it is also important for the Cactoideae subfamily (41% of the species; Mandujano et al. 2010; Camacho-Velázquez et al. 2016). Floral visitors among cactus species correlate with life forms and floral features, which may promote specialized pollination by bats, birds, or insects (Rowley 1980; Anderson 2001; Pimienta-Barrios and Del Castillo 2002). In some genera, like *Rebutia*, pollination is by butterflies, and pollination by birds occurs in many genera whose flowers are zygomorphic (e.g., *Cochemiea*, *Schlumbergera,* and *Cleistocactus*, Rowley 1980; Anderson 2001); butterflies and birds may pollinate despite those functional groups being nectar consumers (Nicolson 2007), while bat pollination is most common in columnar life forms such as *Pachycereus*, *Pilosocereus,* *Stenocereus,* and *Neobuxbaumia* (Fleming et al. 1994; Nassar et al. 1997; Valiente-Banuet et al. 1997; Casas et al. 1999; Anderson 2001; McIntosh 2005).

Overall, clonality in *Grusonia bradtiana* imposes a cost as the species displays high inbreeding depression. Despite this, fruits and seeds are produced in all possible ways of selfing, but outcrossing results in a twofold advantage. Moreover, if pollination or resources are scarce, *G. bradtiana* still has reproductive opportunities, as some fruits and seeds will form, so in unpredictable environments, selfing provides a reproductive insurance for this species. *Grusonia bradtiana* is a very successful species, its longevity, diversity of reproductive strategies, and its high reproductive success partially explain why this cactus is a dominant species of *bajadas* at Cuatro Ciénegas Basin.

Acknowledgments Jordan Golubov, Luis E. Eguiarte, and Francisco Molina-Freaner for comments on previous versions of the chapter. Financial support of project IN205500 PAPIIT-DGPA UNAM to MC Mandujano and PRONATURA provided free stay at La Becerra, Cuatro Ciénegas, during our field trips. Susana Moncada and personnel from the Area de protección de Flora y Fauna Cuatrocienegas provided logistic support. We appreciate the help during field work of Manuel Rosas, Gisela Aguilar Morales and in memoriam of our dearest friend Dolores Rosas Barrera (2018). The first author thanks the help at different stages of the work to M.Sc. Daniel Cervantes.

References

Abrahamson WG (1980) Demography and vegetative reproduction. In: Solbrig OT (ed) Demography and evolution in plant populations. University of California Press, Berkeley, CA, USA, pp 89–106

Anderson EF (2001) The Cactus family. Timber Press, Portland, OR, USA

Bárcenas RT, Yesson C, Hawkins JA (2011) Molecular systematics of the Cactaceae. Cladistics 27:470–489

Barrett SCH (2013) The evolution of plant reproductive systems: how often are transitions irreversible? Proc Royal Soc B 280:1–9

Barrett SCH, Jesson LK, Baker AM (2000) The evolution and function of stylar polymorphisms in flowering plants. Ann Bot 85:253–265

Braam J (2005) In touch: plant responses to mechanical stimuli. New Phytol 165:373–389

Bravo-Hollis H (1978) Las Cactáceas de México Volume 1. Universidad Nacional Autónoma de México, México, DF, Mexico

Bravo-Hollis H, Sánchez-Mejorada H (1991) Las Cactáceas de México Volume 2. Universidad Nacional Autónoma de México, México, DF, Mexico

Camacho-Velázquez A, Ríos-Carrasco S, Vázquez-Santana S (2016) Biología reproductiva de la subfamilia Cactoideae (Cactaceae). Cact Suc Mex 61(4):100–127

Cárdenas-Ramos D, Mandujano MC (2018) Florivory effects on pollinator preference and the reproductive output of a threatened living rock cactus, *Ariocarpus retusus* (Cactaceae). Haseltonia 25:1–7

Carrillo-Ángeles IG, Mandujano MC, Golubov J (2011) Influences of the genetic neighborhood on ramet reproductive success in a clonal desert cactus. Population Ecology 53:449–458

Casas A, Valiente-Banuet A, Rojas-Martínez A, Dávila P (1999) Reproductive biology and the process of domestication of the columnar cactus *Stenocereus stellatus* in Central Mexico. Am J Bot 86:534–542

Cervantes SM (2001) Variación geográfica en el sistema reproductivo de *Pachycereus pringlei*. Bachelor diss., Facultad de Ciencias UNAM, México, DF, Mexico

Charlesworth D, Charlesworth B (1987) Inbreeding depression and its evolutionary consequences. Annu Rev Ecol Syst 18:237–268

Charpentier A (2002) Consequences of clonal growth for plant mating. Evol Ecol 15:521–530

Clark-Tapia R, Molina-Freaner F (2004) Reproductive ecology of the rare clonal cactus *Stenocereus eruca* in the Sonoran Desert. Plant Syst Evol 247:155–164

Cota-Sánchez H, Almeida OJG, Falconer DJ, Choi HJ, Bevan HL (2013) Intriguing thigmonastic (sensitive) stamens in the plains prickly pear *Opuntia polyacantha* (Cactaceae). Flora 208:381–389

Crawley MJ (1993) GLIM for ecologists. Blackwell Scientific Publication, Oxford, UK

Cruden RW (1977) Pollen-ovule ratios: a conservative indicator of breeding systems in flowering plants. Evolution 31:32–46

Dafni A, Firmage D (2001) Pollen viability and longevity: practical, ecological and evolutionary implications. Plant Syst Evol 222:113–132

Del Castillo RF (1988) Fenología y remoción de semillas en *Ferocactus histrix*. Cact Suc Mex 33:5–14

Del Castillo RF (1994) Polinización y otros aspectos de la biología floral de *Ferocactus histrix*. Cact Suc Mex 39:36–42

Del Castillo RF, González-Espinosa M (1988) Una interpretación evolutiva del polimorfismo sexual de *Opuntia robusta* (Cactaceae). Agrociencia 71:185–196

Eckert CG (2000) Contribution of autogamy and geitonogamy to self-fertilization in a mass-flowering, clonal plant. Ecology 81:532–542

Everitt BS (1977) The analysis of contingency tables. Chapman and Hall, New York, NY, USA

Fleming TH, Mauricem S, Buchmann SL et al (1994) Reproductive biology and relative fitness in a trioecious cactus, *Pachycereus pringlei* (Cactaceae). Am J Bot 81:858–867

Flores-Vázquez JC, Rosa Barrera MD, Golubov J, Sánchez-Gallén I, Mandujano MC (2020) Ecological importance of Bajadas in the Chihuahuan Desert. In: Mandujano MC, Pisanty I, Eguiarte LE (eds) Plant diversity and Ecology in the Chihuahuan desert. Springer International, Cham, Switzerland

García-Morales E, Carrillo-Angeles IG, Golubov J et al (2018) Influence of fruit dispersal on genotypic diversity and migration rates of a clonal cactus from the Chihuahuan Desert. Ecol Evol 8:12559–12575

Golubov J, Mandujano MC, Franco M et al (1999) Demography of the invasive woody perennial *Prosopis glandulosa* (honey mesquite). J Ecol 87:955–962

Goulson D (2000) Why do pollinators visit proportionally fewer flowers in large patches? Oikos 91:485–492

Grant V, Grant KA (1979) Pollination of *Opuntia basilaris* and *O. littoralis*. Plant Syst Evol 132:321–325
Grant V, Grant KA (1980) Clonal microspecies of hybrid origin in the *Opuntia lindheimeri* group. Bot Gaz 141:101–106
Grant V, Grant KA, Hurd PD Jr (1979) Pollination of *Opuntia lindheimeri* and related species. Plant Syst Evol 132:313–320
Guzmán U, Arias S, Dávila P (2003) Catálogo de cactáceas mexicanas. Universidad Nacional Autónoma de México- Comisión Nacional para el conocimiento y uso de la biodiversidad, México, DF, Mexico
Handel SN (1985) The intrusion of clonal growth patterns on plant breeding systems. Am Nat 125:367–383
Harder LD, Barrett SCH (1995) Mating cost of large floral displays in hermaphrodite plants. Nature 373:512–515
Harper JL (1977) Population biology of plants. Academic Press, London, UK, 892 pp
Henning T, Weigend M (2013) Beautiful, complicated—and intelligent? Novel aspects of the thigmonastic stamen movement in Loasaceae. Plant Signal Behav 8(6):E24605. https://doi.org/10.4161/PSB.24605
Hoffman MT (1992) Functional dioecy in *Echinocereus coccineus* (Cactaceae): breeding systems, sex ratios, and geographic range of floral dimorphism. Am J Bot 79:1382–1388
Holsinger KE (2000) Reproductive systems and evolution in vascular plants. PNAS 97:7037–7042
Ishii HS, Sakai S (2002) Temporal variation in flower display size and individual floral sex allocation in racemes of *Narthecium asiaticus* (Liliaceae). Am J Bot 89:441–446
Kearns CA, Inouye DW (1993) Techniques for pollination biologists. University Press of Colorado, Boulder, CO, USA
Larson BMH, Barrett CH (2000) A comparative analysis of pollen limitation in flowering plants. Biol J Linn Soc 69:503–520
Leuck EE, Miller JM (1982) Pollination biology and chemotaxonomy of the *Echinocereus viridiflorus* complex (Cactaceae). Am J Bot 69:1669–1672
Liao W, Harder LD (2014) Consequences of multiple inflorescences and clonality for pollinator behavior and plant mating. Am Nat 184:580–592
Mandujano MC, Golubov J (2000) *Opuntia bradtiana* en la zona calcárea del Bolsón de Mapimí México. Cact Suc Mex 45:66–68
Mandujano MC, Montaña C, Eguiarte LE (1996) Reproductive ecology and inbreeding depression in *Opuntia rastrera* (Cactaceae) in Chihuahuan Desert: why are sexually derived recruitments so rare? Am J Bot 83:63–70
Mandujano MC, Carrillo-Angeles IG, Martínez-Peralta C et al (2010) Reproductive biology of Cactaceae. In: Ramawat KG (ed) Desert plants-biology and biotechnology. Springer, Berlin, Germany, pp 197–230
Marsh PC (1984) Biota of Cuatro Ciénegas, Coahuila, México: Preface. JANAS 19:1–2
Martínez-Peralta C, Márquez-Guzmán J, Mandujano MC (2014) How common is self-incompatibility across species of the herkogamous genus *Ariocarpus*? Am J Bot 101(3):1–9
Martínez-Ávalos JG, Martínez-Gallegos R, Guerra-Pérez A, Torres Castillo JA, et al. (2020). Diversity and distribution of cacti species in the Cuatro Ciénegas basin. In: Mandujano MC, Pisanty I, Eguiarte L (eds.) Plant diversity and Ecology in the Chihuahuan desert. Springer International, Cham, Switzerland
McFarland JD, Kevan PG, Lane MA (1989) Pollination biology of *Opuntia imbricata* (Cactaceae) in southern Colorado. Can J Bot 67:24–28
McIntosh ME (2005) Pollination of two species of *Ferocactus*: interactions between cactus-specialist bees and their host plants. Funct Ecol 19:727–734
Michener CD, McGinely RJ, Danforth BN (1994) The bee genera of north and Central America (Hymenoptera: Apoidea). Smithsonian Institution Press, Washington, DC, USA and London, UK

Montiel González C, Bautista F, Delgado C, García-Oliva F (2018) The climate of Cuatro Ciénegas Basin: drivers and temporal patterns. In: Souza V et al (eds) Cuatro Ciénegas Ecology, Natural History and Microbiology, Cuatro Ciénegas Basin: an endangered Hyperdiverse oasis. Springer, Cham, Switzerland

Nagy ES, Strong L, Galloway LF (1999) Contribution of delayed autogamous selfing to reproductive success in mountain Laurel, *Kalmia latifolia* (Ericaceae). Am Midl Nat 142:39–46

Nassar JM, Ramirez N, Linares O (1997) Comparative pollination biology of Venezuelan columnar cacti and the role of nectar-feeding bats in their sexual reproduction. Am J Bot 4:918–927

Neff JL, Simpson BB, Dorr LJ (1982) The nesting biology of *Diadasia afflicta* Cress. (Hymenoptera: Anthophoridae). J Kansas Entomol Soc 53:499–518

Negrón-Ortiz V (1998) Reproductive biology of a rare cactus, *Opuntia spinosissima* (Cactaceae), in the Florida keys: why is seed set very low? Sex Plant Reprod 11:208–212

Nicolson SW (2007) Nectar consumers. In: Nicolson SW, Nepi M, Pacini E (eds) Nectaries and nectar. Springer, Dordrecht, Netherlands, pp 289–342

Ordway E (1987) The life history of *Diadasia rinconis* Cockerell (Hymenoptera:Anthophoridae). J Kansas Entomol Soc 60:15–24

Palleiro N, Mandujano MC, Golubov J (2006) Aborted fruits of *Opuntia microdasys* (Cactaceae): insurance against reproductive failure. Am J Bot 93(4):505–511

Pimienta-Barrios E, Del Castillo RF (2002) Reproductive biology. In: Nobel PS (ed) Cacti: biology and uses. University of California Press, Berkeley, CA, USA, pp 75–90

Piña RH (2000) Ecología reproductiva de *Ferocactus robustus* en el Valle de Zapotitlán Salinas, Puebla. MSc diss., ENCB Instituto Politécnico Nacional, México, DF, Mexico

Piña H, Montaña C, Mandujano MC (2007) Fruit abortion in the Chihuahuan-desert endemic cactus *Opuntia microdasys*. Plant Ecol 193:305–313

Pinkava D (1984) Vegetation and floral of the bolson of Cuatro Cienegas region, Coahuila, México: IV. Summary, endemism and corrected catalogue. JANAS 19:23–47

Pinkava D (2002) On the evolution of continental North American opuntias. Succulent Plant Research 6. David Hunt, Sherborne, UK

Rebman JP (2001) Las suculentas del Islote Toro, Baja California, México. Cact Suc Mex 46:52–55

Ren MX, Tang JY (2012) Up and down: stamen movements in Rutagraveolens (Rutaceae) enhance both outcrossing and delayed selfing. Annals of Botany 110:1017–1025. https://doi.org/10.1093/aob/mcs181

Reyes-Agüero JA, Aguirre JR, Valiente-Banuet A (2006) Reproductive biology of *Opuntia*: a review. J Arid Environ 64:549–585

Rosas Barrera MD, Golubov J, Pisanty I, Mandujano MC (2020) Effect of reproductive modes on the population dynamics of an endemic cactus from Cuatro Ciénegas In: Mandujano MC, Pisanty I, Eguiarte LE (eds) Plant diversity and Ecology in the Chihuahuan Desert. Springer International, Cham, Switzerland

Ross R (1981) Chromosome count, cytology and reproduction in the Cactaceae. Am J Bot 68:463–470

Rowley GD (1980) Pollination syndromes and cactus taxonomy. Cact Succ J (Great Britain) 42:95–98

SAS Institute (1995) JMP statistics and graphics guide. SAS Institute, Cary, NC, USA

Schlindwein C, Wittman D (1997) Stamen movements in flowers of *Opuntia* (Cactaceae) favour oligolectic pollinators. Plant Syst Evol 204:179–193

Schlising RA (1972) Foraging and nest provisioning behavior of the oligolectic bee, *Diadasia bituberculata* (Hymenoptera: Anthophoridae). Pan-Pac Entomol 48:175–188

Simpson BB, Neff JL (1983) Pollination ecology in the arid southwest. Aliso 11:417–440

Snow AA, Spira TP, Simpson R, Klips RA (1996) The ecology of geitonogamous pollination. In: Lloyd DG, Barrett SCH (eds) Floral biology studies on floral evolution in animal – pollinated plants. Chapman & Hall, Toronto, Canada, pp 191–216

Stephenson AG (1981) Flower and fruit abortion: proximate causes and ultimate functions. Ann Rev Ecol Syst 12:253–279

Sutherland S (1987) Why hermaphroditic plants produce many more flowers than fruits: experimental tests with *Agave mckelveyana*. Evolution 41(4):750–759

Trujillo-Argueta S, González-Espinosa M (1991) Hibridización, aislamiento reproductivo y formas de reproducción en *Opuntia* spp. Agrociencia 1:39–58

Valiente-Banuet A, Rojas-Martínez A, Casas A et al (1997) Pollination biology of two winter-blooming giant columnar cacti in the Tehuacan Valley, Central Mexico. J Arid Environ 37:331–341

Wallace RS, Gibson AC (2002) Evolution and systematics. In: Nobel PS (ed) Cacti: biology and uses, Berkeley, CA, pp 1–21

Wyatt R (1983) Pollinator-plant interactions and the evolution of breeding systems. In: Real L (ed) Pollination biology. Academic Press, New York, pp 51–95

Zar JH (1996) Biostatistical analysis. Prentice-Hall Inc, New Jersey

Chapter 6
How Did *Fouquieria* Come to the Chihuahuan Desert? Phylogenetic and Phylogeographic Studies of *Fouquieria shrevei* and *F. splendens* and the Role of Vicariance, Selection, and Genetic Drift

José Arturo De-Nova, Jonás A. Aguirre-Liguori, and Luis E. Eguiarte

Abstract We discuss the radiation and dispersal of *Fouquieria*, the sole genus of Fouquieriaceae, that includes eleven species, all endemic to the warm deserts and dry subtropical regions of North America. Two sister species occur in the Chihuahuan Desert, the broadly distributed *Fouquieria splendens* and the microendemic gypsophilic *F. shrevei* from a few localities in Coahuila, Mexico. Both species coexist in the valleys and mountain ranges west and around Cuatro Ciénegas, but in the Cuatro Ciénegas valley, despite wide availability of gypsum soils, only *F. splendens* is found. Fouquieriaceae represents an ancient lineage that diverged from Polemoniaceae during late Cretaceous, but speciation of *Fouquieria* species occurred during the Mio-Pliocene mainly by vicariance, associated with Neogene orogenesis underlying the early development of regional deserts. The clade formed by *F. shrevei* and *F. splendens* diverged from *F. macdougalii* by a vicariant event that took place during the early Pliocene. However, opposed to the rest of the species, the divergence between *F. shrevei* and *F. splendens*, which grow sympatrically, could have occurred by ecological speciation associated with the colonization of gypsum soils by *F. shrevei*. The patchiness of gypsum deposits in Coahuila has resulted in selection, intense genetic drift, and reduced gene flow among *F. shrevei* populations.

Keywords Cuatro Ciénegas · Fouquieriaceae · Gypsophily · Mexico · Microendemic species · Radiation

---

J. A. De-Nova (✉)
Instituto de Investigación de Zonas Desérticas, Universidad Autónoma de San Luis Potosí, San Luis Potosí, Mexico
e-mail: arturo.denova@uaslp.mx

J. A. Aguirre-Liguori · L. E. Eguiarte
Departamento de Ecología Evolutiva, Instituto de Ecología, Universidad Nacional Autónoma de México, UNAM, Mexico City, Mexico

M. C. Mandujano et al. (eds.), *Plant Diversity and Ecology in the Chihuahuan Desert*, Cuatro Ciénegas Basin: An Endangered Hyperdiverse Oasis,
https://doi.org/10.1007/978-3-030-44963-6_6

Introduction

Fouquieria is a genus comprised of eleven species, all endemic to the warm deserts and dry subtropical regions of North America and it is the only genus of the Fouquieriaceae family (Table 6.1). *Fouquieria* possesses unique characteristics that drastically distinguish it from all other plant lineages, including the mode of spine development, extent of the hard decurrent ridges, anastomosing cortical water-storage tissue, and changes in ovary placentation from flower to fruit (Henrickson 1969a, 1969b, 1972). Flowers are tubular and arranged in terminal clusters, with long stamens and corolla with variable size and color (red to white) commonly associated with specific pollination relationships including ornithophily and entomophily (Henrickson 1972). The xeromorphic adaptations mentioned above, as well as others, make it very difficult to infer its phylogenetic relationships from morphology, and in consequence Fouquieriaceae was previously placed in different orders, including Polemoniales, Tamaricales, and Ebenales (Henrickson 1967, 1972). More recent molecular phylogenetic analyses revealed that Fouquieriaceae is sister to Polemoniaceae inside Ericales, despite contrasting morphological dissimilarities (Bremer et al. 2002; Schönenberger et al. 2005, 2010; Sytsma et al. 2006).

As a lineage almost endemic to the warm deserts in North America, speciation in *Fouquieria* has followed the intricate history associated with the main desert areas in Mexico. The genus was recently used to test hypotheses on how diversification was modeled by vicariant events using a well-resolved phylogeny and molecular clock analyses (De-Nova et al. 2018). Indeed, several studies have shown that vicariance has affected the assembly and diversification of the biota of North American deserts (e.g., Findley 1969; Hubbard 1973; Morafka 1977; Riddle and Hafner 2006; Hafner and Riddle 2011).

Table 6.1 *Fouquieria* species, phylogenetic lineages (see Fig. 6.2) and general distribution (De-Nova et al. 2018)

Species	Lineage	Distribution
Fouquieria burragei Rose	WD	PD
Fouquieria columnaris (C. Kellogg) Kellogg ex Curran	ESG	PD
Fouquieria diguetii (Tiegh.) I.M.Johnst.	WD	PD
Fouquieria fasciculata (Willd. ex Roem. and Schult.) Nash	ESG	HD
Fouquieria formosa Kunth	WD	SW, TCD
Fouquieria leonilae Miranda	WD	SW
Fouquieria macdougalii Nash	WD	SD
Fouquieria ochoterenae Miranda	WD	SW
Fouquieria purpusii Brandegee	ESG	TCD
Fouquieria shrevei I.M.Johnst	WD	CD
Fouquieria splendens Engelm.	WD	CD, HD, PD, SD

WD Woody clade; *ESG* Early succulent grade; *CD* Chihuahuan Desert; *HD* Hidalguense Desert; *PD* Peninsular Desert; *SW* semidesert relicts; *SD* Sonoran Desert; *TCD* Tehuacán-Cuicatlán Desert

The Chihuahuan Desert, including the Southern associated areas of the Mezquital and Tehuacán-Cuicatlán valleys (Shreve 1942; Zavala-Hurtado and Jiménez 2020, this volume), represents an excellent scenario to study vicariant events promoted by the Neogene orogenesis in Mexico that underlies the development of regional arid lands and deserts. Different studies have shown that mainly as a result of vicariant events, the Chihuahuan Desert has high levels of species richness, super and palaeo-endemism, and an elevated phylogenetic diversity (e.g., Rzedowski 1993; Hernández et al. 2001; Hafner and Riddle 2005, 2011; Riddle and Hafner 2006; Hernández and Gómez-Hinostrosa 2011; Sosa and De-Nova 2012; Sosa et al. 2018).

Fouquieria includes two sister species co-occurring in the Chihuahuan Desert, the broadly distributed *Fouquieria splendens* also found in the Hidalguense, Sonoran, and Peninsular deserts, including different infraspecific lineages, and the microendemic gypsophilic *F. shrevei*, dominant in particular gypsum soils from central and southeastern Coahuila (Fig. 6.1). Both species coexist in the valleys and mountain ranges west and around Cuatro Ciénegas, but in the Cuatro Ciénegas Basin (CCB), despite wide availability of gypsum soils (see Ochoterena et al. 2020, this volume), only *F. splendens* is found (personal observations). We hypothesize that, given their sympatric distribution, it is likely that—in contrast with the rest of the family—these species diverged by a process of ecological speciation mediated by natural selection and local adaptation, in which *F. shrevei* colonized gypsum soils. Here we present some recent information about diversification in *Fouquieria* and discuss the implications of the divergence of its species in the Chihuahuan Desert.

The Diversification of an Ancient Lineage

Although Fouquieriaceae includes only one genus and eleven species, recently dated phylogenies show that it represents an ancient lineage that diverged from Polemoniaceae during the late Cretaceous (ca. 75.54 Ma; Magallón et al. 2015; De-Nova et al. 2018; Fig. 6.2). This period corresponds to the moment at which most angiosperm families originated, a diversification process associated with pronounced tectonic and geological activities in the world, as well as an increase in global temperatures (Magallón et al. 2015). Given their anatomical characteristics, Fouquieriaceae or its extinct ancestors were likely pre-adapted to seasonally dry environments; however, it presents now a high taxonomic distinctiveness from other families, particularly with its sister Polemoniaceae, that sets it as a unique family.

Molecular phylogenies show that the lineage of Fouquieriaceae had a long period with no speciation and very low extinction rates (De-Nova et al. 2018). This evolutionary stasis lasted from its origin, in the late Cretaceous, to the diversification of its sole genus *Fouquieria*, during the Mio-Pliocene (De-Nova et al. 2018).

The divergence of most *Fouquieria* species occurred by vicariant events promoted by the development of regional deserts in North America (De-Nova et al. 2018). The recent diversification of *Fouquieria* produced several microendemisms to regional deserts mainly by vicariance and founder effects, like *F. purpusii* in the

Fig. 6.1 *Fouquieria* species in the Chihuahuan Desert: (A) Adult individuals of *F. splendens* (photo by T Lobato de Magalhães); (B) *F. splendens* flowering (photo by MM Salinas-Rodríguez); (C) Adult individuals of *F. sherevei* (photo by LE Eguiarte); (D) *F. sherevei* flowering (photo by E Scheinvar)

Tehuacán-Cuicatlán Valley, *F. ochoterenae* in the Mexican Southwestern semidesert relicts, *F. burragei* in the Peninsular Desert, and *F. shrevei* in the Chihuahuan Desert (Henrickson 1972; De-Nova et al. 2018).

The Recent Speciation of *Fouquieria* in the Chihuahuan Desert

Geographic phylogenetic structure shows that isolation by distance has been the main driver structuring the genetic variation in *Fouquieria* species, with a vicariant-dispersal history (Fig. 6.2), where most restricted or endemic species are older than widespread species (De-Nova et al. 2018).

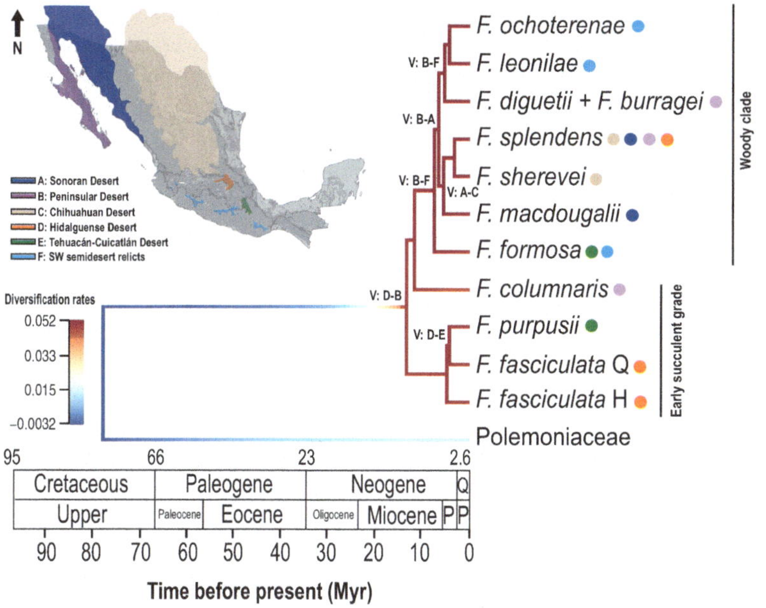

Fig. 6.2 *Fouquieria* diversification dynamics and biogeographic history using plastid DNA sequences based on data from De-Nova et al. (2018). Phylorate plot obtained with BAMM, derived from using a prior on the expected number of shifts equal to 1. "V" indicates vicariance events estimated with BMM in RASP. Color dots indicate distribution areas for the species

An interesting speciation event in the genus, that apparently did not imply vicariance, underlies the divergence between *F. shrevei* and *F. splendens* in the Chihuahuan Desert, which have current overlapping distributions in the Coahuila state (Aguirre-Liguori 2012; De-Nova et al. 2018). This clade diverged by vicariance from *F. macdougalii* from the Sonoran Desert, during the early Pliocene. More recently *F. shrevei* and populations of *F. splendens* diverged sympatrically in the Chihuahuan Desert around 3.99 Ma (De-Nova et al. 2018). As mentioned above, *F. shrevei* is a dominant or common gypsophile plant in some localities with gypsum soils in the Chihuahuan Desert, from central and southeastern Coahuila (Henrickson 1972; Aguirre-Liguori et al. 2014).

Gypsum soils present complex chemical and physical properties, such as a high concentration of sulfur and hard surfaces (Meyer 1986; Ruiz et al. 2003; Palacio et al. 2007), which can promote a diverse assemblage of gypsum adapted endemic plant taxa, in particular in the Chihuahuan Desert (Moore et al. 2014; see Ochoterena et al. 2020, this volume).

Given the restrictive conditions that gypsum soils represent, it is likely that a process of divergent adaptation to these conditions accompanied the divergence of *F. shrevei*. In this scenario the divergence between *F. shrevei* and *F. splendens* could have occurred by a process of ecological speciation, in which barriers to gene flow may have occurred in response to adaptation to contrasting soil conditions (Rundle and Nosil 2005). In addition, it is relevant to mention that while *F. shrevei* has white flowers and is pollinated by insects, *F. splendens* has red flowers and is mainly pollinated by hummingbirds and sometimes bees (Waser 1979; Aguirre-Liguori 2012; McKinney et al. 2012; Aguirre-Liguori et al. 2014). This change of flower coloration and pollinators might contribute to reduced potential gene flow between species and would reduce the presence of hybrids maladapted both to gypsum and non-gypsum soils. Nonetheless as we discuss below, *F. splendens* chloroplast haplotypes have been found in individuals displaying a *F. shrevei* phenotype growing in gypsum soils, indicating that either lineage sorting is incomplete, or that hybrids may occur in low frequencies (Aguirre-Liguori et al. 2014). To determine this, it will be interesting to test the hybridization hypothesis using genome wide markers, and testing soil concentration in populations including hybrids to determine if hybridization is dependent on the selective constraint imposed by gypsum.

The adaptation to gypsum soil is common in the flora of the Chihuahuan Desert (Moore et al. 2014; Ochoterena et al. 2020, this volume) and could be associated with the pluvial and inter-pluvial cycles of the Middle Pleistocene. During the pluvial periods, the xeric flora of the Chihuahuan Desert contracted and remained in southern refugia, while its area was covered by paleolakes and was then expanded during the inter-pluvial periods (Van Devender 1990; Riddle and Hafner 2006; Hafner and Riddle 2011; De-Nova et al. 2018).

Fouquieria splendens is the sole species of Fouquieriaceae that is broadly distributed including infraspecific taxa (Henrickson 1972), implying that dispersal processes occurred from an ancestral distribution in the Sonoran Desert, to the Peninsular, Hidalguense, and Chihuahuan Deserts (1.46–0.25 Ma) (De-Nova et al. 2018). Graham (2010) indicated that the modern version of the North America dry lands vegetation appeared approximately during the Miocene–Pliocene boundary and it became modernized during the dry intervals of the Quaternary period. Graham (2010) highlights the ancestral presence of species pre-adapted to seasonally dry environments, the later spread of these environments in the Miocene, and the increasing edaphic aridity through the generation of coarse soils as some of the main factors that originated in the Mexican dry vegetation. A similar process apparently was involved in the evolution of the complete Fouquieriaceae family and particularly of *F. splendens*.

Genetic Drift and Low Gene Flow in *Fouquieria shrevei*

Given its adaptation to gypsum soil, *F. shrevei* represents an interesting species to understand the importance of gene flow and genetic drift in the differentiation of populations. Gypsum soils in the Chihuahuan Desert show a patchy distribution, generating gypsum islands of different sizes that are also separated by different distances (Moore and Jansen 2007; Moore et al. 2014; Ochoterena et al. 2020, this volume). However, *F. shrevei* produces dry seeds that are wind-dispersed (Henrickson 1972; Humphrey 1974), which could promote long-distance dispersal, as in other *Fouquieria* species (Henrickson 1972; Humphrey 1974; Redfern 2008). In order to understand the dynamics between large dispersal potential (wind dispersal) and reduced geographic size, along with a separated distribution of populations (adaptation to gypsum), demographic, ecological, and microevolutionary processes have been studied in *F. shrevei* (Aguirre-Liguori 2012; Aguirre-Liguori et al. 2014).

Aguirre-Liguori (2012) estimated the census size of *F. shrevei* populations, based on the detailed taxonomic monograph of Henrickson (1972), on the CONABIO database (http://www.conabio.gob.mx/remib), and on fieldtrips. Eighteen populations of the species, found in gypsum "islands," are currently known. These populations are separated by 10–107 km (Fig. 6.3). Aguirre-Liguori (2012) counted the number of individuals in five sampled populations, and based on the size area for each population he estimated that, on average, populations of *F. shrevei* have a density of 61 ind/ha. By averaging the area occupied by five sampled populations and multiplying the number of known populations they also estimated that populations occupy a total area of between 20 and 30 ha. Finally, Aguirre-Liguori (2012) concluded that the total extant census size of *F. shrevei* could be between 1235 and 1835 individuals. This population size is consistent with estimated small populations in

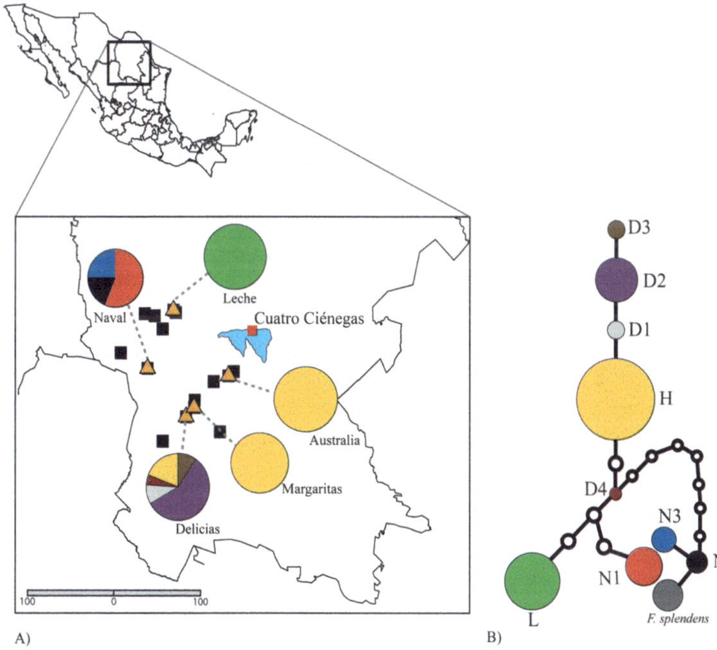

Fig. 6.3 (**A**) Populations of *Fouquieria shrevei* and haplotype frequencies based on data from Aguirre-Liguori et al. (2014). Black squares represent described populations of *F. shrevei*; Yellow triangles represent sampled populations; the blue polygon shows Cuatro Ciénegas; each pie plot shows the haplotype frequency in each sampled population. (**B**) Haplotype network showing the relationship between *F. shrevei* and *F. splendens* (in dark gray) haplotypes. Circle size is proportional to the number of haplotypes

other plants, and this small size would increase the effects of genetic drift (Ellstrand and Elam 1993).

To better understand the effects of gene flow and genetic drift, Aguirre-Liguori et al. (2014) analyzed sequence variation in three chloroplast intergenic spacers from five populations. As found in *Gypsophila struthium*, another gypsum endemic species (Martínez-Nieto et al. 2013), these authors reported that many populations had fixed haplotypes, indicating strong genetic drift (Fig. 6.3). The haplotype diversity in *F. shrevei* is associated with the effective population size, as is expected from theory (Ellstrand and Elam 1993). Some populations, like Las Delicias and La Naval, have currently relatively large population sizes (hundreds of individuals) and have apparently maintained similar sizes for a long term (i.e., no recent bottlenecks) plus they harbor multiple haplotypes. In contrast, other populations with smaller population sizes (e.g., Australia, Leche, and Margaritas) are fixed for a single haplotype (Fig. 6.3). These small populations presumably suffered a bottleneck during the recent colonization of the Pleistocene pluvial lakes where they are found (Aguirre-Liguori et al. 2014).

According to a generalized skyline plot, Aguirre-Liguori et al. (2014) found that effective population sizes of *F. shrevei* have remained unchanged for the past 30–40,000 yr, indicating that the evolutionary history and genetic structure of *F. shrevei* seem to be relatively old, as has been hypothesized for gypsophile species in general (Johnston 1941; Rzedowski 1991; Moore and Jansen 2007; Moore et al. 2014).

Consistent with a static demography maintained for a long time, Aguirre-Liguori et al. (2014) found that the diversity at the total species level is high, yet the majority of the haplotypes are private, i.e., only found in a single population (Table 6.2; Fig. 6.3), even if *F. shrevei* has a narrow total distribution, which is no larger than 200 × 200 km (Fig. 6.3). Interestingly, haplotypes displayed a wide range of mutational steps among them (Aguirre-Liguori et al. 2014), which is concordant with an old species with a prolonged demographic stasis. In the case of *F. shrevei* calibrated phylogenies suggest that the separation of some of its populations occurred a long time ago, during the Pleistocene around 1.39–0.38 Myr (De-Nova et al. 2018).

By performing a PERMUT analyses (Pons and Petit 1996), Aguirre-Liguori et al. (2014) found that genetic similarity between haplotypes was associated with

Table 6.2 Total and within-population diversity in *Fouquieria shrevei*

Locality	N	No. of haplotypes	S	H_d	D (Tajima)
La Sierra de Australia	20	1	0	0	0
Las Delicias	21	5	2	0.648	0.025
Las Margaritas	17	1	0	0	0
La Leche	20	1	0	0	0
La Naval	16	3	8	0.625	2.488[a]
Total	94			0.743	−0.019

N Number of studied individuals; S Number of polymorphic sites; H_d Haplotype diversity
[a]significant values ($P < 0.05$)
Modified from Aguirre-Liguori et al. 2014

geographic proximity between populations, indicating that there is a phylogeographic structure. These authors detected that isolation by distance affects the genetic structure of *F. shrevei*, indicating that there is a significant positive correlation between pairwise F_{ST} (genetic differentiation) and geographic distance, where geographically closer populations share more haplotypes (Aguirre-Liguori et al. 2014).

To test historical processes, Aguirre-Liguori (2012) performed species distribution models (SDM) of *F. shrevei* populations. Comparing the Pleistocene and current SDM, he found that most populations have remained in their ancestral distribution, and only a few areas have been recently established (Fig. 6.4). However, it is interesting to note that the current available potential niche model distribution has increased, although predicted areas have not been colonized yet. We do not know if *F. shrevei* individuals have not had enough time to colonize new sites, or if there are other environmental requirements not used in the analysis of the potential distribution, in particular soil characteristics besides gypsum availability.

Considering haplotype networks (many mutation steps) and distribution of within-population diversity, Aguirre-Liguori et al. (2014) proposed that *F. shrevei* populations have suffered high levels of founder effects associated with the colonization of new "islands," i.e., gypsum patches in the desert, through seed wind dispersal. These results suggest that the colonization of new gypsum rich soils has been an infrequent process, which has not allowed populations to expand to their entire

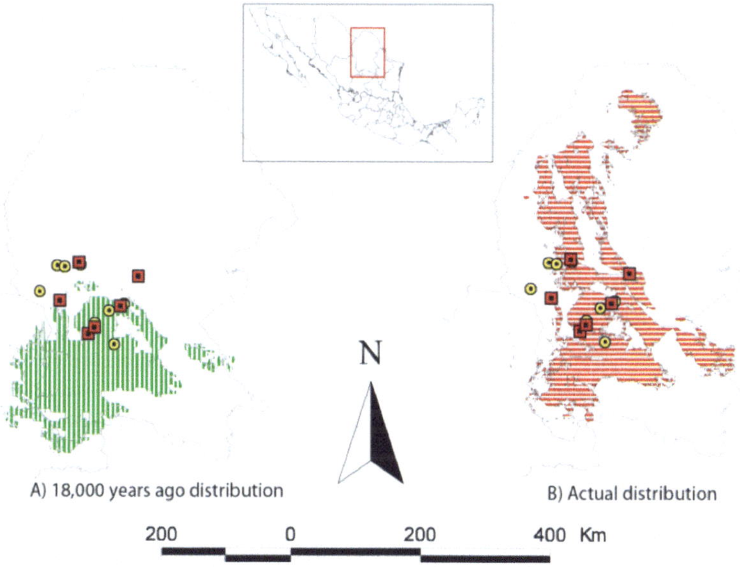

Fig. 6.4 Distribution model of *Fouquieria shrevei*. The green area shows the Pleistocene potential distribution 18,000 years ago. The red area represents the actual potential distribution. Yellow dots represent populations of the species; red dots represent the populations that were used in the analysis. (**A**) Pleistocene distribution model (18,000 years ago); (**B**) Present distribution model

geographic potential distribution (Fig. 6.4). The rare foundation of new populations by seed dispersal and few original individuals is consistent with the reduced diversity found in other populations like La Leche, Las Margaritas, and La Sierra de Australia. The Cuatro Ciénegas valley is an interesting example of how *F. shrevei* has not attained its entire potential distribution where, despite the wide availability of gypsum soils (Ochoterena et al. 2020, this volume), only *F. splendens* is found.

Overall, Aguirre-Liguori et al. (2014) concluded that gypsum adaptation has dominated the genetic structure of *F. shrevei* by imposing limited distribution and increased genetic drift. However, seed dispersal has allowed the establishment of new populations that have been fixed with new haplotypes due to strong founder effects.

On the Origin of *Fouquieria shrevei*

The sympatric speciation of *F. shrevei* from *F. splendens* in the Chihuahuan Desert was estimated by De-Nova et al. (2018) to have occurred 3.99 Myr ago, during the Pliocene. However, as we mentioned above, genetically similar haplotypes were found between individuals of *F. splendens* and some of *F. shrevei* in the La Naval population (Aguirre-Liguori et al. 2014). Haplotype sharing is common in closely related plant species and it has been shown that these patterns could be due either to hybridization between species or, as in these *Fouquieria* species, to ancestral polymorphisms (Avise 2000; Moore and Jansen 2007; Fehlberg and Ranker 2009; Escobar et al. 2011). Individuals of La Naval population have a characteristic *F. shrevei* phenotype, so it is likely that the related haplotypes come from incomplete lineage sorting. Besides, the La Naval population is large, decreasing the impact of genetic drift and increasing the time needed to complete lineage sorting (Avise 2000).

Conclusions

The sympatric *F. shrevei* and *F. splendens* represent the most recent diverging species in Fouquieriaceae, during the Pliocene. Low levels of gene flow between populations, infrequent colonization events, genetic drift, and very likely strong natural selection due to local conditions (not tested yet) could have promoted a rapid differentiation in these species. The use of next generation sequencing methods will allow to better understand the demographic history of *Fouquieria* species and, more specifically, to understand the role that selection has played in the divergence of *F. shrevei*, including the detection of genes involved in gypsum soil adaptation and genomic differentiation. In the long term, the rapid differentiation could promote speciation events and in turn generate new species that would form monophyletic groups of endemic and gypsophile plants (e.g., Moore and Jansen 2007), increasing regional diversity (Aguirre-Liguori et al. 2014).

Similar processes could be occurring in the lineage formed by the sympatric *F. burragei* and *F. diguetii*, from the Peninsular Desert, where phylogenetic boundaries are not totally differentiated. Infraspecific taxa in *F. splendens* could have similar microevolutionary processes promoting a more recent divergence initiated during Pleistocene in different populations throughout the wide distribution range of this species. Interaction between selection, genetic drift, and gene flow could promote speciation events that in turn would increase regional diversity and could explain the high number of narrow endemics associated with the North American deserts.

Acknowledgements The results presented here were partially funded by PAPIIT-UNAM [grant 202310 to S Magallón], CONACyT [grants 2004-C01-46475 to LEE and CB-2014-243454 to JAD-N]. We are deeply grateful to Susana Magallón and Enrique Scheinvar for the supervision of the studies where the data for this chapter came from and to Erika Aguirre-Planter for reviewing the manuscript.

References

Aguirre-Liguori JA (2012) Genética de poblaciones y filogeografía de *Fouquieria shrevei* (Fouquieriaceae). Tesis de Licenciatura, Facultad de Ciencias, Universidad Nacional Autónoma de México, México DF, Mexico

Aguirre-Liguori JA, Scheinvar E, Eguiarte LE (2014) Gypsum soil restriction drives genetic differentiation in *Fouquieria shrevei* (Fouquieriaceae). Am J Bot 101:730–736

Avise JC (2000) Phylogeography: The history and formation of species. Harvard University Press, Cambridge, MA, USA

Bremer B, Bremer K, Heidari N, Erixon P, Olmstead RG, Anderberg AA, Källersjö M, Barkhordarian E (2002) Phylogenetics of asterids based on three coding and three noncoding chloroplast DNA markers and the utility of noncoding DNA at higher taxonomic levels. Mol Phylogenet Evol 24:274–301

De-Nova JA, Sánchez-Reyes L, Eguiarte LE, Magallón S (2018) Recent radiation and dispersal of an ancient lineage: the case of *Fouquieria* (Fouquiericeae, Ericales) in North American deserts. Mol Phylogenet Evol 126:92–104

Ellstrand NC, Elam DR (1993) Population genetic consequences of small population size: implications for plant conservation. Annu Rev Ecol Syst 24:217–242

Escobar PR, Gernandt DS, Piñero D, Garcillán PP (2011) Plastid DNA diversity is higher in the island endemic Guadalupe Cypress than in the continental Tecate Cypress. PLoS One 6:e16133

Fehlberg SD, Ranker TA (2009) Evolutionary history and phylogeography of *Encelia farinosa* (Asteraceae) from the Sonoran, Mojave, and Peninsular Deserts. Mol Phylogenet Evol 50:326–335

Findley JS (1969) Biogeography of southwestern boreal and desert mammals. Misc Pub Univ Kansas Mus Nat Hist 51:113–128

Graham A (2010) Late Cretaceous and Cenozoic history of Latin American vegetation and terrestrial environments. Missouri Botanical Garden Press, St. Louis, MO, USA

Hafner DJ, Riddle BR (2005) Mammalian Phylogeography and evolutionary history of Northern Mexico's deserts. In: Cartron JLE, Ceballos G, Felger RS (eds) Biodiversity, ecosystems, and conservation in Northern Mexico. Oxford University Press, New York, USA, pp 225–245

Hafner DJ, Riddle BR (2011) Boundaries and barriers of North American warm deserts: An evolutionary perspective. In: McGowan A, Upchurch A, Slater PC (eds) Palaeobiogeography biodiversity space time. CRC Press, Systematics Association Special Volume Series. Boca Raton, FL, USA, pp 75–114

Henrickson J (1967) Pollen morphology of the Fouquieriaceae. Aliso 6:137–160

Henrickson J (1969a) Anatomy of periderm and cortex of Fouquieriaceae. Aliso 7:97–126

Henrickson J (1969b) An introduction to the Fouquieriaceae. Cact Succ J (Los Angeles) 41:97–105

Henrickson J (1972) A taxonomic revision of the Fouquieriaceae. Aliso 7:439–537

Hernández HM, Gómez-Hinostrosa C (2011) Areas of endemism of Cactaceae and the effectiveness of the protected area network in the Chihuahuan Desert. Oryx 45:191–200

Hernández HM, Gómez-Hinostrosa C, Bárcenas R (2001) Diversity, spatial arrangement, and endemism of Cactaceae in the Huizache area, a hot-spot in the Chihuahuan Desert. Biodivers Conserv 10:1097–1112

Hubbard JP (1973) Avian evolution in the aridlands of North America. Liv Bird 12:155–196

Humphrey RR (1974) The boojum and its home: *Idria columnaris* Kellogg and its ecological niche. University of Arizona Press, Tucson, AZ, USA

Johnston IM (1941) Gypsophily among Mexican desert plants. J Arnold Arbor 22:145–170

Magallón S, Gómez-Acevedo S, Sánchez-Reyes L, Hernández-Hernández T (2015) A metacalibrated time-tree documents the early rise of flowering plant phylogenetic diversity. New Phytol 207:437–453

Martínez-Nieto MI, Segarra-Moragues JG, Merlo E, Martínez-Hernández F, Mota JF (2013) Genetic diversity, genetic structure and phylogeography of the Iberian endemic *Gypsophila struthium* (Caryophyllaceae) as revealed by AFLP and plastid DNA sequences: connecting habitat fragmentation and diversification. Bot J Linn Soc 173:654–675

McKinney AM, CaraDonna PJ, Inouye DW, Barr B, Bertelsen CD, Waser NM (2012) Asynchronous changes in phenology of migrating Broad-tailed Hummingbirds and their early-season nectar resources. Ecology 93:1987–1993

Meyer SE (1986) The ecology of gypsophile endemism in the Eastern Mojave Desert. Ecology 67:1303–1313

Moore MJ, Jansen RK (2007) Origins and biogeography of gypsophily in the Chihuahuan Desert plant group *Tiquilla* subg. Eddya (Boraginaceae). Sys Bot 32:392–414

Moore MJ, Mota JF, Douglas NA, Olvera HF, Ochoterena H (2014) The ecology, assembly and evolution of gypsophile floras. In: Rajakaruna N, Boyd RS, Harris T (eds) Plant ecology and evolution in harsh environments. Nova Publishers, New York, NY, USA, pp 97–128

Morafka DJ (1977) A biogeographical analysis of the Chihuahuan Desert through its Herpetofauna. Dr. W. Junk B.V. Publishers, The Hague, Netherlands

Ochoterena H, Flores-Olvera H, Gómez-Hinostrosa C, Moore MJ (2020) Gypsum and plant species: A wonder of Cuatro Ciénegas, and the Chihuahuan desert. In: Mandujano MC, Pisanty I, Eguiarte LE (eds) Plants diversity and ecology in the Chihuahuan desert, Cuatro Ciénagas Basin, Chihuahuan desert, Mexico. an endangered hyperdiverse oasis. Springer International Publishing AG, part of Springer Nature, Cham, Switzerland

Palacio S, Escudero A, Monsterrat-Martí G, Maestro M, Milla R, Albert M (2007) Plants living on gypsum: beyond the specialist model. J Bot 99:333–343

Pons O, Petit R (1996) Measuring and testing genetic differentiation with ordered versus unordered alleles. Genetics 144:1237–1245

Redfern JL (2008) Phylogeny and phylogeography of Fouquieriaceae: Evolutionary history of a North American desert plant family. Ph.D. dissertation, University of New Mexico, Albuquerque, NM, USA

Riddle BR, Hafner DJ (2006) A step-wise approach to integrating phylogeographic and phylogenetic biogeographic perspectives on the history of a core North American warm deserts biota. J Arid Environ 66:435–461

Ruiz JM, López-Cantarero I, Rivero RM, Romero L (2003) Sulphur phytoaccumulation in plant species characteristic of gypsipherous soils. Int J Phytoremediation 5:203–210

Rundle HD, Nosil P (2005) Ecological speciation. Ecol Lett 8:336–352
Rzedowski J (1991) El endemismo en la flora fanerogámica mexicana: una apreciación analítica preliminar. Acta Bot Mex 15:47-64
Rzedowski J (1993) Diversity and origins of the Phanerogamic Flora of Mexico. In: Ramamoorthy TP, Bye R, Lot A, Fa J (eds) Biological diversity of Mexico: origins and distribution. Oxford University Press, Oxford, UK, pp 129–148
Schönenberger J, Anderberg AA, Sytsma KJ (2005) Molecular phylogenetics and patterns of floral evolution in the Ericales. Int J Plant Sci 166:265–288
Schönenberger J, von Balthazar M, Sytsma KJ (2010) Diversity and evolution of floral structure among early diverging lineages in the Ericales. Philos T Roy Soc B 365:437–448
Shreve F (1942) The desert vegetation of North America. Bot Rev 8:195–246
Sosa V, De-Nova JA (2012) Endemic angiosperm lineages in Mexico: hotspots for conservation. Acta Bot Mex 100:293–315
Sosa V, De-Nova JA, Vázquez-Cruz M (2018) Evolutionary history of the flora of Mexico: dry forests cradles and museums of endemism. J Syst Evol 56:523–536
Sytsma KJ, Walker JB, Schönenberger J, Anderberg AA (2006) Phylogenetics, biogeography, and radiation of Ericales. Annual BSA Meeting, Botany 2006 Abstract 71, Chico, CA, USA
Van Devender TR (1990) Late quaternary vegetation and climate of the Chihuahuan Desert, United States and Mexico. In: Betancourt JL, Van Devender TR, Martin PS (eds) Packrat middens: the last 40,000 years of biotic change. The University of Arizona Press, Tucson, AZ, USA, pp 104–133
Waser NM (1979) Pollinator availability as a determinant of flowering time in Ocotillo (*Fouquieria splendens*). Oecologia 39:107–112
Zavala-Hurtado JA, Jiménez M (2020). Diversity and uniqueness at its best. Vegetation of the Chihuahuan Desert. In: Mandujano MC, Pisanty I, Eguiarte L (eds.) Plant diversity and Ecology in the Chihuahuan desert. Springer International, Cham, Switzerland

Chapter 7
Between the Arid and the Opulent: Plant Resources of the Mexican Desert

Andrea Martínez Ballesté, Thalía Iglesias Chacón, and María C. Mandujano

Abstract Wild species collected in Mexico are important resources for the subsistence of its inhabitants. Particularly in the arid and semi-arid regions of northeastern Mexico, where agriculture and livestock are limited by climatic conditions, human groups that inhabit these sites depend on the rich diversity of its flora and fauna. The largest desert in North America, the Chihuahuan Desert, is one of the world's biologically richest deserts. To date, extraction of wild resources has been performed sustainably. The farmers that live in this region and depend on the collection of natural resources are mindful of their harvesting practices to secure them as a continuing source of income. The farmers in the area have a wealth of knowledge about the biological characteristics and the most appropriate extraction methods for the native plants. The extraction of candelilla, oregano, mesquite, and lechuguilla remains a very important activity in this region, and the description from farmers of how they harvest these plants is an indication of the sophistication and hard work that their extraction involves.

Keywords Candelilla · Chihuahuan Desert · Cuatro Ciénegas · Lechuguilla · Mesquite · Oregano

---

A. Martínez Ballesté (✉)
Jardín Botánico, Instituto de Biología, Universidad Nacional Autónoma, UNAM de México, Mexico City, Mexico
e-mail: andrea.martinez@ib.unam.mx

T. Iglesias Chacón
Independent editor and historian, Mexico City, Mexico

M. C. Mandujano
Departamento de Ecología de la Biodiversidad, Instituto de Ecología, Universidad Nacional Autónoma de México, UNAM, Mexico City, Mexico

M. C. Mandujano et al. (eds.), *Plant Diversity and Ecology in the Chihuahuan Desert*, Cuatro Ciénegas Basin: An Endangered Hyperdiverse Oasis,
https://doi.org/10.1007/978-3-030-44963-6_7

While raising livestock and practicing agriculture have become the modern-day norms for subsistence for many, the importance of harvesting native plant and animal species has historical importance from both cultural and economics standpoints. Some species have in fact acquired national and international demand, giving them considerable commercial importance, and their commercialization has sustained the development of many rural populations (Eccardi 2006).

In the arid and semi-arid zones of northeastern Mexico, water scarcity limits the development of agricultural and livestock activities. Likewise, managing resources in common agroforestry systems in tropical areas, like family gardens, is not a common practice in arid areas. Due to the lack of water, the diversity of plants that can be cultivated in these sites is very low and ornamental plants tend to predominate. Nonetheless, although it appears that resources are scarce in arid zones, the human groups that have inhabited these sites have found a rich diversity of flora and fauna that they have historically exploited for different purposes (Nabhan 1985; Cervantes-Ramírez 2002).

The state of Coahuila is in the Chihuahuan Desert. It is estimated that about 25% of the known cacti species are found in this desert (Zavala-Hurtado and Jiménez 2020, this volume), beside the largest diversity of bees in the world, 150 butterfly species, 120 lizard species, 260 bird species, and 120 mammalian species. It is the only desert of the world that has permanent wetlands, located in the region of Cuatro Ciénegas, where important populations of endemic fish live. The extraction of natural plant and animal resources in this region continues to be of great economic importance to many local farmers who seek to use these resources in a sustainable way (López-Carrera et al. 2004). Harvesting wild plants requires not only an exhaustive search in the deserts, but also a solid knowledge on the biological characteristics of plants and the most appropriate forms of extraction. In municipalities like Cuatro Ciénegas, the economic activity of many *ejidos* (areas of communal land often used for farming) still depends on the extraction of resources, such as candelilla wax (Rosas Barrera et al. 2020, this volume), oregano, mesquite wood, and *lechuguilla* fibers (Cervantes-Ramírez 2002).

Lechuguilla

L*echuguilla* (*Agave lechuguilla)* is a small agave species that is extracted for its soft but resistant fibers or *ixtle* (Nahuatl name used to refer to agave fibers), which have multiple industrial applications. This plant grows across New Mexico, Arizona, and Texas, in the southwestern United States, crossing into the Mexican states of Chihuahua, Coahuila, Nuevo León, Durango, Zacatecas, and San Luis Potosí, up to the northern portion of the state of Querétaro.

L*echuguilla* is used to make ropes, headbands, brushes, saddle straps, and bridles for yokes, as well as other commonly used objects. The *lechuguilla* fiber has also been used in the elaboration of paper money, denim clothes, and various crafts. Furthermore, during the dry season, it is ground and given to cattle as a complement to their diet. In the Universidad Autónoma de Nuevo León, current research explores the possible use of *lechuguilla* for construction since, when mixed with concrete, it adds flexibility and flex-load capacity.

In addition, the softness, heat resistance, and elongation capacity of *ixtle* make it coveted in the international market. Its fibers are used to make a wide variety of brushes, from those used daily in baths to those used to clean railroad cars. Moreover, because this natural fiber does not become magnetized, it can be used to clean delicate pieces of jewelry (Reyes-Agüero et al. 2000).

However, obtaining the fiber is an arduous process with meager rewards. First, one must collect the plants and remove the buds and central parts of the plant—the best fibers are obtained from the tender leaves. The leaves are then scraped with a knife to separate the *guishe* (local name for the succulent leaf tissue), which has a high content of saponins and can be used as a degreaser and even as a remedy against dandruff. Once cleaned, the fibers are laid out to dry in the sun for a day or two.

The *ixtleros* (people that produce the fiber) are paid 0.73 cents of a dollar per kilo of the fiber of *lechuguilla*. It is estimated that in a full day of work, using manual methods, three to four kilos can be obtained by one person. Machines that offer a higher performance have been developed to do the scraping mechanically, but the fibers obtained from them are of lower quality; hence, many scrapers do not consider the use of machines a good option.

Lechuguilla is better than any synthetic fiber for many applications and has therefore become a commercially important plant. However, little of its commercial value ever reaches the hands of the harvesters and scrapers because of a long value chain that leaves very little profit for producers (Jongitud 2007).

Oregano

What we know as Mexican oregano is a group of 40 species belonging to four botanical families and has nothing to do, taxonomically speaking, with *Origanum* spp. or European oregano; nonetheless, they do share several chemical compounds that give them their distinct aroma. Several studies have shown that the quality of the essential oils of at least two species of Mexican oregano (*Lippia graveolens* and *L. berlandieri*), frequently used in the Chihuahuan Desert, is superior to the European kind (Huerta 1997; Carrera et al. 2005).

Most of us know oregano as an important ingredient for the traditional Mexican soup called *pozole*, but the qualities of oregano go far beyond its flavor. In traditional medicine, when it is prepared as an infusion, it benefits digestion; it helps control asthma, cough and colics; it is employ against intestinal worms; and it is used for the treatment of golden staphylococcus and athlete's foot (Huerta 1997). Oregano's essential oil is used in the food industry because it has antioxidant properties that delay the decomposition of food—it is added to sausages, canned foods, sauces, dressings, and pickles. Because of its antimicrobial qualities, it is also used in the manufacturing of condoms since it inhibits the development of bacteria that damage latex products (Huerta 1997).

Oregano, like many other products of the Mexican countryside, is used and marketed through a production chain that provides little to the harvesters. In northern Jalisco, for example, harvesters were paid 0.3 cents of dollar per kilo, while in local markets the same amount was sold for 1.3 dollars—a 400% increase. Meanwhile, when packaged under three transnational labels and sold in self-service stores, the product can sell for 13 dollars per kilo—more than 4000 % more than the amount the harvester receives, and this for a product that needs little processing prior to its commercialization.

Construction of essential oil extraction plants has been proposed as a potentially more economically promising alternative for oregano producers—one ton of oregano produces 28 liters of oil that can be sold for three thousand pesos each. In the semi-desert of Peñamiller, in the Sierra Gorda of Querétaro, a new project aimed at the sustainable use of this resource and the extraction of its essential oil for commercialization is being carried out (Carrera et al. 2005). Similarly, *ejidatarios* (farmers) could produce and market oregano for medicinal and culinary purposes at a small scale in such a way that profits could go directly to the harvesters and land owners. Thinking of direct benefits for land owners by using the desert without transforming it will translate, in both the short and the long term, to a truly sustainable strategy for the conservation of the Mexican desert.

Mesquite

For thousands of years mesquite has been a source of many benefits for the people of northern Mexico and the southern United States (Nabhan 2018). In the Chihuahuan desert, three species of mesquites are recognized, *Prosopis laevigata*, *P. glandulosa*, and *P. reptans*. These species grow in Coahuila, and the first two are found within the Cuatro Ciénegas region (Standley 1926). Some authors have defended that mesquite is the key species maintaining an innumerable set of important and unique desert birds, such as mockingbirds and roadrunners. Likewise, these plants provide a resource to a great diversity of other species, including an impressive number of insects. They also aid in the establishment of many cacti by preventing soil erosion and providing shade (Golubov et al. 2001).

Components of the mesquite plant have many uses: the pods and seeds for food, the leaves and bark for medicinal purposes, and the wood for fuel and for the construction and manufacture of furniture and crafts (Nabhan 2018). Its distribution was favored during the Spanish occupation of Mexico, due to dispersal by cows and horses introduced by the Spaniards. Álvar Núñez Cabeza de Vaca in 1,555 accounts in *Naufragios* how mesquite was part of the diet of the *Avavares*, an ancient semi-nomadic people of northern Mexico, who are now extinct: "it is a fruit that is very bitter when it is on the tree and it is similar to carob, and it is eaten with soil and with it, it is sweet and good to eat." Followed by a description of the grinding process,

Núñez Cabeza de Vaca explains that after consuming products made with mesquite flour, the bellies of the Indians were "very big and full" (Nuñez Cabeza de Vaca 1985).

Mesquite was also used as a component of the fuel used during mining activities in the north of the country, which was booming in the eighteenth century, reducing its populations considerably. Likewise, during the Mexican Revolution, large mesquite plants were used to hang men and as fuel for the trains.

During the 70s, following a presidential decree, efforts were made to eliminate mesquite to open space for cattle raising. This was done through the application of herbicides, and by direct cutting and burning of the plants. Reducing mesquite is still profitable to ranchers. Nevertheless, mesquite remains a predominant plant of the Mexican semi-desert landscape.

Mesquite wood is valuable as fuel because it is used to produce charcoal and because its reddish color, flexibility, and tolerance for high humidity make it desirable for wooden furniture. The wood is also used for shoe stretchers, flooring, and tool handles and mesquite honey is a favorable treat. This plant is used in Mexico to produce cooking flours, *atole* (a warm drink made from mesquite flour), and it can also be a high-quality forage for cattle (Martínez-Sifuentes web page n.d.). Given its value, there are programs to diagnose mesquite pests and diseases in natural populations based on warnings to phytosanitary authorities notice of twig-girdler and borer outbreaks.

Candelilla Wax

One of the most important non-timber forest activities in the Chihuahuan Desert is the extraction of candelilla wax obtained from the species *Euphorbia antisiphylitica*. This plant grows as a shrub between 20 and 110 cm tall; its aerial part comprises numerous straight stalks with a green glaucous color. Stalks have a waxy layer against water loss, allowing it to remain green despite extreme heat of arid zones. Candelilla grows from Puebla and Hidalgo to the north of Zacatecas, west of Nuevo Leon, east of Durango, Chihuahua, and in almost all the state of Coahuila (Canales-Gutiérrez et al. 2005).

Mexico is the only producer of candelilla wax in the world and the state of Coahuila contributes the majority to production (80%) (Canales-Gutiérrez et al. 2005; Martínez-Ballesté and Mandujano 2013). The candelilla wax route begins in the deserts of northern Mexico, where it is grown and processed, and ends as an ingredient in products all over the world. Because it is harmless to ingest, candelilla wax is used in the food industry to delay maturation and increase the shelf life of products, such as cheese, fruits, and vegetables. In the cosmetic industry, candelilla wax is an attractive ingredient for lotions, lipsticks, and balms, and in makeup and hair care products because it also has moisturizing properties. It is used to add brightness and consistency to medications and confectionary, such as *dragees* (hard shelled sweets) or as part of the base used for chewing gum. Candelilla is also used in crayons because it gives them flexibility and is non-toxic (Tunell 1981). Candelilla wax has many other industrial uses, including as a protective coating against humidity and the corrosion of metals; as an insulator of electronic devices; for ink retention in the manufacture of carbon paper; in the manufacture of matches, wood agglomerates, and refractory bricks; and in the construction industry as a polish, waterproofing layer, and even as a protector against radiation. Wicks and detonators of explosives are also protected with candelilla wax (Tunell 1981).

The demand for candelilla wax worldwide is enormous. Since the beginning of the twentieth century, Mexico has exported candelilla wax to the USA, Japan, Germany, and England. In 1980, 200,000 tons of live plants were harvested, resulting in an annual production of 4,000 tons of the wax. In 1992, annual wax production was 1,672 tons. Since the signing of the North American Free Trade Agreement in 1994, several national and international companies have competed for the commercialization of the wax (Canales-Gutiérrez et al. 2005). By 1993, the FAO estimated the direct economic income to Mexico from the sale of candelilla at three million dollars annually; however, in rural areas, *candelilleros*, those who harvest the plants and extract the wax, realize little of these financial gains. From 100 kilos of candelilla plant, only 2–3 kilos of raw wax are extracted, yielding a meager profit of approximately 1.4 dollar per kilo to those at the production stages. However, this product is further refined and can be worth 20 times this amount, yielding much greater profits to those further along in the distribution stages (Martínez-Ballesté and Mandujano 2013). To regulate the extraction of these desert-plant resources, the ministry of the environment (SEMARNAT) has regulated their use with the official standard NOM-018-RECNAT-1999.

Final considerations

Among the plants harvested from the arid and semi-arid regions of Mexico there is considerable diversity to the potential economic benefits.

Mexico has historically been a producer of cheap raw materials that feed the international industry. As producers, the country reaps only the profits at the bottom of the economic profile for these products, while the greatest gains are to be made toward the later stages of distribution. For Mexico to benefit from the upper part of the economic profile of these products, industry for the manufacture and exportation of finished products must be encouraged.

Alternatively, the number of intermediaries involved in the production-distribution chain could be reduced, allowing producers the opportunity to realize more of the profits from further down the chain. Not only would this scenario create a more just distribution of profits, but it would also ensure that the knowledge acquired during generations that allows for sustainable resource management is conserved.

Acknowledgements We thank the support granted by SEMARNAT and CONACYT (0350) for fieldwork in the San Lorenzo *ejido*, and for the support of CONACYT on the postdoctoral fellowship of the first author. Thank you very much to the members of the San Lorenzo *ejido* in Cuatro Ciénegas for teaching us about the hard work of the *candelilleros*.

References

Canales-Gutiérrez E, Canales-Martínez V, Zamarrón-Rodríguez EM (2005) Candelilla: gotas de cera del desierto mexicano alrededor del mundo. In: López C, Chanfón S, Segura G (eds) La riqueza de los bosques mexicanos: más allá de la madera. Experiencias de comunidades rurales. SEMARNAT, CECADESU, CONAFOR, PROCYMAF, CIFOR, Mexico city, Mexico, pp 101–107

Carrera A, Flores DA, Tejeda de Camilo J, Ocampo Velázquez R (2005) Orégano: oro verde del semidesierto. In: López C, Chanfón S, Segura G (eds) La riqueza de los bosques mexicanos: más allá de la madera. Experiencias de comunidades rurales. SEMARNAT, CECADESU, CONAFOR, PROCYMAF, CIFOR, Mexico city, Mexico, pp 61–65

Cervantes-Ramírez MC (2002) Plantas de importancia económica en las zonas áridas y semiáridas de México. Universidad Nacional Autónoma de México, Instituto de Geografía, Mexico city

Eccardi F (2006) México, valor de origen. Banco Santander, Mexico city, Mexico

Golubov J, Mandujano MC, Eguiarte L (2001) The paradox of mezquites (*Prosopis* spp.): invading species or biodiversity enhancers? Boletín de la Sociedad Botánica de México 69:21–28

Huerta C (1997) Orégano mexicano, oro vegetal. Biodiversitas 3(15):8–13

Jongitud E (2007) Vivir de la lechuguilla, una opción para los habitantes del semidesierto. Revista México forestal, CONAFOR agosto-septiembre.

López-Carrera C, Zazueta-Angulo X, Porras-Andujo CR (2004) Notas sobre el aprovechamiento del medio ambiente noreste: reconocimiento del terreno, inicio de la cacería. Ra Ximhai 1(1):39–59

Martínez-Ballesté A, Mandujano MC (2013) The consequences of harvesting on regeneration of a non- timber wax producing species (*Euphorbia antisyphilitica* Zucc.) of the Chihuahuan desert. Economic Botany 67(2):121–136

Martínez-Sifuentes S. (n.d.) Enamórate del mezquite. http://www.cerritos-slp.com/ARTICULOS/mezquite/mezquite.htm. Accessed 10 Jan 2006

Nabhan GP (1985) Gathering the desert. University of Arizona Press, Tucson, AZ, USA

Nabhan GP (2018) Mesquite. Chelsea Green Publishing, White River Junction, VT, USA

Nuñez Cabeza de Vaca A (1985) Naufragios. Alianza Editorial, Madrid, Spain

Reyes-Agüero JA, Aguirre-Rivera J, Peña-Valdivia C (2000) Biología y aprovechamiento de *Agave lechuguilla* Torrey. Boletín de la Sociedad Botánica de México 67:75–88

Rosas Barrera MD, Golubov J, Pisanty I, Mandujano MC (2020) Effect of reproductive modes on the population dynamics of an endemic cactus from Cuatro Ciénegas. In: Mandujano MC, Pisanty I, Eguiarte L (eds.) Plant diversity and Ecology in the Chihuahuan desert. Springer International, Cham, Switzerland

Standley PC (1926) Trees and shrubs of Mexico. Smithsonian Press, Washington, DC, USA

Tunell C (1981) Wax, men and money. Texas Historical Commission. http://www.texasbeyondhistory.net/waxcamps/techniques.html. Accessed 10 Jan 2006

Zavala-Hurtado JA, Jiménez M (2020). Diversity and uniqueness at its best. Vegetation of the Chihuahuan Desert. In: Mandujano MC, Pisanty I, Eguiarte L (eds.) Plant diversity and Ecology in the Chihuahuan desert. Springer International, Cham, Switzerland

Chapter 8
Ecological Importance of *bajadas* in the Chihuahuan Desert

Juan Carlos Flores Vázquez, María Dolores Rosas Barrera, Jordan Golubov, Irene Sánchez-Gallén, and María C. Mandujano

Abstract *Bajadas* are transition zones between the foothills of the mountainous areas and the valleys in the arid zones of North America; given their microclimatic characteristics (geomorphological composition, high humidity, and low solar radiation) they are probably a refuge for many plant species; however, there are no reports for the *bajadas* in the Cuatro Ciénegas Basin. This work describes the structure, composition, and diversity of the plant community found in *bajadas* in the Cuatro Ciénegas Basin (CCB) in the Chihuahuan desert, where we sampled 0.6 ha. We found 38 plant species, 26 genera, and 11 families. The Shannon diversity index was 3.2, higher than those registered in other arid zones. The Cactaceae family includes the greatest number of species (22 species), seven of them are classified under some threat category; other cacti species had a restricted distribution into the so-called mega Coahuilan region. The most important species based on their density and volume were: *Grusonia bradtiana* (importance value, IV =35.3%), *Larrea tridentata* (IV=15.9%), and *Jatropha dioica* (IV=14.12%), these species might strongly contribute to the composition of *bajadas* since they function as nurse plants for other plant species. As the CCB is one of the most diverse of the Chihuahuan Desert, the knowledge of the diversity, structure, and composition of *bajadas* is relevant to

---

J. C. Flores Vázquez · M. D. Rosas Barrera · M. C. Mandujano (✉)
Departamento d Ecología de la Biodiversidad, Instituto de Ecología, Universidad Nacional Autónoma de México, UNAM, Mexico City, Mexico
e-mail: jcflores@iecologia.unam.mx; mcmandujano@iecologia.unam.mx

J. Golubov
Departmento El hombre y su ambiente, Universidad Autónoma Metropolitana Xochimilco, Mexico City, Mexico

I. Sánchez-Gallén
Laboratorio de Ecología del Suelo, Depto. Ecología y Recursos Naturales, Facultad de Ciencias, Universidad Nacional Autónoma de México, UNAM, Mexico City, Mexico

M. C. Mandujano et al. (eds.), *Plant Diversity and Ecology in the Chihuahuan Desert*, Cuatro Ciénegas Basin: An Endangered Hyperdiverse Oasis,
https://doi.org/10.1007/978-3-030-44963-6_8

define important sites and strategies for the conservation of North America's largest desert.

Keywords Bajadas · Cactaceae · Cuatro Ciénegas · Endemism · Chihuahuan Desert · Diversity

Chihuahuan Desert Region

The Chihuahuan Desert Region (CDR) is the largest arid zone in North America covering an area of approximately 505,000 km^2 (MacMahon 1979; Henrickson and Johnston 2007). The CDR in the USA occupies the southern parts of Texas, New Mexico, and Arizona, while in Mexico CDR covers part of the states of Coahuila, Chihuahua, Durango, Nuevo Leon, San Luis Potosi, Tamaulipas, and Zacatecas (Zavala-Hurtado and Jiménez 2020, this volume). The CDR is one of the most diverse arid zones in the world, with 3,382 species of vascular plants (MacMahon 1979; Henrickson and Johnston 2007; Zavala-Hurtado and Jiménez 2020, this volume). Villarreal-Quintanilla (2001) reported that almost a quarter (24.87%) of the CDR vascular plants are endemic, and the families Cactaceae, Asteraceae, Boraginaceae, and Brassicaceae stand out due to their species richness.

Geologically, most of the CDR is found in the Central Plateau and Chihuahuan-Coahuila Plateau and Ranges morphotectonic provinces, where limestone outcrops dominate (Ferrusquia-Villafranca et al. 2005). Its topography is highly heterogeneous, with altitudes ranging from 700 m in the valleys to more than 3000 meters in the mountain ranges providing adequate conditions for a variety of vegetation types, including (1) pine forest, (2) oak forest, (3) halophilous and gypsophilous scrub, (4) xerophilous scrub, (5) submontane scrub, (6) pasture, and (7) underwater vegetation, to name a few (MacMahon 1979; Henrickson and Johnston 2007).

Out of all the states, both in the USA and in Mexico, Coahuila has the largest proportion (approximately 73% of its territory) covered by the CDR (110 973 km^2). Considering the above, it is very likely that a high proportion of the vascular plants described for Coahuila, (3,207 taxa) belong to this region. This serves as an indicator of the importance of the contribution to richness and diversity (Villarreal-Quintanilla 2001).

Cuatro Ciénegas Basin

In Coahuila, the Cuatro Ciénegas Basin (CCB) is particularly relevant for endemism, habitat, and biological diversity (Ezcurra et al. 2020, this volume). CCB covers more than 2000 km^2 (Pinkava 1984, Souza et al. 2018) where high limestone mountain ranges with peaks covered by coniferous forests surround a desert valley that contains *playas*, streams, *ciénegas,* and other water bodies. Altitudes range from 740 m in the valley to more than 3000 m above sea level in the Menchaca mountain range. These characteristics, together with different soil types, promote a large number of different vegetation types including (1) Sacaton grasslands, in the basin's alkaline floor, (2) aquatic and semi-aquatic habitats, (3) gypsum dunes (Ochoterena et al. 2020, this volume), (4) transition zones with desert scrub in the *bajadas*, (5) chaparral, and (6) oak forests and hills composed of oak-pine vegetation (Pinkava 1984). This makes CCB one of the most diverse sites in the entire CDR (Martínez-Ávalos et al. 2020, this volume), with 879 flora taxa, of which approximately 860 species are vascular plants.

The transition zones between the mountains and the valleys (*bajadas*) are the most biologically diverse areas, probably because of their microclimatic characteristics (geomorphological composition, high humidity, and low solar radiation) that allow the establishment of varied plant growth forms not found in other zones (Montaña 1990). Given these characteristics, the *bajadas* are probably a refuge for many plant species, many of which are endemic cacti. Considering the whole CDR, the CCB harbors a large number of species, with many belonging to Cactaceae (Martínez-Ávalos et al. 2020, this volume). Nevertheless, the *bajadas* in CCB as in other areas of the CDR have not been fully explored but they represent a very diverse area for plant species (Fig. 8.1).

Fig. 8.1 Overview of *bajada* vegetation in San Marcos y Pinos locality, Cuatro Ciénegas Basin (CCB), Coahuila, Mexico, with some notable dominant species such as *Grusonia bradtiana*, *Jatropha dioica*, and *Larrea tridentata*. *(Photograph: Juan Carlos Flores Vázquez)*

Vegetation Census

Six 20 × 50 m plots were established in the different sierras surrounding CCB: two in the San Marcos and Pinos Sierra (SMP), two in Sierra de la Fragua (LA), and two in Sierra de la Madera (LM) (view map in Ochoterena et al. 2020, this volume). In each plot, the taxonomic identity, cover (largest and smallest diameter), and height (cm) of each individual were determined. The ecological diversity of all six plots was calculated with the Shannon uncertainty index (H') and the evenness of each locality (Krebs 1999). Since most species at *bajadas* of CCB have a low frequency, we ran a Bray–Curtis similarity analysis which weighed the dissimilarity between communities through the composition of species and the number of individuals of each community and is not highly affected by the most abundant species (Krebs 1999). Indices and analysis were calculated using EstimateS 9.1.0. (Colwell 2013) and the dendrogram was made with Primer 5.0.

The minimum area curve was obtained until the inflection point, where new species were no longer observed, was reached, following the method proposed by Matteucci and Colma (1982). Density was calculated as the number of individuals per unit area within the plots, then extrapolated to 1 ha. Plant cover (m^2 ha^{-1}) occupied by each species was calculated measuring the major and minor radius and using the formula for the area of an ellipse. The importance value (IV) was obtained as ½ relative density + relative cover, based on the proposal of Curtis and McIntosh (1950). Finally, the percentage of the area within each plot covered by vegetation was calculated by simply comparing the total species cover with respect to the total area of each plot.

According to the minimum area analysis, the accumulated area of the six plots leads to an asymptote. Thus, it could be asserted that by covering this area, a good approximation of the existing vegetation in CCB *bajadas* was recorded, the inflection point was reached at 0.6 ha (Fig. 8.2).

Floristic Composition, Vegetation Structure, and Diversity

In the six plots sampled in the *bajadas* of La Fragua, La Madera, and San Marcos y Pinos mountains in CCB we recorded a total of 38 species, 25 genera belonging to 11 families and three morphospecies, highlighting the dominance of the Cactaceae family, with 13 genera and 22 species, which represent 58% of the total species, followed by the families Agavaceae, Asteraceae, and Mimosaceae with two species each; of the remaining families, only one species was recorded (Table 8.1).

A total of 36.69% of the area we sampled was covered by vegetation. According to the relative importance values (IV), the most important species in terms of density and volume were *Grusonia bradtiana*, Cactaceae (35.36%), *Larrea tridentata*, Zygophyllaceae (15.91%), *Jatropha dioica*, Euphorbiaceae (14.12%), *Agave lechuguilla*, Agavaceae (11.27%), *Suaeda mexicana*, Chenopodiaceae (4.55%), and

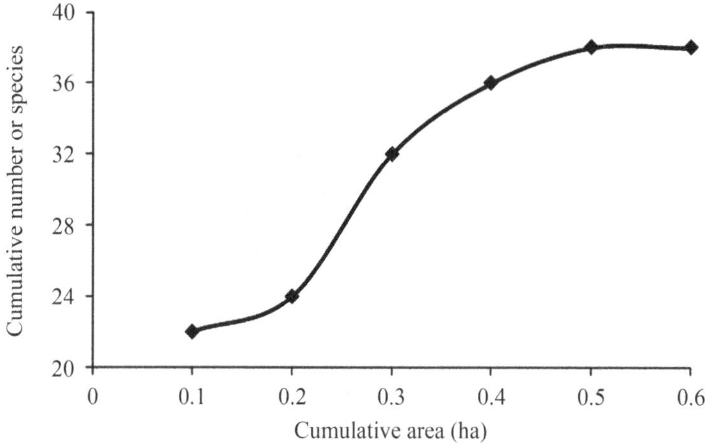

Fig. 8.2 Species accumulation curve, in six localities of *bajadas* in CCB, Coahuila, Mexico

Table 8.1 Number of taxonomical families, genus, and species present in the sampled sites of *bajadas* at Cuatro Ciénegas Basin (CCB), Coahuila, Mexico

Family	Genus	Species
Agavaceae	2	2
Asparagaceae	1	1
Asteraceae	2	2
Boraginaceae	1	1
Bromeliaceae	1	1
Cactaceae	13	22
Chenopodiaceae	1	1
Euphorbiaceae	1	1
Fouquieriaceae	1	1
Mimosaceae	2	2
Zygophyllaceae	1	1
Morphospecies1	1	1
Morphospecies2	1	1
Morphospecies3	1	1

Cordia parviflora, Boraginaceae (3.9%). The remaining species had relative importance values below 4% (Table 8.2). Only four of the 38 species were found in all localities: *A. lechuguilla*, *G. bradtiana*, *J. dioica,* and *L. tridenta*; most species are scarce since they are found only in a few locations, at low frequencies, or both (see Martínez-Ávalos et al. 2020, this volume).

The species richness varied between 8 and 22 for the different plots, with the highest number of species recorded in San Marcos y Pinos (SMP1), with a Shannon diversity index of 2.38, while the lowest number was recorded in Sierra de la Fragua (LF2), with a Shannon diversity index of 0.59. However, density per hectare varied considerably, from 8,610 ind ha^{-1} to 3,020 ind ha^{-1}. The plot with the highest density

Table 8.2 Importance Value (IV), relative density and cover for all species recorded at bajadas of Cuatro Ciénegas Basin (CCB), Coahuila, Mexico. Species with an asterisk are included in the Mexican threatened species list *NOM-ECOL-059, SEMARNAT (2010)

Species	Relative density	Relative cover	IV
Grusonia bradtiana	25.38	45.25	35.30
Larrea tridentata	13.88	17.94	15.91
Jatropha dioica	17.12	11.12	14.12
Agave lechuguilla	16.00	6.55	11.27
Suaeda mexicana	4.36	4.73	4.55
Cordia parviflora	2.18	5.81	3.99
*Epithelantha micromeris**	4.62	0.01	2.31
Cylindropuntia leptocaulis	2.87	1.43	2.15
Echinocactus horizonthalonius	3.38	0.05	1.71
Echinocereus enneacanthus	2.24	0.59	1.41
Euphorbia antisyphilitica	1.43	0.84	1.13
Opuntia streptacantha	0.83	1.35	1.09
Prosopis glandulosa	0.23	1.43	0.83
Morphospecies1	0.46	0.69	0.58
Fouquieria splendens	0.20	0.87	0.54
Mammillaria pottsii	0.69	0.01	0.35
Opuntia phaeacantha	0.23	0.42	0.32
Acacia sp.	0.23	0.37	0.30
*Echinomastus mariposensis**	0.52	0.04	0.28
*Ariocarpus fissuratus**	0.46	0.01	0.23
Coryphantha sp.	0.43	0.01	0.22
Mammillaria heyderi	0.40	0.01	0.20
*Escobaria dasyacantha**	0.40	0.01	0.20
Mammillaria lasiacantha	0.37	0.01	0.19
Flourensia cernua	0.20	0.14	0.17
Parthenium argenteus	0.14	0.14	0.14
Ferocactus hamatacanthus	0.17	< 0.01	0.09
Dasylirion heteracanthum	0.17	0.07	0.12
Mammillaria sp.	0.20	< 0.01	0.10
Agave sp.	0.03	0.07	0.05
Hechtia scariosa	0.03	0.04	0.03
Opuntia microdasys	0.03	< 0.01	0.02
Morphospecies2	0.03	< 0.01	0.01
Morphospecies3	0.03	< 0.01	0.01
Coryphantha echinus	0.03	< 0.01	0.01
*Lophophora williamsii**	0.03	< 0.01	0.01
*Coryphantha pseudoechinus**	0.03	< 0.01	0.01
*Mammillaria plumosa**	0.03	< 0.01	0.01

was the one with the lowest richness and diversity. This reflects the dominance of certain species since the plot with the highest density of individuals also recorded the lowest Shannon index value and the lowest level of evenness (Table 8.3).

Considering the interval of the similarity index, from 0 (similar) to 1 (dissimilar), the similarity analysis of Bray–Curtis denoted a wide heterogeneity in the composition and abundance of species between the localities of *bajadas*. The localities that are geographically close, for example, contiguous localities in La Fragua, have less similarity to each other than locations further away geographically. This was the case in all three mountain ranges, La Fragua, San Marcos y Pinos, and La Madera. In general, two groups are distinguished: site two of Sierra de la Madera and site one of Sierra de la Fragua (LM2 and LF1, respectively) form one group, while the other four locations form a second group (Fig. 8.3).

Species Richness and Diversity

Previous studies mention that *bajadas* are the landscape units in the CDR with the highest species richness (Stein and Ludwig 1979; Pinkava 1984; Montaña 1990). In the present study we recorded a specific richness of 38 species in an area of 0.6 ha, very similar to the 39 species that were recorded in another study conducted in an area of 0.332 ha in the Trans-Pecos subdivision of the CDR, in New Mexico (Stein and Ludwig 1979). Although the area we sampled was larger, it is important to note the study of *bajadas* in New Mexico was done using a different methodology, where several landscape units were covered, i.e., pastures, aquatic vegetation, succulent scrub, and other shrub areas. Some species, mainly shrubs, were recorded in both studies, including *Flourensia cernua, Larrea tridentata,* a very characteristic species of the CDR, *Opuntia phaeacantha*, and *Prosopis glandulosa*, but in general growth forms between study sites were different. Stein and Ludwig (1979) recorded a large amount of annual grasses and forbs (herbaceous flowering plants), but we recorded many species of cacti (several of them under a risk category) at low densities, including *Ariocarpus fissuratus* (Fig. 8.4), *Lophophora williamsii* (Fig. 8.5), and other perennials (Table 8.2).

Table 8.3 Diversity and evenness index, species richness, and individual density for each of the sampled plots of *bajadas* in Cuatro Ciénegas Basin (CCB), Coahuila, Mexico. Localities: San Marcos y Pinos (SMP), La Madera (LM), and La Fragua (LF)

Locality and plot	Shannon index	Evenness	Species richness	Density (ind ha^{-1})
SMP1	2.38	0.53	22	7090
SMP2	2.28	0.57	16	8610
LM1	2.18	0.75	18	3500
LM2	2.02	0.74	16	3020
LF1	1.67	0.65	13	7830
LF2	0.59	0.28	8	4840

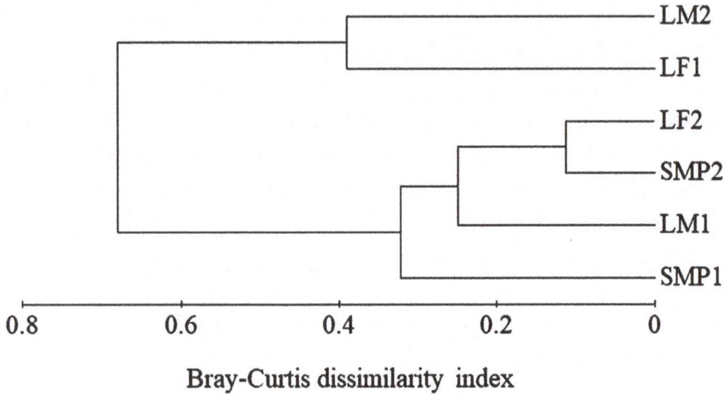

Fig. 8.3 Dissimilarity index between three localities of bajadas in CCB, Coahuila, Mexico. **LM** =Sierra de la Madera, **LF=** Sierra la Fragua, and **SMP=** San Marcos y Pinos; two plots for each locality were sampled

Fig. 8.4 Flowering plant of *Ariocarpus fissuratus* at *bajadas* of San Marcos y Pinos in CCB, Coahuila, Mexico (Photograph: Juan Carlos Flores Vázquez)

Fig. 8.5 Flowering plant of peyote, *Lophophora williamsii* in CCB. The peyote is an endangered cactus endemic the Chihuahuan Desert Region (Photograph: Jorge Jiménez)

If we compare our results with that of other studies conducted in equivalent landscape units in the Sonoran Desert, we can confirm that species richness in CCB *bajadas* is greater than the values reported from Sierra Estrella in Arizona, where 21 species were recorded (Phillips and MacMahon 1978), and from Bahía de Kino, Sonora, with 35 species (Bowers and Lowe 1986). The first of these studies reports a Shannon diversity index, H'= 1.21, lower than the average we recorded in CCB (H'= 1.85). Moreover, although data can be compared globally, there are differences in sampling methods, because in Arizona 20 sampling sites were recorded along two Canfield lines of 500 m in length, which implies that an area-based method was not used. Overall, we can state that plant diversity in CCB is higher.

Structure and Composition

We recorded eleven plant families at *bajadas* of CCB, without considering the three morphospecies which were not identified, because they did not present reproductive structures, flowers, or fruits, during sampling. The Cactaceae family is the best represented, with 13 genera and 22 species. This coincides with other data reported for CCB, which recognizes the area as floristically diverse in comparison to other desert areas (Pinkava 1984; Martínez-Ávalos et al. 2020, this volume). The Cactaceae is particularly diverse, with a high percentage of species which are endemic to the CDR and which have a limited distribution (Hernández et al. 2004; Zavala-Hurtado and Jiménez 2020, this volume). However, of all the cacti species recorded in this study, only *Grusonia bradtiana* had an important IV, both in density (884 individuals) and in frequency, because it was one of the four species found in all sampled sites. In fact, everything seems to indicate that *G. bradtiana* has a restricted distribution to the Cuatro Ciénegas Valley and its adjacent areas (Rosas-Barrera et al. 2020, this volume; Plascencia-López et al. 2020, this volume), which further highlights the importance of the species in the area.

The 21 remaining cacti species presented an IV value below 3%. This indicates they are species with a low density and a low frequency. As mentioned above, some of these species, such as *Ariocarpus fissuratus, Ephitelantha micromeris, Echinomastus mariposensis*, *Lophophora williamsii,* and *Mammillaria plumosa*, to name a few, are included in the Mexican threatened species list (Arias et al. 2005; SEMARNAT 2010). Probably, the abundance pattern of cacti populations at CCB reflects that of the CDR at a larger scale, where it has been reported that many species of cacti have a discontinuous distribution pattern, they are widely distributed throughout the CDR but are infrequent (Hernández et al. 2008, Villareal-Quintanilla et al. 2017; Martínez-Ávalos et al. 2020, this volume). This was observed in 66% of the 71 species of Cactaceae recorded within a 798 km long transect which covered a large part of the CDR (Hernández et al. 2008).

In addition to *G. bradtiana*, other abundant and important species in the *bajadas* of CCB include: *Agave lechuguilla*, *Jatropha dioica,* and *Larrea tridentata*, which unlike *G. bradtiana* have a wide range distribution. *Jatropha dioica* and *A.*

lechuguilla are found in several states of Mexico. The nectar produced by *A. lechuguilla* inflorescences has been recorded as a valuable resource for diverse taxa: bees, hawk moths, and hummingbirds (Silva-Montellano and Eguiarte 2003), throughout its distribution. Additionally, *A. lechuguilla* exhibits biological characteristics that allow it to be widely distributed throughout the CDR, from Zimapán-Metztitlán, in the South to New Mexico in the North (Scheinvar et al. 2017; 2020, this volume). Moreover, it has a high interspecific competitive capacity, as well as a high resistance to herbivory and extreme climatic conditions. It is also a species whose fiber has been used for 8000 years by diverse ethnic groups and a resource used for the livelihood of more than 20,000 families in the northeast of Mexico a few years ago (Reyes-Agüero et al. 2000). This highlights the species biological and social importance in CCB. Finally, *L. tridentata* presents adaptations, mainly in its use of resources, which allow the species to occupy many habitats which results in a wide distribution throughout the CDR. Moreover, it has been recognized as a very important nurse plant for the community structure of the Mapimi area (Silvertown and Wilson 1994; Ezcurra et al. 2020, this volume) because it facilitates the seedling establishment of several species and offers shelter and food to various species of reptiles and birds. It is feasible that *L. tridentata* fulfills the same function in Cuatro Ciénegas, making the species an important plant for structuring the community.

Conclusions

The data obtained in this research indicates that *bajadas* at CCB have high plant species richness compared to the arid zones of the world and of the CDR. In particular CCB is a center of cacti diversity, and many of these species are listed as threatened under some risk category or have a highly restricted distribution, hence the importance of CCB for their conservation. Finally, *bajadas* are an important system for the structure and diversity of deserts and play an important role in shaping desert environments.

Acknowledgments We thank the APFF Cuatro Ciénegas (CONANP) office and the people in charge. We thank the owners of Rancho Orozco, Juan Manuel González and Alfonso González. This contribution is in partial fulfillment of the Graduate Program of Ciencias Biomédicas, Universidad Nacional Autónoma de México (UNAM) of the first author MSc. Juan Carlos Flores Vázquez, who also acknowledges the scholarship and financial support provided by the National Council of Science and Technology (CONACyT), fellowship 176604. To Valeria Souza for her hospitality in Cuatro Ciénegas and Carlos Slim Foundation. To Marco A. Romero Romero for helping with figures edition.

References

Arias S, Guzmán U, Mandujano MC, Galván S et al (2005) Las especies de cactáceas mexicanas en riesgo de extinción. I. Una comparación entre los listados NOM-059-ECOL-2001 México, La Lista Roja (UICN) y CITES. Cact. Suc. Mex 50:100–125

Bowers MA, Lowe CH (1986) Plant-Form Gradients on Sonoran Desert Bajadas. Oikos 46:284–291

Colwell RK (2013) Estimates: Statistical estimation of richness species and shared species from samples. Version 9. Persistent URL purl.oclc.org/estimates
Curtis JT, McIntosh RP (1950) The interrelations of certain analytic and synthetic phytosociological characters. Ecology **31**:434–445
Ezcurra E, Martínez-Berdeja A, Villanueva-Almanza L (2020) The evolution of North American deserts and the uniqueness of Cuatro Ciénegas. In: Mandujano MC, Pisanty I, Eguiarte L (eds.) Plant diversity and Ecology in the Chihuahuan desert. Springer International, Cham, Switzerland
Ferrusquia-Villafranca I, González Gúzman LI, Cartron JL (2005) Northern Mexico's landscape, Part I: The physical setting and constrains on modeling biotic evolution. In: Cartron JL, Ceballos G, Felger RS (eds) Biodiversity, Ecosystems and Conservation in Northern México. Oxford University Press, Oxford, UK, pp 11–38
Henrickson J, Johnston MC (2007) A flora of the Chihuahuan Desert Region. Published by J. Henrickson, Los Angeles, p 1695
Hernádez HM, Goettsch B, Gómez-Hinostrosa C et al (2008) Cactus species turnover and diversity a long a latitudinal transect in the Chihuahuan Desert Region. Bio Conser 17:703–720
Hernández HM, Gómez-Hinostrosa C, Goettsch B (2004) Checklist of Chihuahuan Desert Cactaceae. Harvard Paper Botany 9(1):51–68
Krebs CJ (1999) Ecological Methodology. Addison Wesley Longman, Inc, Menlo Park, CA, USA
MacMahon JA (1979) North American deserts: their floral and faunal components. In: Goodall DW, Perry RA (eds) Arid-Land Ecosystems: Structure, Functioning and Management, IBP 16, vol 1. Cambridge University Press, Cambridge, UK, pp 21–82
Martínez-Ávalos JG, Martínez-Gallegos R, Guerra-Pérez A, Torres Castillo JA, et al. (2020). Diversity and distribution of cacti species in the Cuatro Ciénegas basin. In: Mandujano MC, Pisanty I, Eguiarte L (eds.) Plant diversity and Ecology in the Chihuahuan desert. Springer International, Cham, Switzerland
Matteucci SD, Colma A (1982) Metodología para el estudio de la vegetación Secretaría General de la Organización de los Estados Americanos. Secretaria General de la Organización de los Estados Americanos. Programa Regional de Desarrollo Científico y Tecnológico, Washington, DC, USA
Montaña C (1990) A Floristic-Structural Gradient Related to Land Forms in the Southern Chihuahuan Desert. J Veg Scie. 1:669–674
Ochoterena H, Flores-Olvera H, Gómez-Hinostrosa C, Moore MJ (2020) Gypsum and Plant Species: A Marvel of Cuatro Ciénegas and the Chihuahuan Desert In: Mandujano MC, Pisanty I, Eguiarte LE (eds) Plant diversity and Ecology in the Chihuahuan desert. Springer International, Cham, Switzerland
Phillips DL, MacMahon JA (1978) Gradient analysis of a Sonoran Desert bajada. The South Nat 23:669–680
Pinkava DJ (1984) Vegetation and flora of the Bolson of Cuatro Ciénegas Region, Coahuila, Mexico: IV. Summary, endemism and corrected catalogue. J Ariz Nev Acad Sci 19:23–47
Plascencia-López L, Rojas-Aréchiga M, Mandujano MC (2020) Reproductive Biology of *Grusonia bradtiana* (Cactaceae): a dominant species and endemic clonal Cactus from Cuatro Ciénegas Basin and contiguous areas in the Chihuahuan Desert. In: Mandujano MC, Pisanty I, Eguiarte LE (eds) Plant diversity and Ecology in the Chihuahuan desert. Springer International, Cham, Switzerland
Reyes Agüero JA, Aguirre Rivera JR, Peña Valdivia CB (2000) Biología y aprovechamiento de *Agave lechuguilla* Torrey. Bol. Soc. Bot. México 67:75–88
Rosas-Barrera MD, Golubov J, Pisanty I, Mandujano MC (2020) Effect of reproductive modes on the population dynamics of an endemic cactus from Cuatro Ciénegas In: Mandujano MC, Pisanty I, Eguiarte LE (eds) Plant diversity and Ecology in the Chihuahuan desert. Springer International, Cham, Switzerland
Scheinvar E, Gámez N, Castellanos-Morales G et al (2017) Neogene and Pleistocene history of *Agave lechuguilla* in the Chihuahuan Desert. J Biogeo 44:322–334
Scheinvar E, Gámez N, Moreno-Letelier A, Aguire-Planter E, Eguiarte L (2020). Phylogeography of the Chihuahuan Desert: diversification and evolution over the Pleistocene. In: Mandujano

MC, Pisanty I, Eguiarte L (eds.) Plant diversity and Ecology in the Chihuahuan desert. Springer International, Cham, Switzerland

SEMARNAT (2010) Secretaría de Medio Ambiente y Recursos Naturales. 2010. Norma Oficial Mexicana NOM-059-SEMARNAT-2010. Published on December 30, 2010 in Diario Oficial de la Federación, México, DF, Mexico

Silva-Montellano A, Eguiarte LE (2003) Geographic patterns in the reproductive ecology of *Agave lechuguilla* (Agavaceae) in the Chihuahuan desert. I. floral characteristics, visitors, and fecundity. Amer J Bot 90:377–387

Silvertown J, Wilson JB (1994) Community Structure in a Desert Perennial Community. Ecology 75:409–417

Souza V, Olmedo-Álvarez G, Eguiarte LE (eds) (2018) Cuatro Ciénegas Ecology, Natural History and Microbiology. Springer International, Cham, Switzerland

Stein RA, Ludwig JA (1979) Vegetation and Soil Patterns on a Chihuahuan Desert Bajada. Am Midl Nat 101:28–37

Villareal-Quintanilla JA (2001) Listados Florísticos de México, Flora de Coahuila XXIII. Instituto de Biología UNAM, México, DF, Mexico

Villarreal-Quintanilla JA, Bartolomé-Hernández JA, Estrada-Castillón E, Ramírez-Rodríguez H, Martínez-Amador SJ (2017) El elemento endémico de la flora vascular del Desierto Chihuahuense, The endemic element of the Chihuahuan Desert vascular flora. Act. Bot Mex. 118:56–96

Zavala-Hurtado JA, Jiménez M (2020). Diversity and uniqueness at its best. Vegetation of the Chihuahuan Desert. In: Mandujano MC, Pisanty I, Eguiarte L (eds.) Plant diversity and Ecology in the Chihuahuan desert. Springer International, Cham, Switzerland

Chapter 9
Gypsum and Plant Species: A Marvel of Cuatro Ciénegas and the Chihuahuan Desert

Helga Ochoterena, Hilda Flores-Olvera, Carlos Gómez-Hinostrosa, and Michael J. Moore

Abstract Soils rich in gypsum ($CaSO_4 \cdot 2H_2O$) are both a challenge for plant growth and a trigger for plant evolution. Plants that grow only on this type of soils are known as gypsophiles, but there are also generalists that grow on and off gypsum, known as gypsovags. The Cuatro Ciénegas municipality (CCM) in Coahuila is characterized by a complex mosaic of soils with different gypsum contents, ranging from massive, ancient gypsum evaporite bedrock, crystalline selenite, and anhydrite to very recent formations of secondary evaporites, often mixed with other salts, and gypsum dunes. These exposures of gypsum (gypsum outcrops) host a species-rich gypsophilic flora that has mostly been described in only the last 50 years, and aside from the well-known gypsum deposits of Cuatro Ciénegas Basin (CCB) is still imperfectly known. To date there is no available comprehensive synthesis for where the gypsum outcrops occur within CCM or for the vascular plant species growing on them. In this chapter we use remote sensing techniques to reveal botanically unexplored gypsum outcrops and we present a checklist for the vascular plant species currently known to occur on gypsum outcrops within CCM. We report 297 species in 187 genera and 60 families of vascular plants growing on gypsum outcrops in the CCM, of which 31 species are gypsophiles, five are halogypsophiles (species that grow on soils with a mixture of sodium chloride and gypsum), and three are either gypsophiles or halogypsophiles; 15 are endemic to the municipality. Our results demonstrate that the method presented here for detecting potential gypsum outcrops is powerful, that CCM has an outstanding number of potential gypsum outcrops, many of which are unexplored botanically. The relatively high number of gypsophiles in CCM and the fact that new taxa continue to be discovered

H. Ochoterena (✉) · H. Flores-Olvera · C. Gómez-Hinostrosa
Departamento de Botánica, Instituto de Biología, Universidad Nacional Autónoma de México, UNAM, Mexico City, Mexico
e-mail: helga@ib.unam.mx

M. J. Moore
Department of Biology, Oberlin College, Oberlin, OH, USA

M. C. Mandujano et al. (eds.), *Plant Diversity and Ecology in the Chihuahuan Desert*, Cuatro Ciénegas Basin: An Endangered Hyperdiverse Oasis,
https://doi.org/10.1007/978-3-030-44963-6_9

clearly indicate the need for continued botanical exploration of gypsum environments in the Cuatro Ciénegas region and in Coahuila and the Chihuahuan Desert as a whole.

Keywords Endemism · Gypsum · Gypsophiles · Gypsovags · Halogypsophiles · Checklist of vascular plant species

Introduction

Species Diversity in Coahuila

The state of Coahuila in the North of Mexico has a variety of physiographic, climatic, and edaphic conditions resulting in a diversity of vegetation types and a rich flora, which is calculated to comprise approximately 3,100 taxa. Of these, 350 are endemic to the state and adjacent areas, and 190 are found only within the borders of the state (Villarreal-Quintanilla and Encina-Domínguez 2005). Within Coahuila, the Cuatro Ciénegas area is the most diverse, with the highest endemism at 58 taxa. Among all vegetation types in the state, desert scrub that occurs on soils with high concentrations of sodium chloride or sulfates possesses the fourth highest number of endemic taxa (53), in areas such as Cuatro Ciénegas, Valle Hermanas, and Sierra de Las Delicias (Villarreal-Quintanilla and Encina-Domínguez 2005).

Gypsum in Cuatro Ciénegas

With an area of less than 600 km^2, the Cuatro Ciénegas Basin (CCB) harbors one of the most diverse landscapes in Mexico, including (a) aquatic habitats; (b) desert plains; (c) gypsum dunes and flats; (d) basin grasslands; (e) montane chaparral; (f) oak and oak-pine woodlands; and (g) sky-island conifer forests (Pinkava 1984, Ezcurra et al. 2020, this volume). Surrounded by a complex set of mountain chains (Fig. 9.1), the Cuatro Ciénegas municipality (CCM) includes an impressive array of soils containing high contents of gypsum ($CaSO_4 \cdot 2H_2O$). These include massive evaporite gypsum bedrock in Cretaceous-aged geological formations such as the impressive Acatita Formation of western Coahuila (Lehmann et al. 2000), in which gypsum deposits up to 600 m in elevational extent have been exposed by erosion, as well as the smaller gypsum outcrops of the La Virgen Formation in central Coahuila (Gutiérrez-Alejandro et al. 2017). In contrast to these mountainside outcrops, valleys in CCM typically possess geologically very recent gypseous soils that formed in association with the large salt lakes that once occupied the endorheic basins of

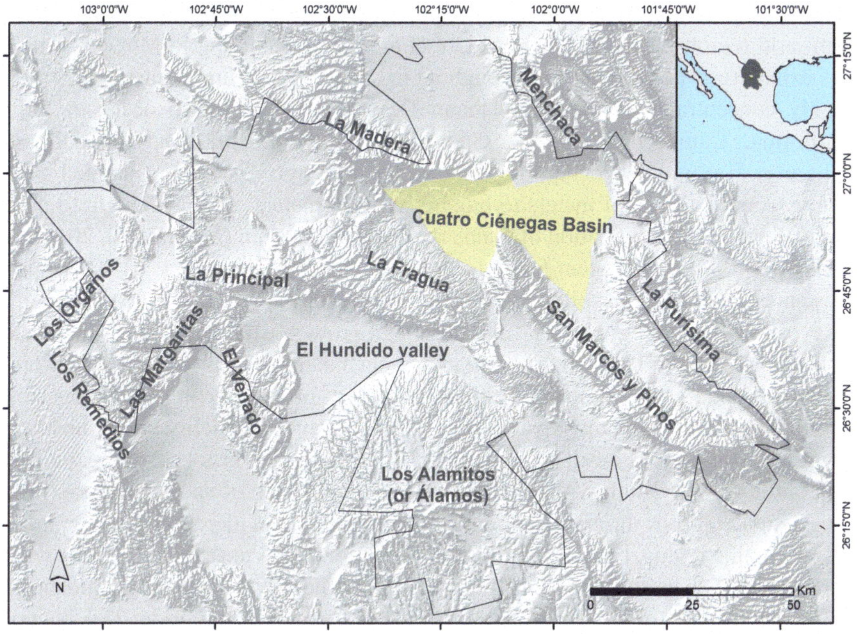

Fig. 9.1 Topography and location of Cuatro Ciénegas municipality (CCM), delinated with a gray line. In the map of Mexico at top right, the state of Coahuila is highlighted in dark gray and CCM is indicated with a dot. On the main map, the protected area within the Cuatro Ciénegas Basin (CCB) is marked in yellow

central and western Coahuila during the Pleistocene. The gypsum in these valley deposits represents reprecipitated gypsum that originated via erosion of the older montane deposits and accumulation in the basins. The most famous formation of gypsum in Mexico—the dunes on the west side of CCB—is derived from such recent, lacustrine gypsum that then formed into dunes via aeolian processes (Minckley 1969). Despite the fact that CCB is so well-known for its gypsum, most gypsum outcrops in the area remain extremely poorly studied biologically. This is largely because biologists are unaware of many such outcrops due to their relatively small size, scattered nature, and inaccessibility via road.

Plant Affinities to Gypsum

Gypsum soils pose significant challenges to plant growth due to severe limitations in certain plant macronutrients (e.g., N, P, K) and because they are typically characterized by hard surface crusts that impede the normal development of seedlings (Moore et al. 2014; Escudero et al. 2015). Nevertheless, plants have become restricted to gypseous soils around the world, forming unique communities that are often highly localized geographically. Gypsum soils are present mainly in arid and semi-arid regions (Parsons 1976; Moore et al. 2014), including in Mexico, which has extensive gypsum outcrops throughout the country. Gypsum outcrops are most highly concentrated in the Chihuahuan Desert region, but also occur in Baja California, Campeche, Chiapas, Guerrero, Oaxaca, Puebla, and Tamaulipas. Wherever gypsum soils occur, they tend to form discrete "islands" separated anywhere from hundreds of meters to hundreds of kilometers. This is thought to promote allopatric diversification of plants restricted to gypsum (Moore et al. 2014).

Different degrees of plant fidelity to gypsum have been recognized. Those species endemic to gypsum soils are known as gypsophiles, those that grow on and off of gypsum are called gypsovags, and those that grow mostly on gypsum are called gypsoclines (Meyer 1986). It is important to note that in the older literature the word gypsophile had a much more flexible connotation, often referring to any species commonly encountered on gypsum, regardless of its overall fidelity to the substrate (e.g., Johnston 1941; Powell and Turner 1977). In general, it is not easy to recognize plant affinity to gypsum without extensive reconnaissance of species distributions on gypsum outcrops throughout a region. Complications in identifying gypsophiles arise from the relatively high solubility of gypsum, which causes it to intermix with surrounding substrates. For example, in the endorheic basins of the Chihuahuan Desert, gypsum is often mixed with other salts, like sodium chloride. This occurs frequently enough that a subset of gypsophile taxa (called halogypsophiles) exist only on such salty gypsum soils.

Despite the obvious effects of gypsum soils on the evolution of plant species, much work remains to understand the ecology, physiology, and evolution of gypsophiles. A necessary basis for such studies is a full understanding of the species assemblages of the gypsophiles themselves in different regions of the world.

Plant Diversity on Gypsum in Cuatro Ciénegas and in the Chihuahuan Desert

The largest known gypsophile flora in the world is found in the greater Chihuahuan Desert region of Mexico and the USA, where at least 230 gypsophile taxa have been documented (Johnston 1941; Powell and Turner 1977; Turner and Powell 1979; Moore and Jansen 2007; Moore et al. 2014). Despite much progress since Johnston's pioneering 1941 study recognizing the existence of gypsophiles throughout the Chihuahuan Desert (e.g., Powell and Turner 1977; Pinkava 1984), a comprehensive floristic list has yet to be produced for the gypsophiles in this region and numerous new gypsophile species continue to be described (e.g., Alexander et al. 2014), including species from CCM (e.g., Turner 2013). Indeed, the CCM is exceptionally important for gypsum biodiversity due to its extensive montane gypsum outcrops and alluvial gypseous soils (most basin-derived soils in the municipality probably have at least some gypsum content) (Fig. 9.2), and due to the fact that it lies at the

Fig. 9.2 Examples of species growing on gypsum soils within the Cuatro Ciénegas municipality (CCM) and nearby areas in the Chihuahuan Desert, representing different lineages of seed plants, life forms, and affinities to gypsum (phylogenetically arranged). (**A**) *Ephedra trifurca* Torr. ex S.Watson (gypsovag). (**B**) *Dasylirion cedrosanum* Trell. (gypsovag). (**C**) *Euphorbia crepitata* var. *longa* M.C.Johnst. (gypsophile). (**D**) *Nerisyrenia* sp. nov. (gypsophile). (**E**) *Coryphantha poselgeriana* (A. Dietr.) Britton & Rose (gypsovag). (**F**) *Suaeda mexicana* (Standl.) Standl. (halogypsovag). (**G**) *Tidestromia rhizomatosa* I.M.Johnst. (gypsophile). (**H**) *Acleisanthes* sp. nov. (gypsophile). (**I**) *Petalonyx crenatus* A. Gray ex S. Watson (gypsophile). (**J**)*Fouquieria shrevei* I.M. Johnst. (gypsophile). (**K**) *Chilopsis linearis* (Cav.) Sweet (gypsovag). (**L**) *Hedyotis teretifolia* (Terrell) G.L. Nesom (gypsophile). (**M**) *Nama stenophylla* A.Gray ex. Hemsl. (gypsophile). (**N**) *Tiquilia gossypina* (Wooton & Standl.) A.T. Richardson (gypsovag). (**O**) *Xanthisma restiforme* (B.L. Turner) D.R. Morgan & R.L. Hartm. (gypsophile). (**P**) *Gaillardia gypsophila* B.L. Turner (gypsophile)

southern end of the Chihuahuan Desert, where species richness and diversity tend to be highest.

Beginning in the 1970s, a number of gypsophiles have been described from the CCM, these mainly centered on CCB itself. These include a diverse array of plant growth forms and families (Fig. 9.2a–p shows a phylogenetic representative sampling), and range from the bizarre succulents *Xanthisma restiforme* (Asteraceae; Fig. 9.2O) and *Tidestromia rhizomatosa* (Amaranthaceae; Fig. 9.2G) to the annual species *Sabatia tuberculata* (Gentianaceae). Some gypsophiles have specialized ecologies, as *Gaillardia gypsophila* (Asteraceae; Fig. 9.2P), which only grows on the gypsum sand dunes in CCB, and *Haploësthes robusta* (Asteraceae), which only grows in the salty gypsum on the east side of CCB, and hence is a halogypsophile (species that grow on soils with a mixture of sodium chloride and gypsum). Other characteristic species of such saline-gypsum soils include halogypsovags such as *Suaeda mexicana* (Fig. 9.2F) and *S. torreyana*, and various species of *Atriplex* (Chenopodiaceae).

More generally, non-salty gypsum soils in CCM are dominated by a diverse array of gypsophiles, gypsoclines, and gypsovags that are largely endemic to Coahuila state, and if not strictly endemic to Coahuila, they are endemic to the Chihuahuan Desert.

Dominant gypsophile taxa include *Fouquieria shrevei* (Fig. 9.2J; De-Nova et al. 2020, this volume), *Nama stenophylla* and *N. constancei* (Namaceae), *Nerisyrenia incana* and *N. castillonii* (Brassicaceae), *Tiquilia turneri* (Ehretiaceae), *Sartwellia mexicana, S. puberula, Haploësthes greggii, Flaveria palmeri*, and *Xanthisma gypsophilum* (Asteraceae), *Petalonyx crenatus* (Loasaceae), *Acleisanthes purpusiana* (Nyctaginaceae), *Drymaria coahuilana* (Caryophyllaceae), *Euphorbia crepitata* (Euphorbiaceae), and *Notholaena bryopoda* (Pteridaceae).

Important gypsocline and gypsovag taxa include *Tidestromia carnosa* (Amaranthaceae), *Xanthisma spinulosum* (Asteraceae), *Tiquilia gossypina* and *T. greggii* (Ehretiaceae), *Anulocaulis eriosolenus* (Nyctaginaceae), *Oenothera boquillensis* (Onagraceae), *Mentzelia mexicana, Eucnide lobata,* and *Cevallia sinuata* (Loasaceae).

Outcrops of montane gypsum have a variety of perennial succulents, including various species of *Agave* and Cactaceae (Flores Vázquez et al. 2020, this volume), including *Coryphantha poselgeriana* (Fig. 9.2E), *Echinocactus horizonthalonius, Echinocereus enneacanthus*, and *Grusonia grahamii.*

Many of the dominant gypsophile taxa noted above belong to clades of gypsophiles whose center of diversity includes CCM (Moore et al. 2014). These include gypsum clades within the genera *Acleisanthes, Nama, Nerisyrenia, Tiquilia, Sartwellia, Haploësthes, Drymaria,* and *Xanthisma* (Moore et al. 2014), all of which have multiple taxa within CCB. In many cases, these congeneric taxa occur in ecologically distinctive sites, most frequently involving montane Cretaceous gypsum deposits vs. recent basin deposits. For example, *Nama constancei* and a recently discovered, undescribed *Acleisanthes* (Fig. 9.2H) are found only on montane deposits, whereas *N. stenophylla* (Fig. 9.2M) and *A. purpusiana* are only known from low-elevation basin deposits. In some cases these species pairs can be found in close proximity, with no obvious signs of hybridization. This and other phylogenetic evidence suggests that the gypsum floras of central and southern Coahuila may have existed for some time, perhaps since the early Pliocene (Moore et al. 2014; De-Nova et al. 2020, this volume). Given this great phylogenetic and ecophysiological diversity of gypsophiles, CCM is an important natural laboratory for plant evolution.

Historical Background of the Cuatro Ciénegas Flora

Johnston (1941) published the earliest list of plants from CCB and nearby areas based on his and earlier botanical explorations. Later, the exploration of the region was notably enhanced by several botanists from Arizona State University, from the Desert Botanical Garden, Phoenix, and, as part of the project on flora of the Chihuahuan Desert from the University of Texas, Austin. Pinkava (1984) published a checklist for this area, including 879 taxa representing 860 species in 456 genera from 114 families of vascular plants. García-Dávila (2000) updated this list to 902 species. According to Pinkava (1984) the CCB is the type locality for 49 plant taxa, of which four where already by then no longer recognized, while 23 were considered endemic, among which Asteraceae is the most prominent. Pinkava (1984) reported 48 species on the gypsum dunes and flats, 33 of which are also in other vegetation types, with 15 restricted to gypsum. Since the publication of Pinkava's list, several new taxa have been described from CCB, as, for example, the gypsophile *Sporobolus coahuilensis* Valdés-Reyna as well as other species endemic to Coahuila (Valdés-Reyna et al. 2015), but which are not known from gypsum soils.

Despite these important efforts, the floristic lists have focused on the CCB itself. The diverse gypsum outcrops of the remaining parts of CCM have been much more poorly explored, leaving important elements of the flora of the municipality overlooked. Since 2011, we have led several expeditions to the Chihuahuan Desert to increase our floristic and evolutionary understanding of plants on gypsum, including to several previously unexplored gypsum sites in CCM, some within 10–20 km of the town of Cuatro Ciénegas itself. All of these sites yielded important new records of plant distributions, and some have yielded new taxa, demonstrating the urgent need for further botanical exploration throughout the municipality and indeed of all of Coahuila.

The aims of this work are to provide a solid basis for the recognition of gypsum outcrops for botanical exploration (focused on CCM) and to produce the first checklist of vascular plant species that grow on gypsum soils in the area.

Methods

Literature searches for species described or collected on gypsum soils of Cuatro Ciénegas were conducted using Pinkava (1984), Villarreal-Quintanilla and Encina-Domínguez (2005), and JStor Global Plants (https://plants.jstor.org/) as starting points. In addition to personal databases (Chihuahuan Desert Gypsophiles by M. J. Moore and Cactaceae by H. M. Hernández), databases for herbarium specimens were downloaded using the key words "Cuatro Ciénegas" and "Cuatrociénegas" (with and without accent on the é) from TEX/LL, MEXU (https://datosabiertos.unam.mx/biodiversidad/), and SEINet (http://swbiodiversity.org/seinet/). The included herbaria cover most of the important botanical collections from the Chihuahuan Desert. Infraspecific taxa were only considered when relevant for gypsum fidelity or endemism. Georeferenced specimens were used to create a common database that included 7,396 records, which after merging duplicates consisted of 5,865 records. This reduced dataset was filtered for CCM using *ArcMap 10.1 (*ESRI 2012*)* resulting in 1,607 collections. Specimens with herbarium labels indicating that the collection was made on gypsum soils were manually identified and used to construct a subset database (Cuatro Ciénegas georeferenced gypsum specimens) that consisted of 527 entries not necessarily restricted to the municipality.

We employed two open-access versions of ArcGIS Landsat imagery that focused on short-wave infrared spectra: (1) Landsat 7 imagery, with channels 7, 4, and 3 (http://www.arcgis.com/home/webmap/viewer.html?services=3ce128 0b32344df499a4cd15a407f2e3) and (2) Landsat 8 imagery (layer 6.1), with channels 7, 6, and 4 (http://imageryworkflows.arcgis.com/LandsatPOI/). Under these channel combinations, gypsum appears turquoise in color (Figs. 9.3 and 9.4), very similar to the color of water ice in the same spectral combination, most likely because gypsum has water molecules trapped within it. Using this imagery, we visually identified potential sites with gypsum outcrops and ground truthed selected sites throughout the Chihuahuan Desert Region to test the Landsat identifications, which confirmed the presence of gypsum at almost all sites (Fig. 9.3 is an example at La Virgen Formation, near the Cuatro Ciénegas town), except for a few cases in New Mexico and west Texas, where the sites proved to be volcanic-derived, ashy deposits (pers.obs.). Using Google Earth (i.e., visible) imagery it is possible to differentiate between these two types of soils; therefore, a comparison of ArcGIS Landsat imagery and Google Earth imagery was used to identify potential gypsum outcrops (Fig. 9.3A and B). Areas identified as potential gypsum outcrops were traced in an approximate fashion using the polygon tool in Google Earth (Fig. 9.3C). These polygons were exported as .kml files and were then imported as shape files (∗.shp) in ArcMap 10.1. To be conservative, we will hereafter refer to the areas so identified as polygons instead of gypsum outcrops, as further soil tests would be required to fully corroborate the actual presence of high concentrations of gypsum.

Fig. 9.3 Example appearance of a gypsum outcrop (in this case in La Virgen Formation, immediately north of Cuatro Ciénegas town) under (**A**) Google Earth visible imagery and (**B**) ArcGIS Landsat 8 (7, 6, 4) imagery. The resulting polygon (**C**) traced with the Google Earth tool broadly corresponds to the gypsum outcrops. (**D**) General view of the outcrop showing correspondence with the Google Earth and ArcGIS Landsat 8 imagery; (**E**–**G**) detailed view of the outcrop showing (**E**) gypsum crystals; (**F**) a sharp boundary between the gypsum outcrops and the xeric vegetation and (**G**) *Nerisyrenia incana* Rollins (gypsophile, endemic to Cuatro Ciénegas Basin) growing on the gypsum outcrops

This method to identify potential gypsum outcrops is powerful, but laborious, as the gypsum concentration results in different blue intensities and areas with gypsum outcrops that are very small or have relatively low gypsum concentrations can be difficult to detect (Fig. 9.4). The CCM georeferenced gypsum specimens (189 collections, without duplicates) were used to complement the visual location of potential gypsum outcrops. All the CCM georeferenced gypsum specimens were mapped

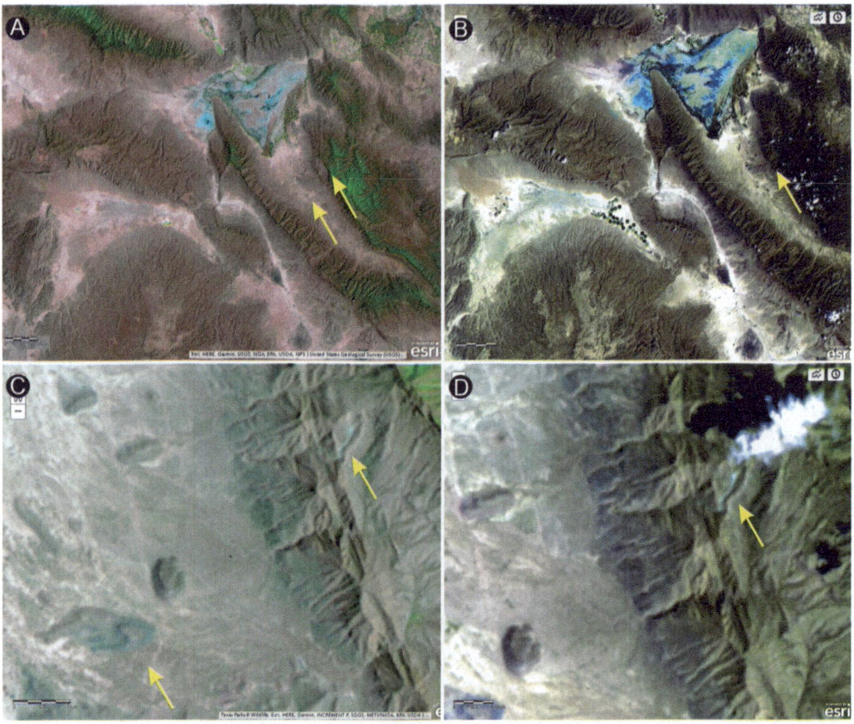

Fig. 9.4 General aspect of the same area using similar scales and different Landsat imagery. (**A**) and (**B**) General view around the Cuatro Ciénegas Basin (CCB). (**C**) and (**D**) Close up of a small potential gypsum outcrop highlighted in the corresponding area with a yellow arrow. Landsat 7 imagery, with channels 7, 4, and 3 (**A** and **C**) vs. Landsat 8 imagery (layer 6.1), with channels 7, 6, and 4 (**B** and **D**)

in Google Earth on top of the already-identified gypsum polygons. Georeferenced localities were occasionally obviously incorrect (the locality did not necessarily match the georeferences) and these were corrected whenever possible, or else discarded. Those entries outside a polygon were checked against the ArcGIS Landsat imagery. If they indeed indicated the potential presence of gypsum, a new polygon was traced. The process was conducted iteratively with every new database that was accessible until no more new polygons were found (Fig. 9.5). A georeference was recorded for each polygon in order to facilitate the use of this information in the future (Table 9.1).

All the polygon shapes were merged into a single layer in ArcMap 10.1, which was used to query the entire CCM collections database. Those entries falling within the municipality and inside a polygon (709 collections) were considered as potential gypsophiles; after that, aquatic and epiphytic species were discarded. Instances in which herbarium collections indicated gypsum on the label, but did not intersect with any known polygons were checked; if the plant

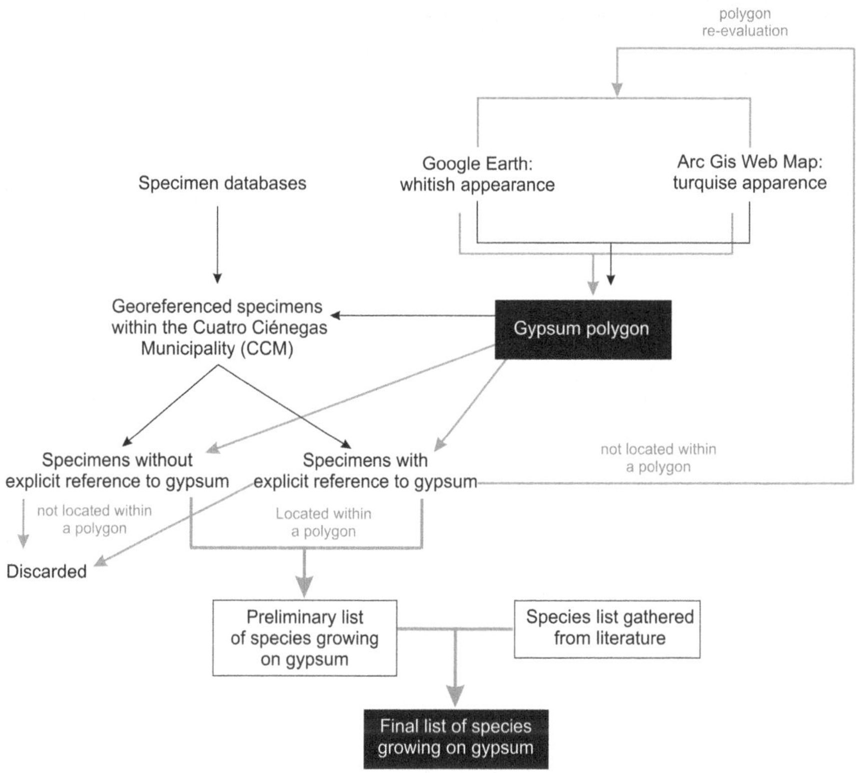

Fig. 9.5 Methodological flowchart for the iterative procedures used to recognize potential areas with gypsum outcrops ("Gypsum polygons") and to produce the checklist of vascular plant species growing on gypsum soils within the Cuatro Ciénegas municipality (CCM). Black boxes indicate the final results; black arrows indicate the initial procedure; gray font and arrows indicate the iterative steps

collections were sufficiently close to a polygon, they were manually added to the list. The preliminary list of species growing on gypsum based on herbarium specimens was then contrasted against literature to produce the final checklist for the plant species growing on gypsum in the CCM (Appendix). Figure 9.5 summarizes the general method as a flowchart.

To assess endemicity we used the Chihuahuan Desert limits sensu Henrickson & Straw (1976); to assess gypsum fidelity and endemicity we used Henrickson and Johnston (1997) as a starting point, complemented by information in TROPICOS (http://www.tropicos.org/) and from specialized literature such as original descriptions, floras, taxonomic treatments, etc.

Table 9.1 List of potential gypsum outcrops (polygons) identified in Cuatro Ciénegas municipality (CCM) as shown in Fig. 9.6, with reference geographic coordinates

Polygon Number	General Location	Reference geographic coordinates
Ecological Reserve and Surroundings		
1	Cuatro Ciénegas Basin (CCB)	26°53’36”N, 102°10’6”W
2	La Virgen	27°04’32”N, 102°02’48”W
3	Sierra Menchaca	27°04’07”N, 101°53’02”W
4	Sierra La Purísima	26°45’11”N, 101°50’45”W
5	Ca. Sierra La Purísima	26°43’26”N, 101°53’18”W
6	Palmira Valley	26°51’07”N, 102°12’38”W
7	Sierra La Madera	26°58’33”N, 102°16’35”W
La Fragua		
8	NE	26°49’37”N, 102°14’06”W
9	N	26°52’10”N, 102°34’23”W
10	N	26°51’23”N, 102°35’03”W
11	W	26°49’08”N, 102°34’41”W
12	W	26°49’04”N, 102°33’40”W
13	W	26°48’04”N, 102°32’48”W
14	SW	26°45’37”N, 102°31’29”W
15	SW	26°43’33”N, 102°26’30”W
16	SW	26°43’21”N, 102°23’27”W
17	SW	26°42’24”N, 102°22’45”W
Valle el Hundido		
18	N	26°40’04”N, 102°25’16”W
19	S	26°32’48”N, 102°30’49”W
20	S	26°31’31”N, 102°29’09”W
Sierra Los Alamitos (Or Álamos)		
21	S	26°16’59”N, 102°27’15”W
Sierra el Venado		
22	N	26°41’28”N, 102°42’37”W
23	N	26°40’15”N, 102°43’12”W
24	N	26°39’07”N, 102°43’24”W
25	Center	26°34’58”N, 102°36’55”W
26	S. El Venado-El Hundido valley	26°31’40”N, 102°34’35”W
Sierra La Principal		
27	S. La Principal-S. Las Margaritas	26°46’30”N, 102°49’16”W
28	S. La Principal-S. Las Margaritas	26°45’52”N, 102°47’30”W
29	S. La Principal-S. Las Margaritas	26°45’28”N, 102°47’07”W
30	N	26°48’06”N, 102°41’44”W
31	N	26°46’38”N, 102°42’38”W
32	S	26°43’37”N, 102°37’41”W
La Esperanza		
33	Northern valley	26°56’12”N, 103°04’10”W
Puerto La Noria (Piedras de Lumbre)		

(continued)

Table 9.1 (continued)

Polygon Number	General Location	Reference geographic coordinates
34	Central valley	26°55’24”N, 102°53’58”W
Sierra Los Órganos		
35	N	26°45’05”N, 103°02’57”W
S. Los Órganos-S. Las Margaritas		
36	Mountains between both sierras	26°45’27”N, 102°57’13”W
Sierra Los Remedios		
37	S. Los Remedios-Valle Buena Vista	26°36’38”N, 103°02’26”W
38	Acatita	26°31’17”N, 103°00’28”W
Sierra Las Margaritas		
39	N	26°44’10”N, 102°50’41”W
40	N	26°43’34”N, 102°47’46”W
41	N	26°42’19”N, 102°50’44”W
42	NW	26°43’36”N, 102°53’31”W
43	NW	26°42’09”N, 102°53’21”W
44	NW	26°41’58”N, 102°54’22”W
45	NW	26°41’41”N, 102°54’58”W
46	NW	26°40’59”N, 102°57’23”W
47	E	26°38’49”N, 102°49’51”W
48	E	26°39’23”N, 102°46’07”W
49	S	26°34’59”N, 102°56’45”W
50	S	26°30’52”N, 102°55’43”W

Results

Gypsum Outcrops

In total, 50 potential gypsum outcrops (polygons) with different sizes and gypsum concentrations were identified in CCM (Fig. 9.6, Table 9.1). All specimen records that fell inside a polygon corresponded to species that are associated with gypsum outcrops (by bibliographic references or personal observations) either as gypsophiles or as gypsovags; hence, we did not find any case that contradicted our method. Nevertheless, it is important to remember that the polygons are traced as general shapes using a relatively large scale, while gypsum rarely is homogeneously distributed across areas. Therefore, we still prefer to be conservative and refer to the area as polygons when they were identified with our method as potential gypsum outcrops, unless they have been confirmed as such.

The most prominent outcrop, as expected, is located inside CCB, near the town, where the dunes with pure gypsum are located. Another three valleys show potentially important amounts of gypsum (Fig. 9.6, Table 9.1): (1) Valle el Hundido (polygons 18 and 19), (2) north side of Valle La Esperanza (polygon 33), (3) parts of a small valley near Puerto La Noria (or Piedras de Lumbre; polygon 34). Most of the sierras within the municipality have potential gypsum outcrops, except for Sierra

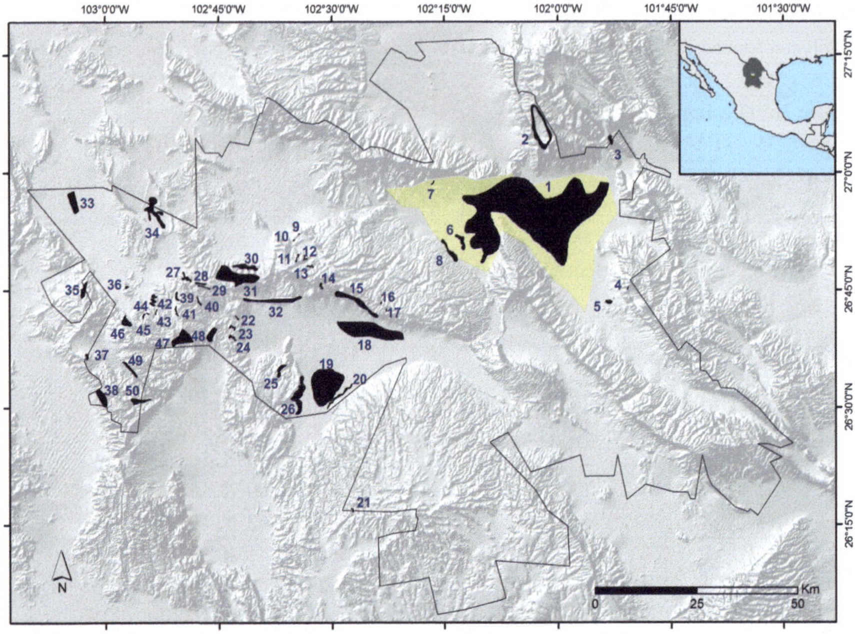

Fig. 9.6 Potential gypsum outcrops (polygons in black) identified in the Cuatro Ciénegas municipality (CCM), delinated with a gray line. In the map of Mexico at top right, the state of Coahuila is highlighted in dark gray, and the location of CCB is indicated with a dot. In the main map, the numbers correspond to the polygons listed in Table 9.1; the Cuatro Ciénegas Basin (CCB) is indicated in yellow

San Marcos y Pinos; nevertheless, the northern tip of this sierra near the Cuatro Ciénegas Basin may have small gypsum outcrops according to herbarium labels, but these were not obvious using the ArcGIS Landsat imagery. The western sierras in the municipality have numerous outcrops of Cretaceous-aged gypsum, especially derived from the Acatita Formation, as in the Sierra Las Margaritas. Perhaps none is more impressive, however, than the massive gypsum outcrops of the Sierra de Los Remedios, which continues along the west side of this sierra well outside the municipality and which forms steep slopes in many places. We cannot rule out the existence of additional small gypsum outcrops/gypseous soils in the northern regions of Sierra La Fragua and the upper southwestern region of Sierra La Purísima, which appear to have some gypsum content as they have a very pale blue appearance when using the ArcGIS Landsat imagery. These were not bright enough to be considered as polygons, however.

Gypsum Outcrops and Plant Collections

Mapping of georeferenced specimens revealed that most of the collecting effort in CCM has been conducted within the basin (CCB) (Fig. 9.7) although the majority of the other valleys and sierras have at least some collections. As expected, most of the gypsum collections come from polygon 1, inside CCB, although there are records for eight other gypsum outcrops (2, 6, 8, 18–20, 22, and 37). Collections were considered to have been made at two additional polygons (21 and 35), even though no herbarium specimens fell strictly within them, because there are specimens collected very nearby with labels that indicate gypsum: we consider these collections to have marginal georeference errors. A few collections were left out from the checklist even though they indicated gypsum in the label, due to the lack of evidence for the presence of gypsum using our method (yellow triangles not associated with a green dot or a polygon in Fig. 9.7).

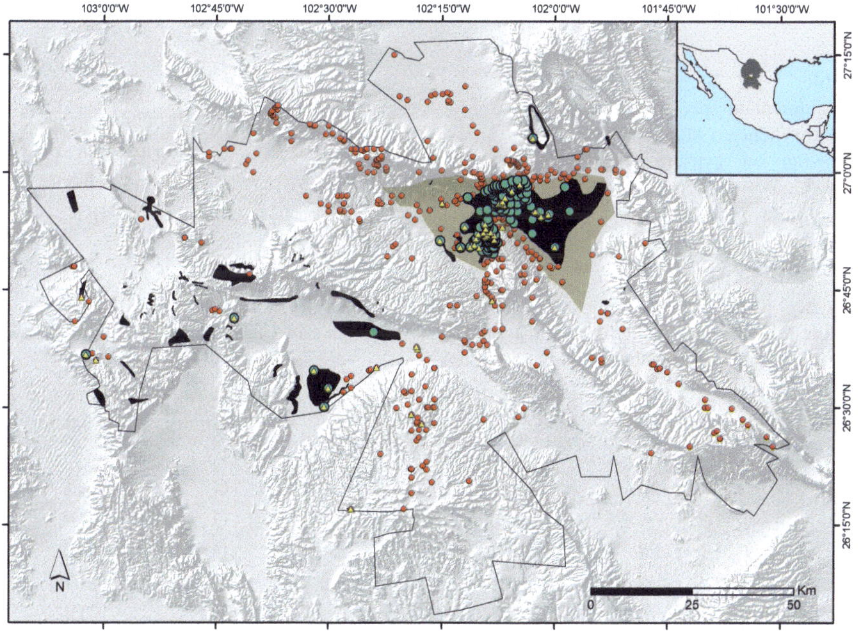

Fig. 9.7 Collecting efforts in the Cuatro Ciénegas municipality (CCM). In the map of Mexico at top right, the state of Coahuila is highlighted in dark gray and CCM is indicated with a dark gray line in the main map. Orange dots correspond to general herbarium collections without indication of gypsum; green dots correspond to collections that intersected with a polygon; yellow triangles indicate collections that have an explicit reference to gypsum on the label. Cuatro Ciénegas Basin (CCB) is indicated in olive-green and the potential gypsum outcrops at CCM are indicated in black

Vascular Plant Diversity on Cuatro Ciénegas Gypsum

We found 297 species in 187 genera and 60 families of vascular plants growing on potential gypsum outcrops in CCM, from which 11 are identified only to genus (Appendix). Among the species growing on gypsum soils in CCM, 31 are gypsophiles (endemic to gypsum), five halogypsophiles (endemic to gypsum mixed with sodium chloride), and three are restricted to soils with gypsum but it can be pure (gypsophiles) or mixed with sodium chloride (halogypsophiles) (Tables 9.2 and 9.3).

The most diverse family (Table 9.4) is Asteraceae (with 54 species), followed by Cactaceae (32 species) and by Poaceae (24 species). At the other extreme, there are 20 families with one species, nine with two, and eight with three species growing on the potential gypsum outcrops of the municipality. Despite the high diversity of Poaceae and Cactaceae, most of their species are gypsovags; in both families, only one species is restricted to gypsum (*Ancistrocactus pinkavanus*, Cactaceae; *Sporobolus spiciformis*, Poaceae; the latter is a halogypsophile). If only gypsophiles and halogypsophiles are considered, Asteraceae is still the family with most species (19 species), followed by Namaceae (three species), Brassicaceae and Euphorbiaceae (two species each) and 13 other families with a single species.

Table 9.2 Number of vascular plant species in the Cuatro Ciénegas Municipality growing on gypsum soils, classified by their fidelity to gypsum

Gypsovags	233
Gypsophiles	31
Halogypsovags	15
Halogypsophiles	5
Either G or HG	3

Table 9.3 Vascular plant species restricted to gypsum soils in Cuatro Cienégas Municipality, arranged by family and their affinity to gypsum

Family	Species		Gypsum affinity	Endemism
Asteraceae	*Dicranocarpus*	*parviflorus*	G	
	Erigeron	*cuatrocienegensis*	G	CCE
	Erigeron	*pinkavii*	G	CDE
	Gaillardia	*candelaria*	G	CDE
	Gaillardia	*gypsophila*	G	CCE
	Gaillardia	*henricksonii*	G	CDE
	Xylothamia	*truncata*	G	CCE
	Haploësthes	*hintoniana*	G	CDE
	Sartwellia	*mexicana*	G	CDE
	Sartwellia	*puberula*	G	CDE
	Solidago	*gypsophila*	G	CCE
	Thymophylla	*gypsophila*	G	CDE
	Varilla	*mexicana* var. *gypsophila*	G	CDE

(continued)

Family	Species		Gypsum affinity	Endemism
	Viguiera	*dentata* var. *longibracteata*	G	
	Xanthisma	*gypsophilum*	G	CDE
	Xanthisma	*restiforme*	G	CCE
Brassicaceae	*Nerisyrenia*	*castillonii*	G	CDE
	Nerisyrenia	*incana*	G	CCE
Cactaceae	*Ancistrocactus*	*pinkavanus*	G	CCE
Caryophyllaceae	*Drymaria*	*coahuilana*	G	CDE
Ehretiaceae	*Tiquilia*	*turneri*	G	CDE
Euphorbiaceae	*Euphorbia*	*crepitata*	G	CDE
	Euphorbia	*fruticulosa* var. *hirtella*	G	CDE
Fouquieriaceae	*Fouquieria*	*shrevei*	G	CDE
Hydrophyllaceae	*Phacelia*	*marshall-johnstonii* var. *marshall-johnstonii*	G	CCE
Loasaceae	*Petalonyx*	*crenatus*	G	CDE
Namaceae	*Nama*	*constancei*	G	CDE
	Nama	*serpylloides*	G, HG	CDE
	Nama	*stenophylla*	G	CDE
Nyctaginaceae	*Acleisanthes*	*purpusianus*	G	
Pteridaceae	*Notholaena*	*bryopoda*	G	
Rubiaceae	*Hedyotis*	*teretifolia*	G	CDE
Amaranthaceae	*Tidestromia*	*rhizomatosa*	G, HG	CCE
Asteraceae	*Haploësthes*	*robusta*	HG	CCE
	Helianthus	*paradoxus* subsp. *cuatrocienegensis*	HG	CCE
	Xanthisma	*johnstonii*	HG	CDE
Chenopodiaceae	*Meiomeria*	*stellata*	HG	CDE
Gentianaceae	*Sabatia*	*tuberculata*	HG	CCE
Poaceae	*Sporobolus*	*spiciformis*	HG	CDE

G= gypsophile; HG= Halogypsophyte. CCE= Endemic to the Cuatro Ciénegas Municipality; CDE= Endemic to the Chihuahuan Desert

The most diverse genus is *Euphorbia* (with 10 species), followed by *Bouteloua*, *Sporobolus*, and *Tidestromia* (five species each), while 10 genera have four species, 10 three, 36 two, and 126 have a single species. Among only gypsophiles and halogypsophiles, *Gallardia*, *Nama,* and *Xanthisma* are the most diverse genera (three species each), followed by *Erigeron*, *Euphorbia*, *Haploësthes*, and *Nerisyrenia* (two species each), and 22 genera with one species.

Several species that fall in Polygon 1 are associated with marshes, wetlands, or meadows near natural pools, as well as a few species in other polygons with springs, such as in Sierra Las Margaritas, near Delicias: *Cirsium coahuilense* (Asteraceae); *Ipomoea sagittata* (Convolvulaceae); *Cladium jamaicense*, *Eleocharis montevidensis*, *E. rostellata*, *Fimbristylis thermalis*, *Fuirena simplex*, *Rhynchospora colorata*, *Schoenus nigricans*, *Scirpus americanus*, *S. maritimus* (Cyperaceae); *Ludwigia*

Table 9.4 Major vascular plant groups and subordinate families with the corresponding number of species growing on gypsum soils in Cuatro Cienégas Municipality

Plant group	Species Number
PTERIDOPHYTES	7
Pteridaceae	5
Selaginellaceae	2
GYMNOSPERMS	2
Ephedraceae	2
ANGIOSPERMS	288
Asteraceae	54
Cactaceae	32
Poaceae	24
Cyperaceae	12
Euphorbiaceae	12
Chenopodiaceae	11
Asparagaceae	9
Fabaceae	9
Amaranthaceae	8
Nyctaginaceae	8
Apocynaceae	6
Loasaceae	6
Malvaceae	5
Polygalaceae	5
Zygophyllaceae	5
Brassicaceae	4
Ehretiaceae	4
Namaceae	4
Rubiaceae	4
Solanaceae	4
Acanthaceae	3
Anacardiaceae	3
Boraginaceae	3
Onagraceae	3
Orobanchaceae	3
Rhamnaceae	3
Rosaceae	3
Verbenaceae	3
Aizoaceae	2
Fouquieriaceae	2
Gentianaceae	2
Krameriaceae	2
Lamiaceae	2
Oleaceae	2
Papaveraceae	2
Plantaginaceae	2
Portulacaceae	2

Table 9.4 (continued)

Plant group	Species Number
Bignoniaceae	1
Caryophyllaceae	1
Celastraceae	1
Convolvulaceae	1
Cucurbitaceae	1
Ebenaceae	1
Ericaceae	1
Fagaceae	1
Hydrophyllaceae	1
Malpighiaceae	1
Oxalidaceae	1
Passifloraceae	1
Polemoniaceae	1
Primulaceae	1
Resedaceae	1
Salicaceae	1
Saururaceae	1
Scrophulariaceae	1
Ulmaceae	1
Violaceae	1

repens (Onagraceae); and *Spartina spartinae* (Poaceae). Although it could be thought that this is a methodological error, herbarium specimens (e.g., Henrickson 23576, TEX) and literature support the potential presence of gypsum in at least some of these wet or semiaquatic habitats.

Endemism

Among the species growing on potential gypsum outcrops in CCM, 79 (26%) are endemic to the Chihuahuan Desert, while 15 (5%; Table 9.5) are restricted to the municipality. Among the endemics of CCM, eight are gypsophiles, three are halogypsophiles, *Tidestromia rhizomatosa* can be a gypsophile or halogypsophile, and three species are gypsovags (Table 9.5).

Other gypsophiles that are not restricted to CCM but that are endemic to the Chihuahuan Desert include: *Erigeron pinkavii*, *Gaillardia candelaria*, *G. henricksonii*, *Haploësthes hintoniana*, *Sartwellia mexicana*, *S. puberula*, *Thymophylla gypsophila*, *Varilla mexicana* var. *gypsophila*, and *Xanthisma gypsophilum* (all from Asteraceae); *Tiquilia turneri* (Boraginaceae); *Nerisyrenia castillonii* (Brassicaceae); *Drymaria coahuilana* (Caryophyllaceae); *Euphorbia crepitata* and *E. fruticulosa* var. *hirtella* (Euphorbiaceae); *Fouquieria shrevei* (Fouquieriaceae); *Petalonyx*

Table 9.5 Taxa that grow on gypsum and that are endemic to CCM with their gypsum fidelity

Taxon		Gypsum fidelity
Ancistrocactus	*pinkavanus*	G
Erigeron	*cuatrocienegensis*	G
Gaillardia	*gypsophila*	G
Nerisyrenia	*incana*	G
Phacelia	*marshall-johnstonii* var. *marshall-johnstonii*	G
Solidago	*gypsophila*	G
Xanthisma	*restiforme*	G
Xylothamia	*truncata*	G
Haploësthes	*robusta*	HG
Helianthus	*paradoxus* subsp. *cuatrocienegensis*	HG
Sabatia	*tuberculata*	HG
Tidestromia	*rhizomatosa*	G, HG
Justicia	*coahuilana*	GV
Nama	*cuatrocienegensis*	GV
Sporobolus	*coahuilensis*	GV

G= gypsophiles; GV= Gypsovags; HG= Halogypsophles

crenatus (Loasaceae); *Nama constancei* and *N. stenophylla* (Namaceae), and *Hedyotis teretifolia* (Rubiaceae).

The following halogypsophiles are endemic to the Chihuahuan Desert: *Xanthisma johnstonii* (Asteraceae), *Meiomeria stellata* (Chenopodiaceae), and *Sporobolus spiciformis* (Poaceae). *Nama serpylloides* (Namaceae) can be considered either a gypsophile or halogypsophile, and is also endemic to the Chihuahuan Desert.

Discussion

While Pinkava (1984) reported 48 species on the gypsum dunes and flats of CCB, and Villarreal-Quintanilla and Encina-Domínguez (2005) reported 35 species for gypsum dunes or soils of CCB, we found 297 species from gypsum soils, including 31 gypsophiles, 5 halogypsophiles, and three either gypsophiles or halogypsophiles in the CCM. Several of the specimens that allowed us to recognize additional species as gypsophiles were already collected at least by the time of the last-mentioned publication; therefore, we believe that the method we developed here is useful for the study of gypsophiles. Differences in the number of reported species are due to differences in covered area (CCB in Pinkava 1984) or in the focus of the research (endemics in Villarreal-Quintanilla and Encina-Domínguez 2005) as well as because of the method here developed, so it is recommended that these results be further evaluated with additional field exploration and soil analyses. In addition, the method provides grounds for further exploration focused on gypsum soils by identifying potential gypsum outcrops in the area.

Our results show that most of the potential gypsum outcrops in CCM identified here are still to be explored. The numbers here presented also could increase when other datasets are incorporated or more specimens are properly georeferenced.

Moore et al. (2014) presented a breakdown of the representation of large angiosperm groups among the world's gypsophiles, and our results are consistent with theirs. We find that Asterids are the most diverse angiosperm group among the gypsophiles and halogypsophiles of CCM (28 species), followed by Caryophyllales (with five species) and Brassicales (with two species).

We emphasize that gypsovag species contribute fundamentally to the flora of gypsum outcrops in CCM (233 spp.). Their study will be just as important to understand the ecophysiological mechanisms that lead to the ability of plants to thrive on gypsum. Likewise, the great ecological diversity of gypsum outcrops in CCM presents an important opportunity to understand how edaphic and climatic factors help shape evolution and community assembly.

Because the Chihuahuan Desert has the largest known flora of gypsophiles in the world, it is an important test bed for studies such as ours. The success of this approach in CCM demonstrates that it will also be valuable in understanding other gypsophile floras that are much more poorly known. Indeed, our approach of synthesizing the available literature and collections from gypsum soils is a necessary first step in any deeper exploration of the evolution and ecology of plants on gypsum. At present there is a developing global interest in gypsum ecosystems, promoted by the collaborative GYPWORLD Project, which is funded by the European Commission under the European Union's Horizon 2020 Programme/MSCARISE-2017.

Conclusions

Our method for identifying potential gypsum outcrops and providing floristic checklists are highly promising techniques for developing similar maps and checklists in other unusual floras, including the world's gypsum associated floras. The analyses presented here demonstrate that the gypsum flora of Cuatro Ciénegas municipality and of the Chihuahuan Desert is globally relevant in terms of diversity and endemism, even though it is still not completely known. Indeed, much work still remains to characterize, understand, and preserve this living laboratory of plant evolution.

Acknowledgments We thank Héctor M. Hernández for allowing us to use his database on Cactaceae and Tom Wendt for downloading and sharing the information from LL and TEX. We thank George Hinton for help with demonstrating how to detect gypsum using online imagery. We thank the National Geographic Society, the US National Science Foundation (DEB 1054539 & DEB 1352907), and Oberlin College for funding. We greatly appreciate the careful revision and suggestions by the editors of this volume.

Appendix

Checklist of the vascular plant species growing on gypsum soils in the Cuatro Ciénegas municipality (CCM) based on georeferenced specimens and literature. Acronyms related to soil fidelity: G = Gypsophile; HG = Halogypsophile; HGV = Halogypsovag; GV = Gypsovag. Acronyms related to endemism: CCE = Endemic to Cuatro Ciénegas municipality; CDE = Endemic to the Chihuahuan Desert (Sensu Henrickson & Straw 1976). For the enlisted specimens, only the first main collector´s lastname is cited to save space. Bibliographic references: [1]= Pinkava (1984); [2]= Villarreal-Quintanilla and Encina-Domínguez (2005); [3]= Nesom (1981); [4]= Urbatsch & Roberts (2004); [5]= García-Morales et al. (2014); [6]= Johnston (1939); [7]= Johnston (1943); [8]= Levin (2002); [9]= Morales García (2018); [10]= Atwood and Pinkava (1977); [11]= Turner (2004).

Pteridophytas

PTERIDACEAE

Astrolepis cochisensis (Goodd.) D.M. Benham & Windham [GV]. Cole 3749A (ASU).

Astrolepis integerrima (Hook.) D.M. Benham & Windham [GV]. Engard 320 (DES).
Cheilanthes hookeri Domin [GV]. Chiang 7635 (LL); Daniel 532B (ASU).
Notholaena bryopoda Maxon **[G]**. Henrickson 12117-5 (LL); Moore 2574 (MEXU, TEX)
Notholaena greggii (Mett. ex Kuhn) Maxon [GV]. Cole 3750 (ASU). **[CDE].**

SELAGINELLACEAE

Selaginella ribae Valdespino [GV]. Moore 1994 (MEXU, TEX)
Selaginella wrightii Hieron. [GV]. Lehto 5497 (ASU).
Selaginella sp. Moore 2573 (MEXU, TEX).

Gymnospermae

EPHEDRACEAE

Ephedra pedunculata Engelm. ex S. Watson [GV]. Henrickson 23183 (TEX); Lehto 5496 (ASU); Meyer s.n. (ASU).
[1] *Ephedra trifurca* Torr. ex S. Watson [GV]. Henrickson 12540 (TEX); Lehto 5264 (ASU).

Angiospermae

ACANTHACEAE

Carlowrightia serpyllifolia A. Gray [GV]. Chiang 7630 (ASU, LL, MEXU); Daniel 218 (MEXU), 530 (ASU, MEXU); Henrickson 20399 (TEX). **[CDE].**
Justicia coahuilana T.F. Daniel [GV]. Johnston 7162 (GH); Lehto 5485 (ASU). **[CCE].**
Ruellia parryi A. Gray [GV]. Henrickson 24432 (TEX). **[CDE].**

AIZOACEAE

Sesuvium sessile Pers. [GV]. Pinkava 5071 (ASU).
[1] *Sesuvium verrucosum* Raf. [HGV]. Chiang 7618 (LL, MEXU), 7634 (LL, MEXU), 7648 (LL), 9156A (LL, MEXU); Henrickson 14279 (TEX), 18742 (MEXU, TEX).

AMARANTHACEAE

Amaranthus blitoides S. Watson [GV]. Henrickson 14311 (TEX); Pinkava P5219A (ASU).
Amaranthus retroflexus L. [GV]. Henrickson 9999 (TEX).
Froelichia arizonica Thornber ex Standl. [GV]. Henrickson 12558 (MEXU); Pinkava 5204 (ASU), 5237 (ASU), 5954 (ASU).
[2] *Tidestromia gemmata* I.M. Johnst. [GV]. **[CDE]**.
Tidestromia lanuginosa (Nutt.) Standl. [GV]. Henrickson 17449 (MEXU, TEX); Pinkava P5051 (ASU).
[1,2] *Tidestromia rhizomatosa* I.M. Johnst. [**G, HG**]. Chiang 7610A (TEX); Flores 1653 (MEXU, TEX), 1815 (MEXU, TEX); Henrickson 14308-3 (TEX), 14450-7 (TEX), 17455-4 (TEX), 20406 (ASU); Moore 1472 (MEXU, TEX), 2857 (MEXU, TEX). **[CCE].**
Tidestromia suffruticosa (Torr.) Standl. [GV]. Pinkava P3757 (ASU). **[CDE].**
Tidestromia tenella I.M. Johnst. [GV]. Henrickson 17438 (TEX), 17453 (TEX), 18741 (TEX). **[CDE].**

ANACARDIACEAE

Pistacia mexicana Kunth [GV]. Pinkava P13086 (ASU).
Rhus virens Lindh. ex A. Gray [GV]. Pinkava P13059 (ASU).
Toxicodendron radicans (L.) Kuntze [GV]. Pinkava 3857 (ASU).

APOCYNACEAE

Asclepias oenotheroides Schltdl. & Cham. [GV]. Henrickson 15921 (MEXU).
Mandevilla macrosiphon (Torr.) Pichon [GV]. Henrickson 24427 (TEX)
Metastelma barbigerum Scheele [GV]. Pinkava 5947 (ASU).
Metastelma pringlei A. Gray [GV]. Cole 3599A (MEXU); Henrickson 20400 (MEXU, TEX).
Metastelma sp. Moore 2567 (MEXU, TEX).

Sarcostemma cynanchoides Decne. [GV]. Henrickson 12554 (MEXU, TEX); Lehto 5269 (ASU).

ASPARAGACEAE

Agave asperrima Jacobi [GV]. Pinkava 5257 (DES).
Agave lechuguilla Torr. [GV]. Pinkava 3758 (ASU).
Dandya purpusii (Brandegee) H.E. Moore [GV]. Henrickson 24434 (TEX). **[CDE].**
Dasylirion cedrosanum Trel. [GV]. Bogler 683 (MO); Engard 293 (DES); Pinkava 3830 (DES), 5003 (DES), 5032 (ASU, DES), 5277 (DES), 5728 (DES).
[1] *Dasylirion leiophyllum* Engelm. ex Trel. [GV]
Hesperaloë funifera (K. Koch) Trel. [GV]. Gentry 23138 (DES).
Yucca thompsoniana Trel. [GV]. García 8116 (MEXU), 8117 (MEXU). **[CDE]**.
Yucca torreyi Schafer [GV]. Leverich 4 (TEX).
[1] *Yucca treculeana* Carrière [GV]. Pinkava 5018 (DES), 5255 (ASU, DES), 5548 (ASU), P5270 (ASU).

ASTERACEAE

[1] *Baccharis neglecta* Britton [GV]. Daniel 680 (ASU).
Baccharis salicina Torr. & A. Gray [GV]. Henrickson 12560 (LL).
[1] *Bahia absinthifolia* Benth. [GV]. Pinkava 5273 (ASU).
Borrichia frutescens (L.) DC. [GV]. Rodríguez 1130 (MEXU).
Brickellia glutinosa A. Gray [GV]. Pinkava 13099P (ASU). **[CDE].**
Chaetopappa pulchella Shinners [GV]. Johnston 7151 (GH). **[CDE].**
Chrysactinia mexicana A. Gray [GV]. Flores 1816 (MEXU, TEX).
Cirsium coahuilense G.B. Ownbey & Pinkava **[GV]**. Henrickson 7957 (ARIZ, LL, MEXU, TEX); Nesom 7045 (NY), 7050 (NY, OBI); Pinkava 5075 (ASU, US, ASU, MEXU), 50667 (MEXU), P5075 (ASU); Wendt 658 (ASU). **[CDE].**
Conoclinium betonicifolium (Mill.) R.M. King & H. Rob. [GV]. García 8119 (MEXU); Henrickson 7966 (ASU, LL, MEXU); Pinkava 3725 (ASU), 5076 (ASU); Wendt 657 (LL).
Dicranocarpus parviflorus A. Gray **[G]**. Moore 1960 (MEXU, TEX), 2559 (MEXU, TEX).
[1,2] *Erigeron cuatrocienegensis* G.L. Nelson **[G]**. Lehto 5511 (NY); Pinkava P3849 (ASU), P5511 (ASU). **[CCE].**
[1,3] *Erigeron pinkavii* Turner [**G**]. **[CDE].**
Flaveria chlorifolia A. Gray [GV]. Brown s.n. (ASU); Chiang 9159 (TEX), 9164 (LL); Flores 1832 (MEXU, TEX); García 8118 (MEXU); Henrickson 7958 (ARIZ, ASU, LL, NMC), 15575 (LL); Villarreal 3193 (MEXU). **[CDE].**
[2] *Flaveria palmeri* J.R. Johnst. [GV]. Henrickson 20413 (MEXU). **[CDE].**
[1] *Flaveria trinervia* (Spreng.) C. Mohr [GV]. Meyer 8 (ASU).
Gaillardia candelaria B.L. Turner var. *mikemoorei* B.L. Turner [**G**]. Moore 2000 (MEXU, TEX). **[CDE].**
[1,2] *Gaillardia gypsophila* B.L. Turner **[HG]**. Henrickson 12544 (LL); Moore 1955 (MEXU, TEX); Turner 6188 (TEX). **[CCE].**
Gaillardia henricksonii B.L. Turner **[G]**. Moore 2575 (MEXU, TEX). **[CDE].**

Gochnatia hypoleuca (DC.) A. Gray [GV]. Moore 2583 (MEXU, TEX).

Xylothamia triantha (S.F. Blake) G.L. Nesom [GV]. Chiang 7615 (LL, MEXU, NY); Iltis 25C (TEX), 25E (MEXU); Pinkava P3645 (ASU). **[CDE].**

[2,4] *Xylothamia truncata* G.L. Nesom **[G]**. Nesom 5254 (TEX). **[CCE]**

Haploësthes greggii A. Gray var. *greggii* [GV]. Moore 1476 (MEXU, TEX). This species prefers gypsum.

Haploësthes hintoniana B.L. Turner **[G]**. Moore 2001 (MEXU, TEX). **[CDE].**

[1,2] *Haploësthes robusta* I.M. Johnst. **[HG]**. Barrie 370 (MEXU); Bogler 100 (MEXU); Engard 289 (DES); Flores 1829 (MEXU, TEX); Henrickson 7959 (ASU, LL, MEXU, NMC), 17440 (ARIZ, MEXU, SRSC, TEX); Moore 2852 (MEXU, TEX); Pinkava 5034 (ASU); Powell 2619 (NY); Wendt 663 (LL, MEXU); White 1923 (ARIZ, GH, MICH). **[CCE].**

Helianthus paradoxus Heiser subsp. *cuatrocienegensis* R.C. Sivinski **[HG]**. Pinkava 4111 (ASU). [**CCE**].

[1] *Isocoma coronopifolia* (A. Gray) Greene [GV]. Engard 294A (DES), 294B (DES); Flores-Olvera 1825 (MEXU, TEX); Henrickson 12539 (LL, MEXU); Leverich 2B (TEX); Pinkava 3727 (ASU).

[1] *Isocoma drummondii* (Torrey & A. Gray) Greene [GV]. Pinkava 5253 (ASU).

Palafoxia texana DC. [GV]. Pinkava 5946 (ASU).

Pectis angustifolia Torr. [GV]. Pinkava 3742 (ASU), 5220 (ASU).

Perityle coahuilensis A.M. Powell [GV]. Pinkava P13099A (ASU). **[CDE].**

Porophyllum scoparium A. Gray [GV]. Chiang 7641 (LL); Pinkava 5488 (ASU).

[1] *Pseudoclappia arenaria* Rydb. [HGV]. Henrickson 15909A (TEX); Pinkava 5066 (ASU); Rodríguez 1353 (MEXU). **[CDE].**

Psilostrophe tagetina (Nutt.) Greene [GV]. Moore 2578 (MEXU, TEX).

Sanvitalia ocymoides DC. [GV]. Rodríguez 1124 (MEXU).

[1,2] *Sartwellia mexicana* A. Gray **[G]**. Chiang 7614 (MEXU, NY); Henrickson 17435A (TEX 86177); Moore 1470 (MEXU, TEX), 1958 (MEXU, TEX); Pinkava 5022 (ASU), 5073 (ASU), 5545 (ASU). [**CDE**]

Sartwellia puberula Rydb. **[G]**. Moore 2582 (MEXU, TEX), 2862 (MEXU, TEX). [**CDE**]

Sidneya tenuifolia E.E. Schill. & Panero [GV]. Pinkava P13096 (ASU).

Simsia calva (A. Gray & Engelm.) A. Gray [GV]. Lehto P5233 (NY); Pinkava 5203 (ASU), 5233 (ASU).

[2] *Solidago gypsophila* G.L. Nesom **[G]**. **[CCE].**

Sonchus oleraceus L. [GV]. Pinkava 5063 (ASU), 5210 (ASU).

Thelesperma cf. *longipes* A. Gray [GV]. Moore 1998 (MEXU, TEX).

Thelesperma megapotamicum (Spreng.) Kuntze [GV]. Pinkava P13081 (ASU).

[2] *Thymophylla gypsophila* (B.L. Turner) Strother **[G]**. Moore 1996 (MEXU, TEX); Turner 6172 (TEX, UC). **[CDE].**

Thymophylla micropoides (DC.) Strother [GV]. Pinkava 5188 (ASU), P5239 (ASU); Pinkava 5952 (DES).

Thymophylla pentachaeta (DC.) Small [GV]. Pinkava 5217 (ASU), 5510 (ASU).

Trixis californica Kellogg. [GV]. Pinkava 5184 (ASU).

[1,2] *Varilla mexicana* A. Gray var. *gypsophila* **[G]**. Pinkava 5260 (ASU), P13105 (ASU); Smith 22157 (TEX); Johnston 10337 (ASU, MEXU); Leverich 1 (MEXU); Marroquín s.n. (MEXU); Moore 2018 (MEXU, TEX), 2551 (MEXU, TEX); Rodríguez 1162 (MEXU). **[CDE]**.

Verbesina encelioides (Cav.) Benth. & Hook. f. ex A. Gray [GV]. Moore 2020 (MEXU, TEX).

Viguiera dentata (Cav.) Spreng. var. *longibracteata* B.L. Turner **[G]**. Moore 2548 (MEXU, TEX).

[1] *Xanthisma gypsophilum* (B.L. Turner) D.R. Morgan & R.L. Hartm. **[G]**. Chiang 9158 (LL), 7617D (LL); Flores 1817 (MEXU, TEX); Henrickson 12542 (LL), 15920 (TEX), 17452 (MEXU, TEX); Moore 1466 (MEXU, TEX), 2579 (MEXU, TEX); Pinkava 5027 (ASU), 5265 (ASU); Turner 6052 (MEXU, TEX,). **[CDE].**

[2] *Xanthisma johnstonii* (S.F. Blake) D.R. Morgan & R.L. Hartm. **[HG]**. **[CDE].**

[1,2] *Xanthisma restiforme* (B.L. Turner) D.R. Morgan & R.L. Hartm. **[G]**. Chiang 7613 (MEXU, TEX), 7653 (LL, MEXU); Flores 1818 (MEXU, TEX); Henrickson 12538 (ARIZ, LL, MEXU, TEX), 17451 (MEXU, TEX); Moore 1471 (MEXU, TEX); Pinkava 3855 (ASU), 5023 (ASU), 5035 (ASU), 5268 (ASU), 13101 (ASU); Turner 6063 (LL); Collector unknown 3806 (NY). **[CCE].**

Xanthisma spinulosum (Pursh) D.R. Morgan & R.L. Hartm. [GV] Henrickson 12546 (LL, MEXU).

Xanthium strumarium L. [GV]. Pinkava 5506 (ASU); Zarate E13 (MEXU).

BIGNONIACEAE

[1] *Chilopsis linearis* (Cav.) Sweet. [GV]. Henrickson 12549 (TEX); Keil 5543 (ASU); Reeves P13106 (ASU); Turner 6194 (TEX).

BORAGINACEAE

Cordia parvifolia A. DC. [GV]. Wendt 7691 (LL, MEXU).

Heliotropium curassavicum L. [HGV]. Henrickson 7952 (ARIZ, ASU, MEXU, TEX); Lehto 5068 (ASU).

Omphalodes aliena A. Gray ex Hemsl. [GV]. Neff 92-3-27-1 (MEXU, TEX).

BRASSICACEAE

Nerisyrenia camporum (A. Gray) Greene [GV]. Fuentes 102 (MEXU); Lehto 196 (ASU), 5070 (ASU).

[1] *Nerisyrenia castillonii* Rollins **[G]**. Fuentes 103 (ARIZ, MEXU); Lehto 5522 (ASU); Moore 1959 (MEXU, TEX), 2869B (MEXU, TEX); Pinkava P5049 (ASU), P13115 (ASU). **[CDE].**

[1,2] *Nerisyrenia incana* Rollins **[G]** Chiang 7616 (LL), 7644 (LL); Flores 1820 (MEXU, TEX), 1830 (MEXU, TEX); Fuentes 101 (ARIZ, MEXU); Henrickson 15919-7 (TEX); Johnston 7130 (GH), 10333 (ASU, LL); Lehto 5267 (ASU); Moore 86 (MEXU, TEX), 88 (MEXU, TEX), 1475 (MEXU, TEX), 2010 (MEXU, TEX), 2552 (MEXU, TEX), 2580 (MEXU, TEX), 2851 (MEXU, TEX), 2856 (MEXU, TEX), 2869A (MEXU, TEX); Neff 92-3-27-3 (TEX); Salywon 73 (DES), 750 (ASU). **[CCE].**

Nerisyrenia sp. Moore 2869C (MEXU, TEX). This is a putative hybrid between *N. incana* and *N. castillonii*.

CACTACEAE

[1] *Ancistrocactus brevihamatus* (Engelm.) Britt. & Rose. [GV].

[5] *Ancistrocactus pinkavanus* García-Mor., Gonz.-Bot. & Rodr.-González **[G]**. Hinton 29472 (GBH); Van Devender s.n. (Greater Good). **[CCE].**

Ariocarpus fissuratus (Engelm.) K. Schum. [GV]. Hernández 2182 (MEXU); Pinkava 3987 (ASU), 5142 (LL), 5500 (ASU), 6139 (ASU); Sánchez Mejorada 4107 (MEXU). **[CDE].**

Astrophytum capricorne (A. Dietrich) Britton & Rose [GV]. Arias 2107 (MEXU). **[CDE].**

Coryphantha echinus (Engelm.) Britton & Rose [GV]. Pinkava 3989 (ASU), 5143 (ASU). **[CDE].**

[1] *Coryphantha macromeris* (Engelm.) Lem. [GV]. Hernández 2183 (MEXU). **[CDE].**

[1] *Coryphantha poselgeriana* (Dietrich) Britt. & Rose [GV]. Chiang 7655 (LL); Hernández 2180 (MEXU); Pinkava 4003 (ASU), 5043 (ASU). **[CDE].**

Coryphantha werdermannii Boed. [GV]. Pinkava 5529a (ASU). **[CDE].**

[2] *Cylindropuntia anteojoensis* (Pinkava) E.F. Anderson [GV]. **[CDE].**

Cylindropuntia imbricata (Haw.) F.M. Knuth [GV]. Pinkava 5958 (ASU).

[1] *Cylindropuntia leptocaulis* (DC.) F.M. Knuth [GV].

Echinocactus horizonthalonius Lem. [GV]. Chiang 7627 (LL); Hernández 2192 (MEXU); Pinkava 4004 (ASU).

[1] *Echinocereus enneacanthus* Engelm. [GV]. Pinkava 5061 (ASU), 5250 (ASU), P5501 (ASU), P5525 (ASU), 5552 (ASU). **[CDE].**

Echinocereus pectinatus (Scheidw.) Engelm. [GV]. Hernández 2185 (MEXU).

Echinocereus stramineus (Engelm.) Seitz [GV]. Hernández 2187 (MEXU). **[CDE].**

Epithelantha bokei Benson [GV]. Chiang 7632 (LL). **[CDE].**

Epithelantha micromeris (Engelm.) Weber [GV]. Gómez 2172B (MEXU); Hernández Macías 2191 (MEXU). **[CDE].**

Escobaria tuberculosa (Engelm.) Britton & Rose [GV]. Hernández 2188 (MEXU); Pinkava 6137 (ASU). **[CDE].**

Escobaria vivipara (Nuttall) Buxb. [GV]. Chiang 7626 (TEX); Pinkava 5080 (ASU).

Ferocactus hamatacanthus (Muehlenpf.) Britton & Rose [GV]. Hernández 2190 (MEXU); Pinkava 5503 (ASU). **[CDE].**

Grusonia bradtiana (J.M. Coult.) Britton & Rose [GV]. Arias 2108 (MEXU); Hernández 2186 (MEXU); Pinkava 3994 (ASU), 4081 (ASU). **[CDE].**

[1,2] *Grusonia moelleri* (A. Berger) E.F. Anderson [GV]. Pinkava 5002A (ASU), 5062 (ASU, MEXU), 5551 (ASU). **[CDE].**

Lophophora williamsii (Lem. ex Salm-Dyck) J.M. Coulter [GV]. Chiang 7643 (LL); Hernández Macías 2189 (MEXU); Pinkava 3759 (ASU). **[CDE].**

Mammillaria grusonii Rünge [GV]. Pinkava 5259 (ASU, RSA), 6138 (ASU). **[CDE].**

Mammillaria heyderi Muehlenpf. [GV]. Pinkava 10430A (ASU, DES), Pinkava PLO430A (DES).

Mammillaria pottsii Scheer ex Salm-Dyck [GV]. Hernández 2181 (MEXU); Pinkava 5505 (ASU). **[CDE].**
Neolloydia conoidea (DC.) Britton & Rose [GV]. Pinkava 5502 (ASU).
Opuntia atrispina Griffiths [GV]. Díaz 153 (MEXU). **[CDE].**
Opuntia macrocentra Engelm. [GV]. Díaz 118 (MEXU); Pinkava P3679 (ASU).
Opuntia phaeacantha Engelm. [GV]. Pinkava 5960 (ASU).
Sclerocactus mariposensis (Hester) N.P. Taylor [GV]. Gay 2962 (POM). **[CDE].**
Sclerocactus uncinatus (Galeotti) N.P. Taylor [GV]. Hernández 2184 (MEXU); Pinkava 5747 (ASU).

CARYOPHYLLACEAE

[1,2] *Drymaria coahuilana* (I.M. Johnst) B.L. Turner **[G].** Chiang 7646 (MEXU); Flores 1822 (MEXU, TEX); Johnston 9154 (LL, MEXU); Moore 1469 (MEXU, TEX), 1954 (MEXU, TEX), 2015 (MEXU, TEX); Pinkava 5542 (ASU), P4344 (ASU). **[CDE].**

CELASTRACEAE

Mortonia scabrella A. Gray [GV]. Moore 2569 (MEXU, TEX).

CHENOPODIACEAE

Allenrolfea occidentalis (S. Watson) Kunze [GV]. Moore 2850 (MEXU, TEX).
[2] *Atriplex acanthocarpa* (Torr.) S. Watson subsp. *stewartii* (I.M. Johnst.) Henr. [GV] Chiang 7621 (TEX); Henrickson 1986 (ASU), 14274 (ASU), 14312 (ASU, TEX), 14301 (ASU, TEX), 20397 (MEXU, TEX, UCR). **[CDE].**
[1] *Atriplex canescens* (Pursh) Nutt. [GV]. Henrickson 12553-2 (TEX), 14304-12 (TEX); Pinkava 5263 (ASU).
[2] *Atriplex prosopidium* I.M. Johnst. [HGV]. Chiang 7612 (MEXU, TEX); Henrickson 14303 (TEX), 14307 (ASU); Pinkava 4343 (ASU), 5077 (ASU), 5251 (ASU). **[CDE].**
Atriplex texana S. Watson [GV]. Henrickson 14305 (ASU)
[2] *Meiomeria stellata* (S. Watson) Standl. **[HG]. [CDE].**
Sarcocornia utahensis (Tidestr.) A.J. Scott [HGV]. Henrickson 710 (TEX).
Suaeda mexicana (Standl.) Standl. [HGV]. Flores 1654 (MEXU, TEX); Henrickson 20407 (ASU, MEXU); Moore 2859 (MEXU, TEX).
Suaeda nigra (Raf.) J.F. Macbr. [HGV]. Flores 1827 (MEXU, TEX); Henrickson 7953 (MEXU, TEX); Johnston 7128 (GH).
[2] *Suaeda palmeri* (Standl.) Standl. [GV]. Brown s.n. (ASU); Chiang 7624 (TEX); Henrickson 7968 (TEX), 14299 (ASU); Johnston 10338 (MEXU, TEX); LaBounty 5957 (ASU); Rodríguez 1139 (MEXU). **[CDE]**.
Suaeda sp. Flores 1630 (MEXU, TEX).

CONVOLVULACEAE

Ipomoea sagittata Poir. [GV]. Engard 318 (DES); Henrickson 7963 (TEX), 15371 (TEX).

CUCURBITACEAE

[1] *Ibervillea tenuisecta* (A. Gray) A. Gray [GV]. Meyer s.n. (ASU); Pinkava 5254 (ASU).

CYPERACEAE

Carex pringlei L.H. Bailey [HGV]. Cole 3737 (MICH); Daniel 683 (MICH); Henrickson 15367 (MEXU, TEX); Reeves P13034 (DES). **[CDE].**

Cladium jamaicense Crantz [GV]. Daniel 684 (MICH); Henrickson 7956 (ASU, LL, MEXU), 15369 (ARIZ); Johnston 7629 (LL, MEXU).

Eleocharis caribaea (Rottb.) S.F. Blake [GV]. Pinkava 3853 (ASU).

Eleocharis cellulosa Torr. [GV]. Henrickson 23180 (TEX).

Eleocharis montevidensis Kunth [GV]. Boke 146 (MICH); Henrickson 23179 (TEX).

Eleocharis rostellata (Torr.) Torr. [GV]. Harvey 1235 (MICH); Jones 7178 (MICH).

Fimbristylis thermalis S. Watson [GV]. Daniel 677 (MICH), 678 (MICH), 682 (MICH); Henrickson 7967 (LL, MEXU), 14294D (LL); Kral 25771 (MICH).

Fuirena simplex Vahl [GV]. Harvey 1236 (MICH); Henrickson 7962 (ASU, LL); Jones 7176 (MICH); Pinkava P3841 (ASU).

Rhynchospora colorata (L.) H. Pfeiff. [GV]. Pinkava 3854 (ASU).

Schoenus nigricans L. [GV]. Henrickson 7954 (ASU, LL), 15372 (TEX), 23190 (TEX), 23191 (TEX); Meyer s.n. (MICH).

Scirpus americanus Pers. [HGV]. Daniel 685 (MICH); Harvey 1238A (MICH); Henrickson 23192 (TEX); Thompson 589 (MICH).

Scirpus maritimus L. [GV]. Pinkava P3633 (ASU).

EBENACEAE

Diospyros texana Scheele [GV]. Pinkava P13082 (ASU).

EHRETIACEAE

[1] *Tiquilia gossypina* (Woot. & Standl.) Richardson [GV] Moore 2571 (MEXU, TEX). **[CDE].**

[1] *Tiquilia greggii* (Torrey & A. Gray) A.T. Richardson [GV]. Flores 1823 (MEXU, TEX); Moore 1995 (MEXU, TEX), 2547 (MEXU, TEX), 2576 (MEXU, TEX), 2861 (MEXU, TEX). **[CDE].**

Tiquilia mexicana (S. Watson) A.T. Richardson [GV]. Moore 2545 (MEXU, TEX). **[CDE].**

[1,2] *Tiquilia turneri* Richardson **[G]**. Flores 1819 (MEXU, TEX); Henrickson 12541 (LL, MEXU); Lehto P5262 (ASU); Moore 87 (MEXU, TEX), 89 (MEXU, TEX), 90 (MEXU, TEX), 1467 (MEXU, TEX), 2011 (MEXU, TEX); Reeves P13108A (ASU); Richardson 1595 (ARIZ, ASU, MEXU, MICH, RM, UC, US). **[CDE].**

ERICACEAE

Arbutus xalapensis Kunth [GV]. Engard 311 (DES).

EUPHORBIACEAE

Croton incanus Kunth [GV]. Pinkava 5182 (ASU); P13077 (ASU).

Euphorbia antisyphilitica Zucc. [GV]. Moore 2866 (MEXU, TEX); Pinkava 5200 (ASU).

[1] *Euphorbia astyla* Engelm. ex Boiss. [GV]. Chiang 7617C (LL), 7619 (LL, MEXU), 9157 (LL, MEXU); Engard 292 (DES); Pinkava 3627 (ASU), 5037 (ASU); Wendt 866 (TEX). **[CDE].**

[2] *Euphorbia crepitata* L.C. Wheeler. **[G]**. Moore 2557 (MEXU, TEX); Pinkava P5479 (ASU). **[CDE].**

Euphorbia exstipulata Engelm. [GV]. Pinkava P13050 (ASU).

[1,2] *Euphorbia fruticulosa* Engelm. ex Boiss. var. *hirtella* M.C. Johnst. **[G]**. Henrickson 12548 (LL, TEX); Pinkava P13104 (ASU); Wendt 864 (TEX). **[CDE].**

Euphorbia lata Engelm. [GV]. Moore 2558 (MEXU, TEX), 2867 (MEXU, TEX).

[2] *Euphorbia scopulorum* Brandegee [GV]. Chiang 7642 (LL, MEXU); Henrickson 7970 (ARIZ, MEXU, TEX); Pinkava 3753 (ASU), 5245 (ASU), 5477 (ASU), P5948 (ASU); Rodríguez 1143 (MEXU). **[CDE]**.

Euphorbia serpyllifolia Pers. [GV]. Pinkava P5219B (ASU).

Euphorbia stictospora Engelm. [GV]. Pinkava P5222 (ASU).

Euphorbia sp. Moore 1477 (MEXU, TEX), 2017 (MEXU, TEX), 2855 (MEXU, TEX), 2868 (MEXU, TEX).

Tragia amblyodonta (Müll. Arg.) Pax & K. Hoffm. [GV]. Pinkava 5235 (ASU), 5186A (ASU), P13048 (ASU).

FABACEAE

Acacia berlandieri Benth. [GV]. Lehto 5181 (NY); Minckley 7237 (ASU).

Acacia farnesiana (L.) Willd. [GV]. Pinkava 5520 (ASU).

Acacia rigidula Benth. [GV]. Pinkava 5471 (ASU).

Acacia roemeriana Scheele [GV]. Pinkava 5468 (ASU).

Dalea wrightii A. Gray [GV]. Pinkava 5244 (DES).

Mimosa unipinnata B.D. Parfitt & Pinkava [GV]. Henrickson 24433 (TEX). **[CDE].**

Prosopis glandulosa Torr. [GV]. Henrickson 12551 (LL).

Senna monozyx (H.S. Irwin & Barneby) H.S. Irwin & Barneby [GV]. Chiang 7633 (TEX). **[CDE].**

Senna pilosior (B.L. Rob. ex J.F. Macbr.) H.S. Irwin & Barneby [GV]. Pinkava 5494 (ASU). **[CDE].**

FAGACEAE

Quercus pungens Liebm. [GV]. Pinkava P13060 (ASU), P13088 (ASU). **[CDE].**

FOUQUIERIACEAE

[2,6] *Fouquieria shrevei* I.M. Johnst. **[G]**. Moore 1997 (MEXU, TEX), 2555 (MEXU, TEX), 2864 (MEXU, TEX). **[CDE].**

[1] *Fouquieria splendens* Engelm. in Wisliz. [GV]. Flores 1826 (MEXU, TEX); Henrickson 14315-2 (TEX); Henrickson 18863 (MEXU); Lehto 5278 (ASU).

GENTIANACEAE

Eustoma exaltatum (L.) Salisb. ex G. Don [GV]. Chiang 7645 (LL), Flores 1831 (MEXU, TEX); Henrickson 15373 (TEX), 15373A (NMC), 15373B (MEXU) 19742 (MEXU, SRSC, UCR); Iltis 28Y (MEXU); Lehto 5031 (ASU), 5041 (ASU).

[1,2,9] *Sabatia tuberculata* J.E. Williams **[GH]**. Chiang 9163 (LL); Daniel 681 (ASU, MEXU); Henrickson 7965 (TEX), 12562 (LL, MEXU), 15366 (TEX), 19744 (MEXU, TEX); Pinkava 3852 (ASU). **[CCE].**

HYDROPHYLLACEAE

[1,2,10] *Phacelia marshall-johnstonii* Atwood & Pinkava var. *marshall-johnstonii* **[G]**. Chiang 7649 (LL); Flores 1835 (MEXU, TEX); Henrickson 15910 (TEX); Johnston 10334 (LL); Moore 2853 (MEXU, TEX). **[CCE].**

KRAMERIACEAE

Krameria erecta Willd. [GV]. Reeves P13053 (ASU).

Krameria grayi Rose & J.H. Painter [GV]. Pinkava P5526 (ASU).

LAMIACEAE

Hedeoma sp. Moore 2570 (MEXU, TEX).

Mentha rotundifolia Huds. [GV]. Pinkava 5515 (ASU); Reeves P13116 (ASU).

LOASACEAE

[1] *Cevallia sinuata* Lag. [GV]. Chiang 7654 (LL); Moore 2013 (MEXU, TEX).

Eucnide floribunda S. Watson [GV]. Henrickson 7971 (ARIZ, MEXU, NMC, TEX). **[CDE].**

Eucnide lobata (Hook.) A. Gray [GV]. Moore 2012 (MEXU, TEX).

[1] *Mentzelia mexicana* H.J. Thomps. & Zavort. [GV]. Chiang 7652 (LL, MEXU); Moore 1473 (MEXU, TEX); Pinkava 5274 (ASU). **[CDE].**

Mentzelia sp. Moore 1956 (MEXU, TEX).

[1,2] *Petalonyx crenatus* A. Gray ex S. Watson **[G]**. Chiang 7650 (LL, MEXU); Flores 1824 (MEXU, TEX); Minckley s.n. (ASU); Moore 2553 (MEXU, TEX), 2581 (MEXU, TEX); Pinkava 4345 (ASU), 5261 (ASU), 5541 (ASU, DES); Smith 4 (MEXU), 22158 (TEX). **[CDE].**

MALPIGHIACEAE

Cottsia gracilis (A. Gray) W.R. Anderson & C. Davis [GV]. Daniel 524 (ASU).

MALVACEAE

Abutilon malacum S. Watson [GV]. Pinkava 5953 (ASU).

Abutilon pinkavae Fryxell [GV]. Pinkava 5481 (ASU). **[CDE].**

Ayenia microphylla A. Gray [GV]. Pinkava 5201 (ASU).

Gossypium sp. Moore 2566 (MEXU, TEX).

[1] *Sida longipes* A. Gray [GV]. Chiang 7616 (MEXU), 7617 (LL).

NAMACEAE

Nama constancei J.D. Bacon **[G]**. Moore 2554 (MEXU, TEX). **[CDE].**
[2] *Nama cuatrocienegensis* G.L. Nesom. [GV]. **[CCE].**
[1,7] *Nama serpylloides* Hemsl. [**G, HG**]. Chiang 7647 (LL); Flores 1834 (MEXU, TEX); Johnston 7126 (GH); Lehto 5067 (ASU); Moore 1483 (MEXU, TEX), 2854 (MEXU, TEX). **[CDE].**
[1] *Nama stenophylla* A. Gray ex Hemsl. **[G]**. Chiang 7609 (LL, TEX), 9160 (LL); Flores 1828 (MEXU, TEX); Jackson 107 (TEX); Lehto 5045 (ASU), 5212 (ASU); Moore 1474 (MEXU, TEX), 2014 (MEXU, TEX), 2858 (MEXU, TEX); Wendt 655 (LL). **[CDE].**
Nama sp. Iltis 27 (MEXU).

NYCTAGINACEAE

Acleisanthes angustifolia (Torr.) R.A. Levin [GV]. Cole 3744 (ASU); Henrickson 20402 (MEXU, TEX).
Acleisanthes longiflora A. Gray [GV]. Chiang 7640 (LL).
[1,2,8] *Acleisanthes purpusiana* (Heimerl) R.A. Levin [**G**]. Flores 1821 (MEXU, TEX); Henrickson 12552 (MEXU, NMC), 18861 (ARIZ, MEXU); Lehto 5544 (ASU); Moore 1468 (MEXU, TEX), 2016 (MEXU, TEX), 2863 (MEXU, TEX); Reeves P13107 (ASU).
Acleisanthes undulata (B.A. Fowler & B.L. Turner) R.A. Levin [GV]. Daniel 533 (ASU); Henrickson 23195 (TEX).
Allionia choisyi Standl. [GV]. Cole 3773 (ASU); Henrickson 20405 (MEXU, NMC, TEX).
Allionia incarnata L. [GV]. Cole 3755 (ASU); Lehto 5215 (ASU); Pinkava 5059 (ASU).
Anulocaulis eriosolenus (A. Gray) Standl. [GV]. Moore 1957 (MEXU, TEX), 2019 (MEXU, TEX), 2556 (MEXU, TEX), 2865 (MEXU, TEX); Rodríguez 1158 (MEXU).
Boerhavia anisophylla Torr. [GV]. Moore 2002 (MEXU, TEX).

OLEACEAE

Menodora scabra A. Gray [GV]. Pinkava 5192 (ASU).
Menodora scoparia Engelm. ex A. Gray [GV]. Daniel 526 (ASU).

ONAGRACEAE

Ludwigia repens J.R. Forst. [HGV]. Lehto 5508 (ASU).
Oenothera boquillensis (P.H. Raven & D.P. Greg.) W.L. Wagner & Hoch [GV]. Moore 2546 (MEXU, TEX). This species seems to prefer gypsum.
[1] *Oenothera macrosceles* A. Gray [GV]. Cole P3850 (ASU); Lehto 5509 (ASU).

OROBANCHACEAE

Castilleja lanata A. Gray [GV]. Moore 2549 (MEXU, TEX).
Castilleja sp. Moore 2577 (MEXU, TEX).

[1] *Orobanche ludoviciana* Nutt. [GV]. Pinkava 5178 (ASU), 5271 (ASU).

OXALIDACEAE

Oxalis alpina (Rose) Rose ex R. Knuth [GV]. Pinkava P13098 (ASU).

PAPAVERACEAE

Argemone sanguinea Greene [GV]. Pinkava 5945 (ASU), P5208 (ASU).
Argemone sp. Moore 2568 (MEXU, TEX).

PASSIFLORACEAE

Passiflora tenuiloba Engelm. ex A. Gray [GV]. Daniel 531 (ASU).

PLANTAGINACEAE

Maurandya antirrhiniflora Humb. & Bonpl. ex Willd.subsp. *hederifolia* (Rothm.) Elisens [GV]. Rodríguez 141 (MEXU).
Penstemon barbatus (Cav.) Roth [GV]. Engard 315 (DES).

POACEAE

Aristida glauca (Nees) Walp. [GV]. Pinkava 5465 (ASU).
Aristida purpurea Nutt. [GV]. Rodríguez 1146 (USU).
Bouteloua curtipendula (Michx.) Torr. [GV]. Harvey 1208 (MICH); Pinkava 5464 (ASU).
Bouteloua eriopoda (Torr.) Torr. [GV]. Harvey 1207 (MICH).
Bouteloua ramosa Scribn. ex Vasey [GV]. Daniel 503 (MICH).
Bouteloua trifida Thurb. ex S. Watson [GV]. Boke 134 (MICH).
Bouteloua sp. Daniel 505 (MICH).
Cynodon dactylon (L.) Pers. [GV]. Pinkava 5078 (ASU).
Dasyochloa pulchella (Kunth) Willd. ex Rydb. [GV]. Rodríguez 1147 (USU).
Digitaria patens (Swallen) Henrard [GV]. Harvey 1211 (MICH).
Disakisperma dubia (Kunth) P.M. Peterson & N. Snow [GV]. Harvey 1226 (MICH).
Distichlis spicata (L.) Greene [HGV]. Chiang 7651 (LL), 9156 (LL); Harvey 1234 (MICH); Henrickson 7951 (ARIZ, ASU, NMC), 14310 (ASU); Johnston 10334A (LL).
Monanthochloë littoralis Engelm. [HGV]. Chiang 7625 (LL); Henrickson 7969 (LL), 14302 (LL).
Muhlenbergia asperifolia (Nees & Meyen ex Trin.) Parodi [GV]. Pinkava 5065 (ASU), 5512 (ASU).
Muhlenbergia lindheimeri Hitchc. [GV]. Wendt s.n. (ASU).
Panicum hallii Vasey [GV]. Rodríguez 1136 (MEXU).
Spartina spartinae (Trin.) Merr. ex Hitchc. [HGV]. Henrickson 7955 (ASU, LL); 15368 (TEX).
Sporobolus airoides (Torr.) Torr. [GV]. Chiang 9155 (LL, MEXU); Henrickson 7964 (LL), 14309 (LL), 14316 (ASU, LL); Pinkava 5079 (ASU), 5547 (ASU).
[2, 11] *Sporobolus coahuilensis* Valdés-Reyna. [GV]. **[CCE].**
Sporobolus flexuosus (Thurb. ex Vasey) Rydb. [GV]. Henrickson 12555 (LL).

[1,2] *Sporobolus spiciformis* Swallen **[HG]**. **[CDE].**
Sporobolus wrightii Munro ex Scribn. [HGV]. Henrickson 14287 (LL).
Tragus berteronianus Schult. [GV]. Pinkava P13047A (ASU).
Trichachne californica (Benth.) Chase [GV]. Pinkava 5955 (ASU).

POLEMONIACEAE

Giliastrum stewartii (I.M. Johnst.) J.M.Porter [GV]. Johnston 10335 (LL, MEXU); Lehto 5210 (ASU).

POLYGALACEAE

Hebecarpa barbeyana (Chodat) J.R. Abbott [GV]. Chiang 7617A (TEX); Reeves P13099 (ASU).
Polygala alba Nutt. [GV]. Chiang 7617B (TEX).
Polygala turgida Rose [GV]. Chiang 7639 (MEXU, TEX); Cole 3851 (ASU); Daniel 679 (ASU); Engard 288 (DES); Henrickson 7960 (ASU, TEX), 18860 (MEXU, TEX); Johnston 9162 (MEXU).
Rhinotropis lindheimeri (A. Gray) J.R. Abbott [GV]. La Bounty 5956 (ASU).
Rhinotropis nudata (Brandegee) J.R. Abbott [GV]. Moore 1993 (MEXU, TEX).

PORTULACACEAE

Portulaca oleracea L. [GV]. Henrickson 20403 (RSA).
Portulaca pilosa L. [GV]. Henrickson 20404 (MEXU).

PRIMULACEAE

[2] *Samolus ebracteatus* Kunth var. *coahuilensis* Henrickson [HGV]. Chiang 7638 (LL), 9165 (LL), 9160A (LL); Flores 1833 (MEXU, TEX); Henrickson 7930 (TEX), 18859 (ARIZ), 18859H (TEX), 18859M (MEXU); Iltis 28X (MEXU); Lehto 5040 (ASU), 5069 (ASU); Rodríguez 1132 (MEXU). **[CDE].**

RESEDACEAE

Oligomeris linifolia (Vahl) J.F. Macbr. [GV]. Lehto 5055 (ASU); Pinkava 10528 (ASU).

RHAMNACEAE

Condalia warnockii M.C. Johnst. [GV]. Rodríguez 1119 (MEXU).
Karwinskia humboldtiana (Schult.) Zucc. [GV]. Lehto 5470 (ASU), P5195 (ASU); Pinkava 5793B (DES); Reeves P13049 (ASU).
Ziziphus obtusifolia A. Gray [GV]. Leverich 8 (TEX)

ROSACEAE

Amelanchier denticulata (Kunth) K. Koch [GV]. Gentry 23248 (DES).
Purshia ericifolia (Torr. ex A. Gray) Henrickson [GV]. Reeves P13091 (ASU).
Vauquelinia corymbosa Bonpl. [GV]. Pinkava P6020 (DES).

RUBIACEAE

Bouvardia ternifolia (Cav.) Schltdl. [GV]. Reeves P13094 (ASU).
Hedyotis intricata Fosberg [GV]. Reeves P13071 (ASU).
Hedyotis teretifolia (Terrell) G.L. Nesom **[G]**. Moore 1999 (MEXU, TEX), 2550 (MEXU, TEX). **[CDE].**
Randia pringlei A. Gray [GV]. Chiang 7608 (LL, MEXU), 7637 (LL, MEXU); Henrickson 20398 (LL); Moore 2544 (MEXU, TEX); Pinkava 5472 (ASU). **[CDE]**.

SALICACEAE

Salix nigra Marshall [GV]. Cole 3926 (LL).

SAURURACEAE

Anemopsis californica Nutt. ex Hook.& Arn. [GV]. Henrickson 15574B (MEXU).

SCROPHULARIACEAE

Buddleja scordioides Kunth [GV]. Moore 2560 (MEXU, TEX).

SOLANACEAE

Chamaesaracha villosa Rydb. [GV]. Chiang 7628 (LL); Henrickson 12565 (LL); Johnston 10336A (LL); Neff 92-3-27-2 (MEXU, TEX); Pinkava 5033 (ASU); Rodríguez 1156 (MEXU).
Lycium berlandieri Dunal [GV]. Henrickson 23182 (TEX).
[1,2] *Lycium parishii* A. Gray var. *modestum* (I.M. Johnston) Chiang [GV]. Chiang 7636 (LL); Henrickson 12545 (LL, MEXU); Pinkava 5266 (ASU); Wendt 661 (LL, MEXU). **[CDE]**.
Solanum elaeagnifolium Cav. [GV]. Pinkava 5074 (ASU). **[CDE].**

ULMACEAE

Celtis pallida Torr. [GV]. Pinkava 5474 (ASU).

VERBENACEAE

Bouchea spathulata Torr. [GV]. Chiang 7610 (ASU, LL, MEXU). **[CDE].**
Lippia graveolens Kunth [GV]. Pinkava 5202 (ASU), P13051 (ASU).
Phyla nodiflora (L.) Greene [GV]. Pinkava 5514 (ASU).

VIOLACEAE

Hybanthus verticillatus (Ortega) Baill. [GV]. Pinkava 5236 (ASU), 5951 (ASU).

ZYGOPHYLLACEAE

Guaiacum angustifolium Engelm. [GV]. Pinkava P5173 (ASU).
Larrea tridentata (DC.) Coville [GV]. Pinkava 3743 (ASU), 5207 (ASU).
Peganum mexicanum A. Gray [GV]. Pinkava 5042 (ASU).
Sericodes greggii A. Gray [GV]. Daniel 523 (ASU). **[CDE].**
Tribulus terrestris L. [GV]. Pinkava 3770 (ASU).

References

Alexander PJ, Douglas NA, Ochoterena H, Flores Olvera H, Moore MJ (2014) Recent findings on the gypsum flora of The Rim of the Guadalupe Mountains, New Mexico: a new species of *Nerisyrenia*, a new state record, and an updated checklist. J Botan Res Instit Texas 8:383–393

Atwood ND, Pinkava DJ (1977) A new gypsophilous species of *Phacelia* (Hydrophyllaceae) from Coahuila, Mexico. Madrono 24:212–214

De-Nova JA, Aguirre-Liguori JA, Eguiarte L(2020). How did *Fouquieria* come to the Chihuahuan Desert? Phylogenetic and phylogeogrphic studies of *Fouquieria shrevei* and *F. splendens*, and the role of vicariance, selection and genetic drift. In: Mandujano MC, Pisanty I, Eguiarte LE (eds) Plant diversity in the Chihuahuan Desert. Springer, Cham, Switzerland

Escudero A, Palacio S, Maestre FT, Luzuriaga AL (2015) Plant life on gypsum: a review of its multiple facets. Biol Rev 90:1–18

ESRI (Environmental Scientific Research Institute) (2012) ArcGIS 10.1. Redlands, CA, USA

Ezcurra E, Martínez-Berdeja A, Villanueva-Almanza L (2020) Plants in the desert: diversity and life-forms. In: Mandujano MC, Pisanty I, Eguiarte LE (eds) Plant diversity in the Chihuahuan Desert. Springer, Cham, Switzerland

Flores Vázquez JC, Rosas Barrera MD, Golubov J, Sánchez-Gallén I, Mandujano MC (2020). Ecological Importance of Bajadas in the Chihuahuan Desert. In: Mandujano MC, Pisanty I, Eguiarte LE (eds) Plant diversity in the Chihuahuan Desert. Springer, Cham, Switzerland

García-Dávila A (2000) Las Plantas del Valle de Cuatro Ciénegas, Coahuila, México. Tesis Ingeniero Agrónomo Zootecnista. Universidad Autónoma Agraria Antonio Narro. Saltillo, Coahuila, Mexico

García-Morales LJ, González-Botello MA, González GRH (2014) *Ancistrocactus pinkavanus* (Cactaceae), a new species from northern Mexico. Acta Succulenta 2(1):27–44

Gutiérrez-Alejandro AG, Chacón-Baca E, Rosales-Domínguez C, Ramírez-Peña CF, Girón P, Grajales-Nishimura JM (2017) A clastic-evaporitic deposit from the Cretaceous of northeastern Mexico: La Mula-La Virgen transition. J S Am Earth Sci 80:411–421

Henrickson J, Johnston MC (1997) A flora of the Chihuahuan Desert Region. Plant Resources Center, University, Austin, TX, USA

Henrickson J, Straw RM (1976) A gazetteer of the Chihuahuan Desert Region. California State University, Los Angeles, CA

Johnston IM (1939) New Phanerogams from Mexico. J Arnold Arbor 20: 238–239

Johnston IM (1941) Gypsophily among Mexican desert plants. J Arnold Arbor 22:145–170

Johnston IM (1943) Noteworthy species from Mexico and adjacent United States, I. J Arnold Arbor 24:227

Lehmann C, Osleger DA, Montañez I (2000) Sequence stratigraphy of Lower Cretaceous (Barremian-Albian) carbonate platforms of northeastern Mexico: regional and global correlations. J Sediment Res 70:373–391

Levin RA (2002) Taxonomic Status of *Acleisanthes*, *Selinocarpus*, and *Ammocodon* (Nyctaginaceae). Novon 12(1):58–63

Meyer SE (1986) The ecology of gypsophile endemism in the eastern Mojave Desert. *Ecology* 67:1303–1313

Minckley WL (1969) Environments of the Bolsón of Cuatro Ciénegas, Coahuila, Mexico. Science Series No. 2. El Paso: Texas Western Press, University of Texas at El Paso, TX, USA

Moore MJ, Jansen RK (2007) Origins and biogeography of gypsophily in the Chihuahuan Desert plant group *Tiquilia* subg. *Eddya* (Boraginaceae). Syst Bot 32:392–414

Moore MJ, Mota JF, Douglas NA, Flores Olvera H, Ochoterena H (2014) The ecology, assembly and evolution of gypsophile floras. In: Rajakaruna N, Boyd RS, Harris TB (eds) Plant ecology and evolution in harsh environments. Nova Science, Hauppauge, NY, USA, pp 97–128

Morales García ED (2018) La familia Gentianaceae en México. Tesis Maestría en Ciencias Biológicas, Instituto de Biología, UNAM, Mexico City, Mexico

Nesom GL (1981) Five new species of Mexican *Erigeron* (Asteraceae). Madrono 28(3):143–144

Parsons RF (1976) Gypsophily in plants—a review. Am Midl Nat 96:1–20
Pinkava DJ (1984) Vegetation and flora of the bolsón of Cuatro Ciénegas Region, Coahuila, México. J Ariz Nev Acad Sci 19:23–47
Powell AM, Turner BL (1977) Aspects of the plant biology of the gypsum outcrops of the Chihuahuan Desert. In: Wauer RH, Riskind DH (eds) Transactions of the symposium on the biological resources of the Chihuahuan Desert region, United States and Mexico, Sul Ross State University, Alpine, Texas, 17–18 October 1974. U.S. Department of the interior, National Park Service Transactions and Proceedings Series, Number 3, Washington, DC, USA, pp. 315–325.
Turner BL (2004) *Sporobolus coahuilensis* (Poaceae): a new record for the U.S.A. from Trans-Pecos, Texas. Sida 21(1):455–457
Turner BL (2013) Taxonomic overview of the Mexican species of *Salvia* sect. *Flocculosae* (Lamiaceae). Phyton 36:1–11
Turner BL, Powell AM (1979) Deserts, gypsum and endemism. In: Goodin JR, Northington DK (eds) Arid Land Plant Resources. Proc. Int. Arid Lands Conf. Pl. Res., Texas Tech. Univ., Lubbock, TX, USA, pp. 96–116
Urbatsch LE, Roberts RP (2004) New combinations in the genus *Gundlachia* and four new genera of Astereae (Asteraceae) from northern Mexico and the southern United States. Sida 21(1):243–257
Valdés-Reyna J, Villaseñor Ríos JL, Encina-Dominguez JA, Ortiz E (2015) The grass family (Poaceae) in Coahuila, Mexico: diversity and distribution. Botan Science 93(1):119–129
Villarreal-Quintanilla JÁ, Encina-Domínguez JA (2005) Plantas vasculares endémicas de Coahuila y algunas áreas adyacentes, México. Acta Bota Mexican 70:1–46

Chapter 10
The Ages of Life: The Changing Forms and History of *Coryphantha werdermannii* Throughout Its Development

Carlos Martorell and Rosa Maricel Portilla-Alonso

> *The waves of the sea, the little ripples on the shore, the sweeping curve of the sandy bay between its headlands, the outline of the hills, the shape of the clouds, all these are so many riddles of form, so many problems of morphology.*
>
> *D'Arcy W. Thomson, 1917.*

Abstract Form reflects the problems that plants face throughout their life, and the way these problems are solved. We discuss the development and form of a globose cactus, *Coryphantha werdermannii*, endemic to the Cuatro Ciénegas area, Coahuila, Mexico. Its spheroidal form reduces the surface of the plant, diminishing the amount of water lost through it. Nevertheless, in a small plant, water loses are larger compared with the exiguous amounts of stored water, leading to huge mortality rates. As the plant develops, white spines that may reflect sunlight cover the plant, modulating the extreme temperatures of the desert. Form changes again as adulthood approaches. A spiny pouch lined in wool that will protect reproductive organs is formed in the plant's apex. This shady pouch, altogether with other spines, reduces photosynthesis and growth rates. Every year, flowers are produced and open almost exactly on the same day for a few hours. The large numbers of seeds produced allow at least some of them to reach safe sites. Unlike other cacti that require the shade of

C. Martorell (✉)
Facultad de Ciencias, Universidad Nacional Autónoma de México, UNAM, Mexico City, Mexico
e-mail: martorell@ciencias.unam.mx

R. M. Portilla-Alonso
Comisión Nacional para el Conocimiento y Uso de la Biodiversidad, Conabio, Mexico City, Mexico
e-mail: maricel.portilla@conabio.gob.mx

M. C. Mandujano et al. (eds.), *Plant Diversity and Ecology in the Chihuahuan Desert*, Cuatro Ciénegas Basin: An Endangered Hyperdiverse Oasis,
https://doi.org/10.1007/978-3-030-44963-6_10

shrubs to survive, seedlings prefer growing below loose gravel. Gravel is more common in eroded areas, such as those produced by livestock, so this plant's populations are denser where people are active. Nevertheless, extreme livestock rates kill large plants, which would cause population decline.

Keywords Ecomorphology · Development · Endangered species · Conservation

Nature's shapes possess intriguing beauty. It is no coincidence that Sir D'Arcy Thomson's classic *On Growth and Form* (1917) has been revered as the most refined literary work that science has ever produced: its author studied nature's forms because he had a profound sense of esthetics. He was amazed by the enigmatic connection between the golden section, which the Greeks reputed as the base for everything beautiful, the spirals found in a sunflower's inflorescence or a nautilus's shell and a series of numbers discovered in medieval Italy by Leonardo Fibonacci (Nobel 1994). He synthesized his profound sentiment for esthetics and the natural world by exclaiming "For the harmony of the world is made manifest in Form and Number, and the heart and soul and all the poetry of Natural Philosophy are embodied in the concept of mathematical beauty" (Thompson 1917).

The habit of searching for reasons behind the shapes of vegetation in mathematics and in physics has its origin with the biologists that study deserts. This is where plants have the greatest collection of architectural shapes (Cody 1991). Apart from mundane-looking herbs and bushes, in deserts one can find oddly shaped plants such as prickly pears, candelabra-like cacti, and agaves (Fig. 10.1). Many authors have found that the peculiar shapes of these plants help them cope with the severe environment that the desert imposes on survival (Ezcurra et al. 2020, this volume).

However, everyone knows that the environment, and the way that we cope with it, changes throughout our lifetime. It is the same for plants, and that is why their shape and form change when they mature and grow old. In this chapter, we describe the morphological changes along the life cycle of *Coryphantha werdermannii* Boed. and discuss the extensive variation of growth forms of this endemic cactus from the Cuatro Ciénegas region.

Coryphantha werdermannii (Fig. 10.2) is one of Mexico's most beautiful cacti and this situation has made it very attractive to collectors. A strong looting event was reported in the 1980s, leading to its protection in Appendix I of the Convention on International Trade in Endangered Species of Wild Fauna and Flora (CITES 2017). It is also protected by Mexican law (NOM-059-ECOL-2010, SEMARNAT-2010) as an endangered species. However, one only needs to search on the Internet to find many sites that sell seeds or specimens, sometimes even openly admitting that they were illegally obtained from one or another Mexican locality. Other economic activities have also contributed to the degradation of areas where this cactus is found. Therefore, we shall also discuss its conservation.

Fig. 10.1 Some plants with unusual shapes from Mexican deserts. (**a**) *Agave applanata* (Asparagaceae). (**b**) A cholla with a characteristic cylindrical stem, *Cylindropuntia imbricata* (Cactaceae). (**c**) The candelabriform *Polaskia chende* (Cactaceae). Photographs: Carlos Martorell

Of Spiny Bubbles

Coryphantha werdermannii does not have a common name in Spanish other than *biznaga*, which also applies to hundreds of other small to medium sized cacti. This term is derived from the Nahuatl word *huitznahuac* which means surrounded by spines and applies to any spherical, small barrel, or globose cactus, as botanists would say. This last term is doubly correct because globes and *biznaga* cacti are

Fig. 10.2 An adult plant of *Coryphantha werdermannii* in habitat. Photographs: Carlos Martorell

round for the same exact mathematical reason: they both try to enclose the greatest amount of volume with least amount of surface.

We should first note that the volume a globe can hold increases with its surface. In a balloon, for example, the rubber (the surface) must expand enough to hold certain amount of volume when it is inflated. Whoever has played with a balloon knows that when it is stretched, a balloon may adopt any form. Thus, we might think that a balloon may take the form of a pyramid or a cube, but they are typically spherical. If we could pinch a blown-up balloon or stretch it into the shape of a pyramid, it would go back to its spherical form when released. This happens because the rubber tends to reduce its size: because the rubber of a balloon in the shape of a pyramid would be more stretched out than a globe holding the same volume (which remains constant because the amount of air in the balloon stays the same—at least approximately, because air is a gas and as such it may be compressed), it naturally adopts a more spherical shape when it shrinks back. Soap bubbles are perfect spheres for the same reason. Technically, it is said that a sphere is the geometric shape that has the least amount of surface-area per unit of volume. Simply put, it has the smallest surface-area-to-volume ratio possible (Nobel 1994, 2003).

The green epidermis of a cactus is covered in very small pores that open and close at different moments of the day called stomata. Atmospheric oxygen and CO_2 for photosynthesis pass through them. However, when they open, water is inevitably lost in the form of vapor. Cacti have adapted to hold their breath during the day and to only open their stomata at night when temperatures are lowest, and evaporation is reduced to the minimum. Even if evaporation is six times lower at nighttime than during daytime (Nobel 1994), the loss of moisture through the surface of a cactus is still very onerous, considering that water loss is a luxury one cannot afford in the desert. Thus, this cactus's survival depends on having great volume to store water but a small surface through which little water can escape to the atmosphere. The solution to this problem is that of a balloon: to adopt the form of a sphere and in doing so acquire the minimum surface-area-to-volume ratio (Nobel 1994, 2003; Ezcurra et al. 2020, this volume).

Size also has a strong impact on the surface-area-to-volume ratio. A smaller sphere has a larger surface relative to its volume than a larger sphere. For instance, a sphere with a radius of 1 mm (more or less the size of a *C. werdermannii* seedling) has a surface-area-to-volume ratio of $0.13\ cm^2/0.00419\ cm^3 = 31.03\ cm^{-1}$. For a 4 cm radius sphere (equivalent to a large adult), the ratio diminishes drastically to $201.06\ cm^2/268.08\ cm^3 = 0.75\ cm^{-1}$. This means that a small globose cactus may lose all of its water $31.03/0.75 = 41$ times faster than adults, leading to high mortality. Therefore the first weeks of a cactus's life are maybe the most precarious of its entire existence.

Difficult Beginnings

In order to reduce their surface to the maximum, seedlings of *C. werdermannii* that have only just germinated have evolved into a spherical form with just a pair of barely perceptible remnant embrionary leaves (Fig. 10.3a). However, an adult may measure up to 7 cm in diameter while a seedling may only measure 1 mm. As a result of these differences in its surface-area-to-volume ratio we can expect seedlings to lose water reserves 70 times faster than an adult cactus. Furthermore, a seedling's radicular system is minuscule. To try to compensate for the accelerated loss of water it is only able to barely reach the moisture that is near the soil surface, which tends to quickly evaporate after each rain. The first months of life for *C. werdermannii* are probably its thirstiest.

The highest temperatures recorded near the desert soil surface reach up to 70 °C at noon. The soil surface may freeze during the severe winters of the Chihuahuan Desert, reaching frigid temperatures several times a year (Nobel et al. 1986). In order to survive, minuscule newborn plants must be able to overcome this microclimate of bitter cold winters and scorching hot summers. And, as if this were not enough, *C. werdermannii* seedlings must also confront other enemies within their

Fig. 10.3 *Coryphantha werdermannii* throughout its life cycle. (**a**) A carpet of seedlings growing below gravel. Some of them have their first spines. (**b**) Small plants (<1 cm in diameter). (**c**) A juvenile with the first, black spines near the apex. (**d**) Adult plant with well-developed black spines. (**e**) A flowering individual. Photagraphs: Carlos Martorell

ecosystem. Ants and arthropods are in a continuous search for seedlings because of the water stored in their tiny stems (Brown et al. 1979; Anderson and MacMahon 2001). As a result of extreme temperatures, aridity and natural enemies, only 1.7 out of a thousand *C. werdermannii* seeds that find their way to the soil in the summer will survive until the winter (Portilla-Alonso 2010; Portilla-Alonso and Martorell 2011).

A Long Childhood

Perhaps the most characteristic trait of cacti are the areoles, pillow-like organs where spines develop. Cactus spines are just leaves that have evolved to be very thin, minimizing their surface and reducing moisture loss (Mauseth 2006). Particularly, in plants from the genus *Coryphantha*, areoles are found at the tip of a protrusion called a tubercle (Bravo-Hollis 1978; Mauseth 2006).

When *C. werdermannii* germinates it lacks tubercles, but it produces them as it grows (Fig. 10.3b). A new tubercle can only grow at the apex—that is at the "tip"—of a plant. A few weeks after germination this species already has an open tuft of tubercles with spines that probably function as parasols (Fig. 10.3a). Eventually, the production of new tubercles compresses the uncovered part of the stem into the ground. From this moment forth, a comb-like layer of spines completely covers the photosynthetic stem of the cactus (Fig. 10.3a, b).

Because *C. werdermannii* spines are white, one could speculate that these spines help reduce the temperature of the stem by reflecting great quantities of light away from it, as it happens in other cacti (Nobel 1983). There are no direct measurements of this happening in *C. werdermannii*. However, there is another globose cactus in the south of Mexico called *Mammillaria pectinifera*, with stems covered in spines similar to those found on young *C. werdermannii* cacti. Experiments in which spines have been removed from this cactus have shown that there is hardly any change on temperature, which means that these spines do no help temperature regulation. However, it was observed that spine removal resulted in rapid water loss, perhaps because the plant's spine cover maintains a humid environment around the plant reducing water loss through the stomata (Rodríguez-Ortega and Ezcurra 2000). It is not possible to say for certain, but the white spiny crown of young *C. werdermannii* may also help to protect it against desiccation. Having prevented death from dehydration, these small cacti may now dedicate all their strength to growing, and they are specialists in doing so. In this life stage they are able to increase their diameter 2 mm every year, which may not seem like much but is the fastest growth rate recorded for this species over its life cycle (Portilla-Alonso and Martorell 2011).

The Changing Times of Adolescence

Species of the genus *Coryphantha* typically have two types of spines: Radial spines, which are thin, numerous and grow from areoles like the spokes of a wheel; and central spines, which are typically sparse, much more robust and project outward from the plant ready to prick anyone who would dare eat its juicy stem (Bravo-Hollis and Sánchez-Mejorada 1991). During its first 13 years of life *C. werdermannii* produces only white radial spines (Fig. 10.4a), but it suffers a metamorphosis at this age that foretells that sexual maturity is near. Small central spines begin to appear and grow from the youngest areoles (Fig. 10.3c). As the tubercles that bear them appear at the apex of the plant, central spines grow quickly. Radial spines also become longer, especially those with tips pointing upwards (Fig. 10.4b,c).

Because areoles can only produce spines when they are very young, the lower part of the plant conserves its juvenile appearance for many years, while younger tubercles at the end of the plant look like a crown of shinning black scimitars (Fig. 10.3c, d). This duality confers *C. werdermannii* exceptional beauty among cacti. Furthermore, if form divulges function, a new form should correspond to a novel function. To understand what this is about, we should look at the parts of the stem that are now protected from the outside world by this cactus's new spiny shell.

Even though they are almost invisible from the outside, tubercles that sport central spines have almost tripled in size in order to be able to accommodate the largest of spines. However, their growth is not immediate, elongation occurs only after other young tubercles have been produced near the apex. When this happens, the longer tubercles form a circle around smaller ones creating a cavity in the upper part of the cactus. The largest of radial spines form a cover that encloses the entrance to this opening (Fig. 10.3d), making it almost imperceptible from outside. The inside of this hidden sanctuary is covered in a thick wool-like yellow cover.

We had already mentioned that spines reflect sunlight preventing damage to the stem from overheating, but when spines are so big and close together, this effect can be spectacular. In some cases, for instance, in *Opuntia bigelovii* (= *Cylindropuntia bigelovii*) where the spine cover is so dense, spines can reduce the stem's temperature up to 10 °C (Nobel 1983). This may be how *C. werdermannii*'s wool-like cover protects the top portion of the plant from excessive heat, preparing it for its final morphological transformation.

In cacti, the only cells that can give rise to new organs are in the areoles (Bravo-Hollis 1978; Mauseth 2006). However, in the genus *Coryphantha*, areoles differentiate into two portions early on in their development: the lower half of the areole originates spines and remains in the middle of the tubercle and the upper half begins to migrate through the tubercle's surface until it arrives at the tubercle's base where it gives rise to the reproductive tissue that will produce flowers. During this migration, the meristem leaves a deep ridge behind on the tubercle, which lets us easily distinguish species of *Coryphantha* from other cacti (Bravo-Hollis 1978).

One explanation for these profound morphological changes is that the tissue that will give rise to flowers migrates in order to protect itself between the tubercles and

Fig. 10.4 Areoles and spines of *Coryphantha werdermannii*. (**a**) Typical areole of infantile individuals with radial spines only. (**b**) Areole of juveniles and adults with a thick, dark central spine. Upper radial spines are longer than the rest; two or three becoming thicker. (**c**) Adult with both types of areoles. The entangled, upper radial spines of adult areoles protect the plant's apex. Below them there is a chamber where the flower buds and fruits are sheltered. Photagraph: Carlos Martorell

the thick wool-like cover, gaining cool temperatures and a protective shield of spines. The appearance of the first tubercle with a ridge tells us that the woolly cover at the apex of the cactus is about to turn into a nuptial den and an incubator where *C. werdermannii* flowers (Fig. 10.3e), fruits, and seeds will develop protected from predators and the hostile environment. The appearance of the crown of thorns is the prelude to a crown of flowers, or, translated to Greek, a *koryfantha*.

The Long Wait

Deserts are systems that march to the pace of the unpredictable. It does not rain much, but when it does, it rains in heavy downpours that are difficult to anticipate (Noy-Meir 1973). Sometimes, after years of no significant precipitation, rainstorms occur without previous warning, marking an opportunity that no one in the desert can waste. It is during these events that the vulnerable *C. werdermannii* seedlings can germinate and grow enough to survive the upcoming drought.

If *Coryphantha* individuals reproduced at random times and perhaps just once in their lifetime, only a small fraction of them would have been lucky enough to reproduce on a rainy year and be successful. The rest of them would have lived only to squander their single opportunity to procreate on a dry year that would annihilate their entire progeny. However, *C. werdermannii* is a long-lived species that reproduces each year, with the hope that at least one daughter seed will have enough water to join the population. Such strategy is apparently common in desert plants (Wilbur and Rudolf 2006). Because of this, the species' persistence is more feasible if adults can live for a long time: the growth and perpetuation of *C. werdermannii* populations depend more on adult survival than on any other component of the life cycle (Portilla-Alonso and Martorell 2011).

In *C. werdermannii* growth becomes slower as time goes by. After it reaches 6 cm in diameter it practically remains the same size for the rest of its life. Individuals that grow larger frequently shrink back (Portilla-Alonso and Martorell 2011). As a result, the longevity of adults is unknown. What we do know is that this cactus takes almost 20 years to reach 6 cm in diameter after having reached puberty. Many wild individuals are probably more than 40 or 50 years old and surely have been avidly waiting for the appropriate year for their seeds to be successful (Portilla-Alonso and Martorell 2011).

The notable reduction in the velocity at which larger plants grow may occur precisely because of the changes in their size and appearance. The surface-area-to-volume ratio of adults has greatly reduced, which decreases evaporation of water that is precious. However, their now vast water reserve is contained in living cells that consume sugars to stay alive. These sugars come from photosynthesis, a process that only occurs on the surface of the plant. Because the surface of the plant has become very small compared to its volume, the amount of light a plant is able to us for photosynthesis is just barely enough for its survival, but is no longer enough to enable plant growth. Although it may seem surprising, many cacti are limited by the amount of light they can use for photosynthesis even though they live in sunlight-saturated deserts (Nobel 1980; Zavala-Hurtado et al. 1998). Succulence (the capacity to store water) has a cost: there is a small green surface that must provide for large number of cells that are deeply immersed in the dark plant's body and thus cannot photosynthesize and feed themselves.

This is one of the unbreakable laws of nature: there is no benefit without a cost. Growth in order to reduce the surface-area-to-volume ratio reduces water evaporation, but also reduces the capture of light usable for photosynthesis. A crown of spines and wool reduce extreme temperatures but can also produce a heavy shadow over photosynthetic tissue. A great reserve of water aids survival through long droughts but is also costly to maintain.

A Day of Frenzy

Each year, after months of slow growth and patiently confronting extreme temperatures and droughts, the first flower buds begin to form. Their growth is so slow that it seems they may never mature. However, one day in spring, an unpredictable rainstorm informs every adult that the time has finally come. The melody's tempo suddenly changes from an *adagio grave* to a *prestissimo appassionato*. Within a few hours flower buds grow, they emerge from their protective apical chamber and almost simultaneously produce an explosion of flower color (Fig. 10.3e). Commonly there is one, but sometimes there are two flowers per adult plant. Nearly every plant flowers on the same day in a synchronous event. After all those months of waiting, some distracted individuals inadequately read the times and flower on the wrong day. This asynchronous flowering notably reduces their chances of reproducing sexually with their neighbors, probably imposing strong selection towards the synchrony of reproductive efforts. For instance, asynchronic flowering reduces by one half the number of seeds produced in *Hybanthus prunifolius* (Augspurger 1981). It seems that there may be two or three more reproductive events per year, or "multiple bangs" (sensu Gentry 1974), seemingly confined to Spring.

During the day of reproductive frenzy, the yellow flowers begin to open when the first rays of sun touch them. Little by little they modify their form, and the first insects responsible for their fertilization start to visit them. Different species of wasps and bees arrive at the flower and submerge themselves into the stamens. They move constantly, covering themselves in pollen and quickly fly away. In general, insect visits are short, lasting a few seconds. One or two hours after dawn the style begins to become receptive, indicating that the flower is ready to be fertilized by the pollen that the insects gently rub them with. Little after mid-day, the petals close. If during this short display a flower was not pollinated, all this effort has been in vain, for it shall open no more: the flowers of *C. werdermannii* open for a single day. Such short-lived flowers are not unusual among cacti. For instance, the flowers of the canelabriform *Polaskia chende*, the prickly pear *Opuntia rastrera*, and even the primitive tree-like *Pereskia guamacho* also live for only one day (Mandujano et al.

2010). The flowers that were pollinated develop into fruits that will be ripe in just a few weeks. On average, a 6-cm adult may produce a bit less than 4 fruits per year (Portilla-Alonso and Martorell 2011).

The fruits will develop within the crown of central spines, protected from predators. Each mature fruit will have more than 600 brown seeds that are less than 1 mm long. Most of them are capable of germinating and are possible candidates to recruit into the population if they are able to establish and survive in the most hostile of conditions (Portilla-Alonso and Martorell 2011).

Searching for a New Beginning

It seems that after being released form the fruit, seeds may remain in the soil for a long time waiting for the best moment to germinate. At least in the laboratory, we have not observed any reductions in seed viability after storing the seeds for over a year. The seeds of this species may form a transient seed bank (Thompson and Grime 1979; Walck et al. 2005). Same as the adults, seeds may live for a long time in the search of a window of opportunity. In deserts, some of the limiting factors for cacti seed germination and establishment are temperature, light, and most importantly, the amount of water (Rojas-Aréchiga and Vázquez-Yanes 2000; Rojas-Aréchiga et al. 2013). After the first stimulus provided by rain, seeds of *C. werdermannii* in the ground take only three days to germinate. Other cacti in the Cuatro Cienégas region germinate even faster—in just one day or few days (Mandujano et al. 2020, this volume).

But seeds do not only look for opportunities by patiently waiting for the correct moment; they also explore their surroundings. Not just any place is good for them. Seedlings of many species of cacti may only survive when a bush or a large rock gives them shade (Valiente-Banuet et al. 1991; Filazzola and Lortie 2014), but *C. werdermannii* seems to be as successful in direct sunlight, similar to *Ariocarpus fissuratus* (Mancilla-Ramírez et al., this volume). Adults are even found more frequently than expected by chance in places that are totally missing shrubs and rocks. However, seedlings are frequently found bellow gravel, which may be providing a more humid environment protected from sunlight. Survival and germination in this microenvironment are much larger than anywhere else (Portilla-Alonso 2010).

An individual seed cannot move at will from one place to another in search for gravelly cover. What happens is that the mother plant produces over 5000 seeds per year so, when they disperse, it is very likely that at least one of them will arrive at a suitable site. Supposing that these plants have a reproductive life of 25 years, we could expect that an adult *C. werdermannii* plant could produce an astounding number of 126,000 seeds in its lifetime. Thus, it is almost inevitable that some of them end up in the right place at the right time in order to successfully incorporate themselves into the population. However, if the population remains stable, it is likely that only one of these 126,000 seeds may reach the adult stage. Undoubtedly, life in the great Chihuahuan desert is very tough even for its most resistant and spiny inhabitants.

New Neighbors with Strange Customs

The genus *Coryphantha* probably appeared in the Chihuahuan desert about 5 million years ago (Hernández-Hernández et al. 2014). *Coryphantha werdermannii* has been exposed to the natural limitations of its environment for thousands of years and evolution has prepared it to overcome them even if it is with difficulty. However, the presence of humans has modified these conditions. When mankind brings changes into natural ecosystems such as the introduction of livestock, vegetation cover removal, construction of roads and other structures, illegal removal of specimens, and other activities, we are talking about anthropogenic disturbance (Martorell and Peters 2005).

Anthropogenic disturbance is normally associated to changes as serious as desertification, species extinction, erosion, or the reduction of water availability among other damage to the ecosystem (Martorell and Peters 2005). However, *C. werdermannii*'s populations increase in size in the presence of disturbance. The reason for this is simple: human beings reduce vegetation cover of shrubs and promote slight soil erosion, which leaves gravel exposed from the earth. These are ideal condition for the incorporation of new individuals of *C. werdermannii* into the population (Portilla-Alonso 2010).

However, at sites where disturbance is too strong it is no longer possible to find this cactus. We had already said that adult survival is crucial in order to maintain this species because they are the individuals that are capable of living for many years waiting for the appropriate events to reproduce. Nevertheless, livestock bite and step on adults damaging their apexes, which results in the death of some of the individuals, and leaves behind individuals that are sterile from damage to their apex, which is where they develop flowers (Portilla-Alonso and Martorell 2011). Many other globose cacti are favored by intermediate or even high intensities of disturbance, but in general they are hampered when human disturbance becomes too severe (Martorell and Peters 2009; Martorell et al. 2012).

Climate change seems to be even more dangerous than anthropogenic disturbance. Using methods such as "niche modeling" it is possible to identify the appropriate climate conditions for *C. werdermannii*, as well as geographic regions in which we may find these conditions. From niche modeling and projected future changes to the planet's climate it is possible to identify areas that would be appropriate for a species in the future. This procedure suggests that there will be no remaining areas adequate for *C. werdermannii* in the state of Coahuila before the end of the twenty-first century (Martorell et al. 2015).

Pending Homework

Since the last century, man has gained consciousness of the effects it has on its surroundings and has begun to conceive of different strategies to conserve what we have left. Thus, *Coryphantha werdermannii* is protected by CITES (2017), and because of the illegal commerce of its seeds and of individuals it is classified as an endangered species by Mexican laws (SEMARNAT 2010). However, this has not been enough to successfully conserve it.

It is fundamental to have adequate estimates of the amount of disturbance *C. werdermannii* may tolerate. The good news is that the preservation of this cactus is not in conflict with human activities. Not only is it possible to conserve *C. werdermannii* in places where productive activities are held, it is desirable to do so. However, data suggests that excessive and irresponsible use of the environment may result in the speedy local extinction of this species (Martorell et al. 2015).

In general, the situation confronted by *C. werdermannii* is unsettling: it is a very beautiful plant, with a complicated life cycle. It has been able to modify its form at different moments of its life to confront the natural adversities the ecosystem imposes. However, this may imply that its form and life cycle may not be adequate to confront new pressures caused by changes to the climate. To conserve cacti, it is necessary to understand that most species only grow in Mexico (Anderson 2001), and because of that they are part of Mexico's natural and cultural richness. It is thus in Mexico's best interest, but also a responsibility of the international community, to avoid the loss of *C. werdermannii* and populations of many other species that are in the same situation. Conservation of a species is not only important for the species itself, it is also important for the organisms that depend on it such as the ones that eat their fruit or pollinate it. If this was not enough reason to conserve them, we should also consider the fact that cacti are also a great pleasure to look at.

Acknowledgements We are indebted with D Montañana, D Rábago, C Almaza, D García-Meza, A Martínez-Blancas and MA Romero.

References

Anderson E (2001) The cactus family. Timber, Portland, OR, USA

Anderson CJ, MacMahon JA (2001) Granivores, exclosures, and seed banks: harvester ants and rodents in sagebrush-steppe. J Arid Environ 49(2):343–355. https://doi.org/10.1006/jare.2000.0781

Augspurger CK (1981) Reproductive synchrony of a tropical shrub: experimental studies on effects of pollinators and seed predators in *Hybanthus prunifolius* (Violaceae). Ecology 62(3):775–788. https://doi.org/10.2307/1937745

Bravo-Hollis H (1978) Las cactáceas de México, vol I, 2nd ed. Universidad Nacional Autónoma de México, México, DF, Mexico

Bravo-Hollis H, Sánchez-Mejorada H (1991) Las cactáceas de México, vol III. Universidad Nacional Autónoma de México, México, DF, Mexico

Brown JH, Reichman O, Davidson DW (1979) Granivory in desert ecosystems. Ann Rev Ecol Syst 10(1):201–227. https://doi.org/10.1146/annurev.es.10.110179.001221

CITES (2017) Appendices I, II and III. https://cites.org/sites/default/files/eng/app/2017/E-Appendices-2017-04-04.pdf. Accessed 1 Feb 2019

Cody M (1991) Niche theory and plant growth form. Vegetatio 97(1):39–55. https://doi.org/10.1007/BF00033900

Ezcurra E, Martínez-Berdeja A, Villanueva-Almanza L (2020) The evolution of North American deserts and the uniqueness of Cuatro Ciénegas. In: Mandujano MC, Pisanty I, Eguiarte L (eds.) Plant diversity and Ecology in the Chihuahuan desert. Springer International, Cham, Switzerland

Filazzola A, Lortie CJ (2014) A systematic review and conceptual framework for the mechanistic pathways of nurse plants. Global Ecol Biogeog 23(12):1335–1345. https://doi.org/10.1111/geb.12202

Gentry AH (1974) Flowering phenology and diversity in tropical Bignoniaceae. Biotropica 6(1):64–68. https://doi.org/10.2307/2989698

Hernández-Hernández T, Brown JW, Schlumpberger BO, Eguiarte LE, Magallón S (2014) Beyond aridification: multiple explanations for the elevated diversification of cacti in the New World succulent biome. New Phytol 202(4):1382–1397. https://doi.org/10.1111/nph.12752

Mancilla-Ramírez MR, Flores-Vázquez JC, Martínez-Peralta C, Aguilar-Morales G, Mandujano MC (2020) Cuatro Cienégas as a refuge for the living rock cactus, *Ariocarpus fissuratus*: demographic and conservation studies. In: Mandujano MC, Pisanty I, Eguiarte LE (eds.) Plant diversity and Ecology in the Chihuahuan desert. Springer International, Cham, Switzerland

Mandujano MC, Carrillo-Angeles I, Martínez-Peralta C, Golubov J (2010) Reproductive biology of Cactaceae. In: Ramawat KG (ed) Desert plants. Springer, Heidelberg, Germany, pp 197–230

Mandujano MC, García Naranjo A, Rojas-Aréchiga M, Golubo J (2020). Conservation status, germination and establishment of the divine cactus, *Lophophora williamsii* (Lem, Ex Salm-Dyck). J. M. Coult. At Cuatrocienegas. In: Mandujano MC, Pisanty I, Eguiarte L (eds.) Plant diversity and Ecology in the Chihuahuan desert. Springer International, Cham, Switzerland

Martorell C, Peters EM (2005) The measurement of chronic disturbance and its effects on the threatened cactus *Mammillaria pectinifera*. Biol Conserv 124(2):199–207. https://doi.org/10.1016/j.biocon.2005.01.025

Martorell C, Peters E (2009) Disturbance-response analysis: a method for rapid assessment of the threat to species in disturbed areas. Cons Biol 23(2):377–387. https://doi.org/10.1111/j.1523-1739.2008.01134.x

Martorell C, Garcillán P, Casillas F (2012) Ruderality in extreme-desert cacti? Population effects of chronic anthropogenic disturbance on *Echinocereus lindsayi*. Pop Ecol 54(2):335–346. https://doi.org/10.1007/s10144-012-0307-8

Martorell C, Montañana DM, Ureta C, Mandujano MC (2015) Assessing the importance of multiple threats to an endangered globose cactus in Mexico: cattle grazing, looting and climate change. Biol Conserv 181:73–81. https://doi.org/10.1016/j.biocon.2014.10.035

Mauseth JD (2006) Structure–function relationships in highly modified shoots of Cactaceae. Ann Bot 98(5):901–926. https://doi.org/10.1093/aob/mcl133

Nobel PS (1980) Interception of photosynthetically active radiation by cacti of different morphology. Oecologia 45(2):160–166. https://doi.org/10.1007/bf00346455

Nobel P (1983) Spine influences on PAR interception, stem temperature, and nocturnal acid accumulation by cacti. Plant Cell Environ 6(2):153–159. https://doi.org/10.1111/j.1365-3040.1983.tb01888.x

Nobel PS (1994) Remarkable agaves and cacti. Oxford University Press, Oxford, UK

Nobel PS (2003) Environmental biology of agaves and cacti. Cambridge University Press, Cambridge, UK

Nobel P, Geller G, Kee S, Zimmerman A (1986) Temperatures and thermal tolerances for cacti exposed to high temperatures near the soil surface. Plant Cell Environ 9(4):279–287. https://doi.org/10.1111/1365-3040.ep11611688

Noy-Meir I (1973) Desert ecosystems: environment and producers. Ann Rev Ecol Syst 4(1):25–51. https://doi.org/10.1146/annurev.es.04.110173.000325

Portilla-Alonso RM (2010) Causas de la ruderalidad en cactáceas globosas: reproducción y establecimiento de *Coryphantha werdermannii* (Cactaceae) en zonas perturbadas y conservadas. MSc Thesis, Universidad Nacional Autónoma de México, México, DF, Mexico

Portilla-Alonso RM, Martorell C (2011) Demographic consequences of chronic anthropogenic disturbance on three populations of the endangered globose cactus *Coryphantha werdermannii*. J Arid Environ 75(6):509–515. https://doi.org/10.1016/j.jaridenv.2011.01.015

Rodríguez-Ortega CE, Ezcurra E (2000) Distribución espacial en el hábitat de *Mammillaria pectinifera* y *M. carnea* en el valle de Zapotitlán Salinas, Puebla, México. Cact Suc Mex 45(1):4–14

Rojas-Aréchiga M, Vázquez-Yanes C (2000) Cactus seed germination: a review. J Arid Environ 44(1):85–104. https://doi.org/10.1006/jare.1999.0582

Rojas-Aréchiga M, Mandujano MC, Golubov J (2013) Seed size and photoblastism in species belonging to tribe Cacteae (Cactaceae). J Plant Res 126(3):373–386. https://doi.org/10.1007/s10265-012-0526-2

SEMARNAT (2010) Secretaría de Medio Ambiente y Recursos Naturales. 2010. Norma Oficial Mexicana NOM-059-SEMARNAT-2010. Published on December 30, 2010 on Diario Oficial de la Federación, México, DF, Mexico

Thompson DW (1917) On growth and form. Cambridge University Press, Cambridge, UK

Thompson K, Grime JP (1979) Seasonal variation in the seed banks of herbaceous species in ten contrasting habitats. J Ecol 67(3):893–921. https://doi.org/10.2307/2259220

Valiente-Banuet A, Bolongaro-Crevenna A, Briones O, Ezcurra E, Rosas M, Núñez H, Barnard G, Vázquez E (1991) Spatial relationships between cacti and nurse shrubs in a semi-arid environment in Central Mexico. J Veg Sci 2(1):15–20. https://doi.org/10.2307/3235893

Walck JL, Baskin JM, Baskin CC, Hidayati SN (2005) Defining transient and persistent seed banks in species with pronounced seasonal dormancy and germination patterns. Seed Sci Res 15(3):189–196. https://doi.org/10.1079/SSR2005209

Wilbur HM, Rudolf VH (2006) Life-history evolution in uncertain environments: bet hedging in time. Am Nat 168(3):398–411. https://doi.org/10.1086/506258

Zavala-Hurtado JA, Vite F, Ezcurra E (1998) Stem tilting and pseudocephalium orientation in *Cephalocereus columna-trajani* (Cactaceae): a functional interpretation. Ecology 79(1):340–348. https://doi.org/10.2307/176887

Chapter 11
Cuatro Ciénegas as a Refuge for the Living Rock Cactus, *Ariocarpus fissuratus:* Demographic and Conservation Studies

M. Rosa Mancilla-Ramírez, Juan Carlos Flores-Vázquez, Concepción Martínez-Peralta, Gisela Aguilar Morales, and María C. Mandujano

Abstract *Ariocarpus fissuratus* is a subglobose cactus, endemic of the Chihuahuan Desert, from San Luis Potosí, Mexico to south of Texas, restricted to limestone soils. Its stem resembles the soil surface where it grows; therefore, it is commonly known as "living rock" or *falso peyote*. Under harsh conditions, its taproot shrinks, enhancing its survival by making the plant less visible to predators and reducing its exposure to high temperature and water loss. Due to its intriguing morphology and the beauty of their flowers it is highly appreciated by illegal collectors, that along with recent habitat loss have driven the species near extinction. We studied the population ecology with data obtained during the years 2005–2009 of the species in three populations with contrasting densities and distribution of size classes in the Cuatro Ciénegas Basin. We found that the populations are stable (neither increasing nor decreasing in numbers) and Cuatro Ciénegas Basin contains the largest populations of *A. fissuratus* along its distribution. Moreover, the populations have variable densities, and differences in plant sizes, individual growth rates and survival probabilities, flowering, fruit and seed production, and pollinator behavior. Individual reproductive success, estimated as fruit and seed production, decreases at low population densities, in comparison to reproductive success of individuals located at high population densities, suggesting that a large population size is required to ensure

M. R. Mancilla-Ramírez · J. C. Flores-Vázquez · G. Aguilar Morales · M. C. Mandujano (✉)
Departamento de Ecología de la Biodiversidad, Instituto de Ecología, Universidad Nacional Autónoma de México, UNAM, Mexico City, Mexico
e-mail: mcmandu@iecologia.unam.mx

C. Martínez-Peralta
Departamento de Ecología de la Biodiversidad, Instituto de Ecología, Universidad Nacional Autónoma de México, UNAM, Mexico City, Mexico

Escuela de Estudios Superiores del Jicarero, Universidad Autónoma del Estado de Morelos, Cuernavaca, Mexico

M. C. Mandujano et al. (eds.), *Plant Diversity and Ecology in the Chihuahuan Desert*, Cuatro Ciénegas Basin: An Endangered Hyperdiverse Oasis,
https://doi.org/10.1007/978-3-030-44963-6_11

efficient pollination. Overall, Cuatro Ciénegas Basin is the most important area for this living rock conservation, because it offers protection to the populations and its interacting species, especially its pollinators, which are solitary bees.

Keywords Chihuahuan desert · Density · Life cycle · Population dynamics · Threats

The Living Rock Cactus

Ariocarpus fissuratus is a subglobose cacti (Fig. 11.1), commonly known as *chautle, falso peyote*, or *living rock* due to the morphology of its stem, which consists of triangular, flattened, and fissured tubers that protrude just above the ground, resembling the rocks found in its habitat (Bravo-Hollis and Sánchez-Mejorada 1991; Anderson 2001). These morphological characteristics make it particularly beautiful and rare, making the species an attractive plant for collectors and thus, for illegal harvesting. Moreover, the deterioration of its habitat (mainly due to land use change) and its biological characteristics (slow growth and high mortality in juvenile stages) have contributed to the threatened status of this species. *Ariocarpus fissuratus* is included in the Appendix I of CITES (Anderson 2001) and listed as "endangered" in Mexico's threatened species list NOM-059-ECOL-2010 (SEMARNAT 2010; Aguilar-Morales et al. 2011).

In addition, *A. fissuratus* is also known for containing alkaloids. Some ethnic groups, such as the Tarahumara, consider it as powerful as peyote (*Lophophora williamsii*, Mandujano et al. 2020, this volume). They attribute magical properties to the species and use it as a narcotic to cure fevers and rheumatic pains. In contrast, it

Fig. 11.1 A flowering specimen of *Ariocarpus fissuratus*, wool in the apex is a secondary character indicating the plant is potentially reproductive. Photo: Flores-Vázquez

is considered a "bad" cactus by the Huicholes because it brings bad thoughts or madness (Batis and Rojas 2002).

Ariocarpus fissuratus is an endemic species of the Chihuahuan desert with a wide distribution including sections of the Mexican states of Chihuahua, Coahuila, Durango, Nuevo Leon, San Luis Potosí, Tamaulipas, Zacatecas and, in the USA, portions of Texas. Of the seven species in the *Ariocarpus* genus, *A. fissuratus* has the widest geographical distribution (Aguilar-Morales et al. 2011). Although there is a high density of individuals within *A. fissuratus* in most of the known populations, they are restricted to sites with specific soil characteristics (Martínez-Peralta 2007), limestone soils in xerophilous scrub vegetation types (Aguilar-Morales et al. 2011). Cuatro Ciénegas Basin is one of the localities where the greatest number of individuals has been reported; hence, it is a site of great importance for the conservation of this species (Mancilla-Ramírez 2012).

Despite its distribution, little is known about the ecology of *A. fissuratus.* Some of these studies have focused on the species reproductive biology (Martínez-Peralta and Mandujano 2011; Martínez-Peralta et al. 2020, this volume) or other ecological aspects, for example, limited geographical distribution, population density and conservation status and life history characteristics (López González and García Ponce 2004; Mancilla-Ramírez 2012; Valencia et al. 2016). Increasing the knowledge base for this species is very important for the implementation of management and conservation plans.

Populations Status and Demographic Traits of *Ariocarpus fissuratus*

Coahuila is one of the states of Mexico with the widest known and potential distribution of *A. fissuratus* (Fig. 11.2, Aguilar-Morales et al. 2011), it is present in the municipalities of Cuatro Ciénegas, Ocampo, General Cepeda, Parras de la Source, and Viesca (Villavicencio et al. 2006); nonetheless, the conservation status of these populations is not optimal in most of the regions reported due to the illegal collection of individuals and the disturbance of their habitat (López González and García Ponce 2004, Valencia et al. 2016).

López González and García Ponce (2004) conducted a study in 21 sites in Coahuila where *A. fissuratus* is found, to evaluate the conservation status of the species in each site. Their criteria were based on the level of disturbance and the population density; these were used to group populations into three categories: *bad* (looted sites with anthropogenic disturbance and densities less than 25 individuals per 100 m^2), *good* (little anthropogenic disturbance and a density between 24 and 36 ind/100 m^2), and *excellent* (sites without disturbance and densities greater than 25 ind/100 m^2). They found that 42.85% of the sites were deteriorated, mainly due to human activities (looting and extraction of materials for construction) and by disturbance due to wild and domestic fauna. A contrasting population density was

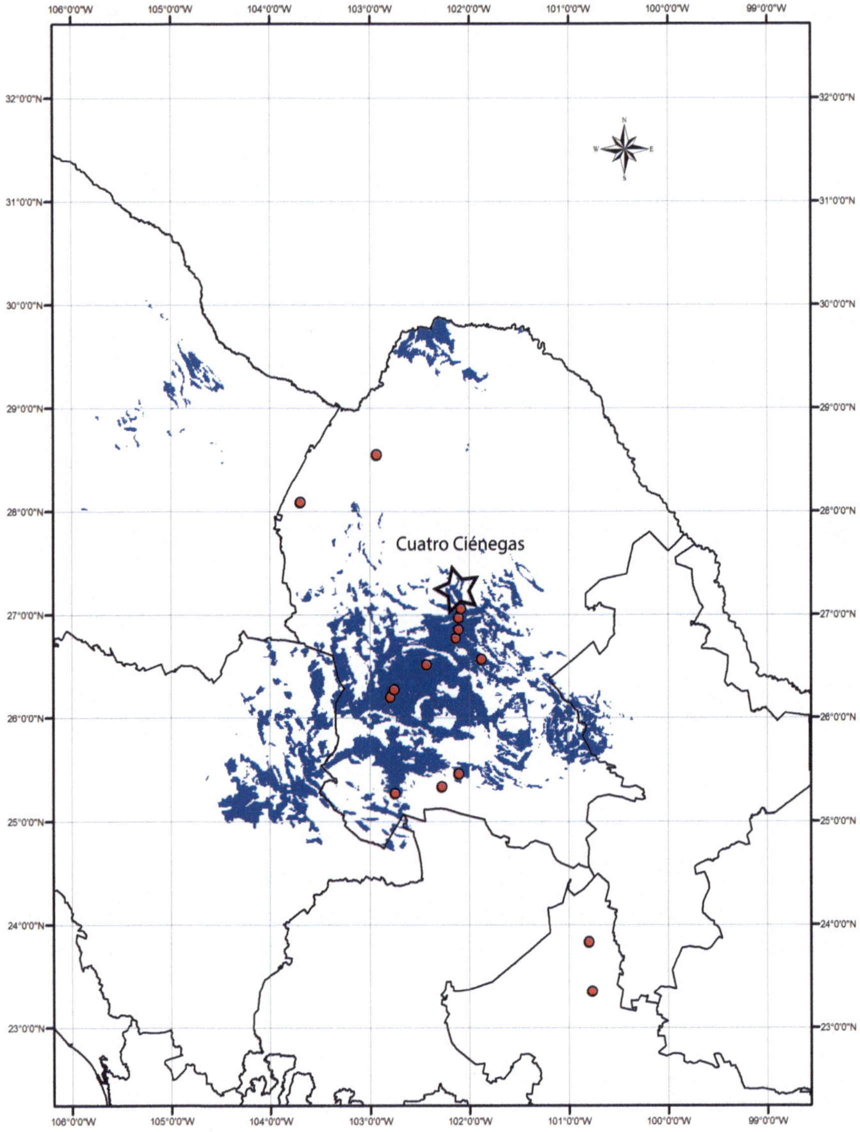

Fig. 11.2 Known (red) and potential (blue) distribution of *Ariocarpus fissuratus* in Mexico. Source: Aguilar-Morales et al. (2011). (Reproduced with permission of Cactáceas y Suculentas Mexicanas)

registered in two of the study sites in Cuatro Ciénegas area; in Sierra La Cuchilla a high density of 64 ind/100 m^2 was reported, while in the Sierra San Marcos there was a low density of 9 ind/100 m^2. In the latter, extraction of materials for construction and extraction of fiber from *Agave lechuguilla* are activities commonly practiced in the area. They conclude that *A. fissuratus* is in a steady state of conservation.

Another study, conducted by Valencia et al. (2016), which evaluated the demographic aspects of the living rock in three different sites of the Sierra and Cañón de Jimulco Municipal Ecological Reserve in Torreón, Coahuila, Mexico, found a low density of individuals (between 0.006 and 0.049 ind/m^2). They also reported a high rate of illegal extraction of individuals, evidenced by the presence of "extraction wells" which are cavities in the soil caused by extraction of plants. The sites also presented a high mortality of plants and the exploitation in the studied sites of other species, such as *A. lechuguilla,* which increases the level of disturbance. All this results in a bad conservation status of *A. fissuratus* populations in the Jimulco reserve.

Mancilla-Ramírez (2012) conducted a demographic study of three *A. fissuratus* populations with different densities within Cuatro Ciénegas Basin in Rancho Orozco (Fig. 11.3): high (H; located at 26° 54′ 43.14″ N 102° 7′ 21.54″ W), low (L; 26° 54′ 40.02″ N 102° 7′ 37.39″ W), and medium (M; 26° 54′ 38.27″ N 102° 7′ 18.6″ W), with data obtained during the years 2005–2009. Individuals were classified into six categories: seeds plus five size categories based on plant diameter (Fig. 11.4). The life cycle of the species was determined, the intrinsic population growth rate (λ) was estimated (Caswell 2001) and the most important categories for population maintenance and conservation were evaluated.

Fig. 11.3 (**a**) Study sites of *Ariocarpus fissuratus* in Rancho Orozco, Cuatro Ciénegas, Coahuila, Mexico. (**b**) Individuals tagged in the study. (**c**) The population density of each site is shown: high (H), medium (M), and low (L) density. Plots are scaled in meters for the three sites

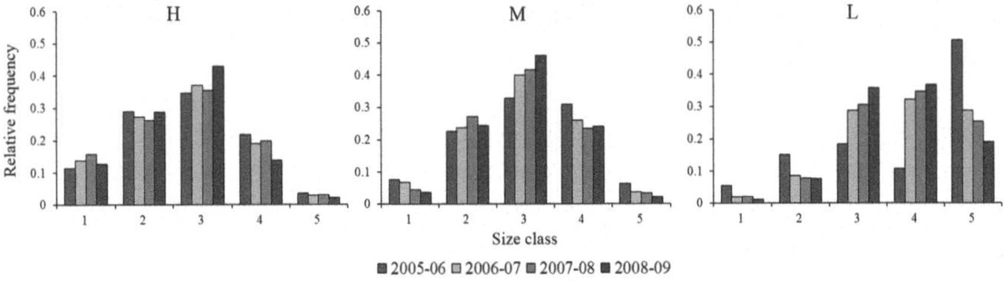

Fig. 11.4 Population structure of *Ariocarpus fissuratus* in the Cuatro Ciénegas Basin observed in each population (Density: H= high, M= medium, L= low) and each annual period. The seed category is not included due to its high expected values. (Size class (mm): 1 = 0-20, 2 = >20-40, 3= >40-60, 4=>60-80, 5 = >80)

The population structure of *A. fissuratus* (i.e., the relative frequency of individuals in each size category, which add up to one) was different between the three sites (Fig. 11.4). It was observed that in populations H and M the individuals of intermediate categories prevail (classes 3, 2, and 4), while in population L there is a greater frequency of larger individuals (classes 4, 3, and 5). The categories of bigger size individuals are the ones that contribute most to the population growth in all sites. This pattern, in which categories corresponding to adults contribute more to population growth than seedling and juvenile stages, has been found in other species of cacti (Godínez-Álvarez et al. 2003), such as *Mammillaria crucigera, Opuntia rastrera*, and *Astrophytum myriostigma* (Contreras and Valverde 2002, Mandujano et al. 2001, López-Flores 2012).

Ariocarpus fissuratus has a complex life cycle (Fig. 11.5) in which four important demographic processes occur: stasis (permanence in the same size category), growth, fecundity (number of average seeds produced per fruit), and regression or shrinking (size reduction). The demographic process of growth is when a plant reaches the next size category or when more favorable conditions exist, when a plant grows to even larger sizes, reaching a maximum diameter of around 150 mm (Table 11.1). Regarding fertility, it was observed that individuals begin to be reproductive as soon as they reach a 20 mm diameter and the probability of contributing to reproduction is greater as plants size increases.

Stasis or survival is the most important demographic process for *A. fissuratus*, which is typical of organisms with a great longevity (Silvertown et al. 1993; Godínez-Álvarez et al. 1999; Mandujano et al. 2001; Esparza-Olguín et al. 2002; Rosas-Barrera and Mandujano 2002; Jiménez-Sierra et al. 2007; Martorell and Portilla-Alonso 2020, this volume). In the particular case of *A. fissuratus*, where environmental conditions are extreme and plants are often subjected to various stressful conditions, the importance of surviving is even greater. Because survival is affected by different biotic and abiotic factors, sometimes it is not possible for plants to increase their size or reproduce, and instead, they remain in the same size category or even decrease their size if conditions are unfavorable. This size reduction can happen through different mechanisms, one of them is the loss of tubers

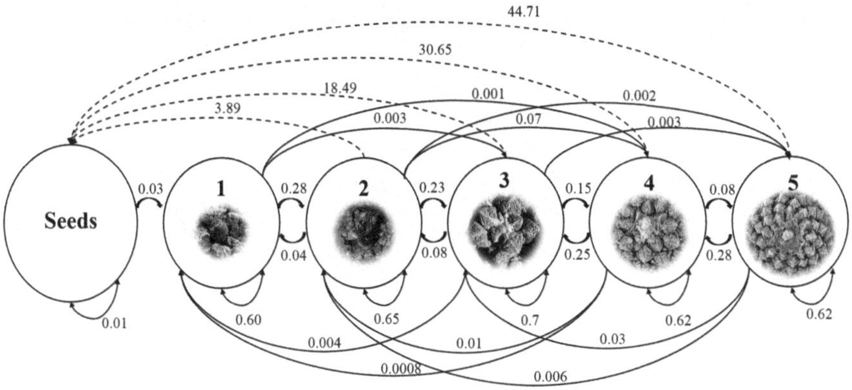

Fig. 11.5 Average life cycle of *Ariocarpus fissuratus* for the three study sites (High, Medium, and Low population density) in the Cuatro Ciénegas Basin, Coahuila, Mexico

Table 11.1 Demographic characteristics of *Ariocarpus fissuratus* in three populations with different densities in Cuatro Ciénegas, Coahuila, Mexico

Characteristics	H	M	L
Sampled area (m^2)	600	800	2500
# of individuals	987.6 (SD ± 62.25)	359.4 (SD ± 34.09)	354 (SD ± 24.20)
Density (ind/m^2)*	1.64 (SD ± 0.10)	0.45 (SD ± 0.04)	0.14 (SD ± 0.01)
Average diameter (mm)*	45.31 (SD ± 18.86)	50.24 (SD ± 17.54)	65.78 (SD ± 20.81)
Mortality (average % of dead individuals)	4.73 (SD ± 4.96)	6.97 (SD ± 4.96)	3.67 (SD ± 4.96)
Proportion of reproductive individuals	0.07 (SD ± 0.03)	0.16 (SD ± 0.20)	0.28 (SD ± 0.17)
Minimum reproductive size (mm)	20.3	26.8	36.6
Maximum size (mm)	131	103.5	157.5

H= high, M= medium and L= Low population density.* denotes a siginifcant difference between sites

(mainly those at the edge of the plant, Fig. 11.6), as has also been reported for *Ariocarpus scaphirostris* (Mandujano et al. 2007a).

Another shrinking mechanism is through root contraction, which allows plants to reduce their exposure to high temperatures (Garrett et al. 2010), which can exceed 70 °C on the soil surface (Nobel et al. 1986). This mechanism allows the plant to position part of its structure below the soil surface. This helps to reduce water loss (Garrett et al. 2010), as has been demonstrated in the study of two stone plants, *Lithops* (Aizoaceae) species (Eller and Ruess 1982). Regression is a process occurring in various species of cacti; it has been associated with unfavorable environmental conditions or as a result of some type of damage (e.g., herbivory) (Mandujano

Fig. 11.6 Individual of *Ariocapurs fissuratus* with dead tubers around the plant's edge. Dead tubers are yellow while live tubers have a grey color like that of the surrounding rocks. Photograph: R. Mancilla-Ramírez

et al. 2001, 2007a, b, Contreras and Valverde 2002, Jiménez-Sierra et al. 2007, López-Flores 2012).

The magnitude and combination of these demographic processes, together with environmental conditions, determine the status of populations, that is, we can infer whether populations are increasing or decreasing. Mancilla-Ramírez (2012) reports that the population growth rate of *A. fissuratus* is close to the numerical balance ($\lambda \approx 1$). This has been observed in several species of Cactaceae regardless of their growth form; for example, in the columnar *Escontria chiotilla* (Ortega-Baes 2001) and *Neobuxbaumia macrocephala* (Esparza-Olguín et al. 2002, 2005), the prickly pear *Opuntia rastrera* (Mandujano et al. 2001), the globose *Mammillaria crucigera* (Contreras and Valverde 2002) and *Mammillaria magnimamma* (Valverde et al. 2004), the barrel *Echinocactus platyacanthus* (Jiménez-Sierra et al. 2007) and the geophytic *Ariocarpus scaphirostris* (Mandujano et al. 2007a, b).

There are different factors that limit the population growth of *A. fissuratus*; for example, the low recruitment rate and high mortality during the first stages of the life cycle, which also occurs in other species of cacti (Godínez-Álvarez et al. 1999; Esparza-Olguín et al. 2002; Jiménez-Sierra et al. 2007; Valverde et al. 2004; Mandujano et al. 2007a, b; Martorell and Portilla-Alonso 2020, this volume). In arid and semi-arid environments, these demographic processes can be significantly affected by adverse weather conditions, such as sharp temperature fluctuations, scarce rainfalls, and prolonged droughts, as well as predation (Nobel 1989, Mandujano et al. 1998). These conditions can easily change a population in numerical balance (i.e., $\lambda \approx 1$) into one with an accelerated decline (i.e., λ less than 1). Long-term studies show there are clear demographic responses to unfavorable environmental conditions where populations decrease (Mandujano et al. 2001).

In *Ariocarpus fissuratus,* we performed numerical simulations using the species average life cycle (Fig. 11.5, Mancilla-Ramírez 2012) to evaluate the impact on the population growth rate of various possible scenarios. Independent numerical simulations were used to evaluate the effect of recruitment, survival, and mortality by looting on the rate of population increase (λ). A greater recruitment of individuals or a small increase in the number of seedlings results in the population increasing its growth rate ($\lambda = 1.84$), while reducing the survival of reproductive adults, equivalent to illegal looting, results in a decreased growth rate ($\lambda = 0.96$).

In addition to environmental factors, population density also has an effect at the individual and population level (Mancilla-Ramírez 2012). The populations of *A. fissuratus* of Cuatro Ciénegas Basin with contrasting population densities (H, M, and L, 164, 45 and 14 ind/100 m^2, respectively), differ significantly in several characteristics (Table 11.1). For example, the average plant diameter is significantly higher in the low-density population (L), while in population H individuals have, on average, smaller sizes and reproduce earlier, at smaller size. These relationships between density and demography may be the result of resource competition, i.e., when there is a greater number of individuals, the availability of resources for every plant diminishes; therefore, the individual growth is lower (Harper 1977; Antonovics and Levin 1980; Mancilla-Ramírez 2012), or it could also be that a greater recruitment of individuals occurs in the populations. It was also observed that the proportion of individuals reproducing annually is greater in site L (even though there are a smaller number of individuals), in this case it may be due to the benefit of a greater availability of resources in low-density scenarios that allows for more plants to present a reproductive event.

The presence of adult individuals in a population is very important, because they are the plants that contribute to reproduction. The reproductive values reported for *A. fissuratus* increase when size increases. This is the case for plants in general and has been documented in other cactus species (e.g., *Ariocarpus scaphirostris* (Mandujano et al. 2007a); *Echinocactus platyacanthus* (Jiménez-Sierra et al. 2007); *Mammillaria crucigera* (Contreras and Valverde 2002); *Neobuxbaumia macrocephala* (Esparza-Olguín et al. 2002); *Opuntia macrocentra* (Mandujano et al. 2007b)), because larger individuals are more likely to survive and can invest a greater amount of resources in reproduction (Stearns 1992).

Mortality was greater at site M, followed by sites H and L. There are no certain causes for the mortality of individuals. Some individuals had insect larvae inside, which possibly caused the death of some plants. In other cases, the spots where individuals were marked were found empty; this suggests looting, given that "extraction wells" were observed where the cacti were supposed to be found (Valencia et al. 2016).

Density may have an indirect influence within populations of *A. fissuratus*, rather than between them, and may not necessarily cause negative effects, for example, the intrinsic population growth rate (λ) between populations is similar and indicates population maintenance (i.e., $\lambda \approx 1$), however there are differences in individual growth rate, mortality, and fecundity (Mancilla-Ramírez 2012). There are demographic processes that are apparently favored by a higher population density, for example reproductive success, that benefits from a greater number of individuals which increases the probably to attract pollinators, favoring pollen transfer between individuals and increasing fruit and seed set (Martíez-Peralta et al. 2020, this volume). However, the effect of other factors such as environmental conditions, availability of resources, physicochemical characteristics of the soil should not be excluded, since these factors can affect the intra- and inter-specific interactions of individuals and cause changes in the demographic characteristics of *Ariocarpus fissuratus*.

Ariocarpus fissuratus and Its Association with Other Species and with Rocky Soils

The association of *A. fissuratus* with other species is very important, particularly for its reproduction, because this species depends on pollination for seed production. It has been reported that the species has a bee pollination syndrome (Martínez-Peralta and Mandujano 2011; Martínez-Peralta et al. 2020, this volume).

In arid environments, it is common to find plants associated with other species for protection against unfavorable environmental conditions and possible predation (Ezcurra et al. 2020, this volume). Nonetheless, *A. fissuratus* is a plant that usually grows in open spaces, free of vegetation (see Martorell and Portilla-Alonso 2020, this volume), but can nevertheless be found associated with *Larrea tridentata*, *Agave lechuguilla*, *Agave striata*, and several other cactus species such as *Grusonia bradtiana*, *Echinocactus horizonthalonius*, and *Opuntia* spp. (Villavicencio et al. 2006).

Because *A. fissuratus* grows in very stony soils, it is suggested that rocks can have a protective function and act as abiotic nurses. Rocks create microhabitats that dampen extreme environmental conditions by reducing solar radiation and temperature and by decreasing loss of soil moisture. These microhabitat conditions favor seed germination, as well as the establishment and growth of seedlings (Valiente-Banuet and Ezcurra 1991; Peters et al. 2008; Martorell and Portilla-Alonso 2020, this volume). The nursing function of rocks has some advantages over that of nurse plants, because the association with rocks does not result in negative interactions, like competition for limited resources, and rocks also do not attract herbivores or pathogens (Haussmann et al. 2010). Peters et al. (2008) conducted a review on several cactus species associated with rocks, among the studied species were *Carnegiea gigantea, Stenocereus thurberi, Cereus calcirupicola* and *Astrophytum asterias*, and some *Mammillaria* species.

Importance of Cuatro Ciénegas Basin for Conservation of *Ariocarpus fissuratus*

Considering our data, the status of the populations and the ecological characteristics of *A. fissuratus* reported by Mancilla-Ramírez (2012), and taking into account the criteria of López González and García Ponce (2004) in relation to the density of individuals and the level of disturbance, the conservation status of the populations within Cuatro Ciénegas is "excellent," in contrast to the population status of the species outside this area.

There is clear evidence on why Cuatro Ciénegas is an area of great importance for the conservation of the living rock. Within this area the levels of anthropogenic disturbance are much lower compared with other regions where resource exploitation occurs indiscriminately, and human settlements have a great impact on biodiversity. For example, although populations reported by Valencia et al. (2016) are found within the Municipal Reserve of Sierra y Cañón de Jimulco, their proximity to large cities, such as Torreón, makes the populations of *A. fissuratus*, as well as that of other species and their pollinators, more vulnerable.

The characteristics of the evaluated populations of *A. fissuratus* in Cuatro Ciénegas reflect the plasticity of the species when facing different ecological and environmental conditions. For example, the fact that in some places there is more recruitment of new individuals than in others, or that in some populations there is a big proportion of larger individuals. These ecological characteristics can ensure the maintenance of healthy populations of the species in Cuatro Ciénegas.

Moreover, natural protected areas are important for biodiversity conservation. For instance, within the Flora and Fauna Protection Areas of Cuatro Ciénegas and in the Sierra Maderas del Carmen (including the Sierras de La Madera and San Marcos) a total of 108 plant taxa can be found, representing 30.1% of the total endemic plants reported for the state of Coahuila (Villarreal-Quintanilla and Encina-Domínguez 2005); many of these taxa are included in Mexico's threatened species list NOM-059-ECOL-2001 (SEMARNAT 2010).

We can conclude that the recipe to avoid the extinction of *Ariocarpus fissuratus* (and many other species of plant and animals) is very simple: conserve the Cuatro Ciénegas Basin populations, and this will ensure the long-term conservation of the living rock and many other organisms.

References

Aguilar-Morales G, Martínez-Peralta C et al (2011) Distribución geográfica del género *Ariocarpus* Scheidweiler (Cactaceae). Cact Suc Mex 56:49–63

Anderson EF (2001) The cactus family. Timber Press Inc. Portland, OR, USA

Antonovics J, Levin DA (1980) The ecological and genetic consequences of density-dependent regulation in plants. Annu Rev Ecol S 11:411–452

Batis A, Rojas M (2002) El peyote y otros cactus alucinógenos de México. Biodiveristas 40:12–17

Bravo-Hollis H, Sánchez-Mejorada H (1991) Las cactáceas de México, vol II. Universidad Nacional Autónoma de México, México, DF, Mexico

Caswell H (2001) Matrix population models. Sinauer Associates Inc Publishers, Sunderland, MA, USA

Contreras C, Valverde T (2002) Evaluation of the conservation status of a rare cactus (*Mammillaria crucigera*) through the analysis of its population dynamics. J Arid Environ 51:89–102

Eller BM, Ruess B (1982) Water relations of *Lithops* plants embedded into the soil and exposed to free air. Physiol Plant 55:329–334

Esparza-Olguín L, Valverde T, Vilchis-Anaya E (2002) Demographic analysis of a rare columnar cactus (*Neobuxbaumia macrocephala*) in the Tehuacan Valley, Mexico. Biol Conserv 103:349–359

Esparza-Olguín L, Valverde T, Mandujano MC (2005) Comparative demographic analysis of three *Neobuxbaumia* species (Cactaceae) with differing degree of rarity. Popul Ecol 47:229–245

Ezcurra E, Martínez-Berdeja A, Villanueva-Almanza L (2020) The evolution of North American deserts and the uniqueness of Cuatro Ciénegas. In: Mandujano MC, Pisanty I, Eguiarte L (eds.) Plant diversity and Ecology in the Chihuahuan desert. Springer International, Cham, Switzerland

Garrett TY, Huynh CV, North GB (2010) Root contraction helps protect the "living rock" cactus *Ariocarpus fissuratus* from lethal high temperatures when growing in rocky soil. Am J Bot 97:1951–1960

Godínez-Álvarez H, Valiente-Banuet A, Valiente-Banuet BL (1999) Biotic interactions and the population dynamics of the long-lived columnar cactus *Neobuxbaumia tetetzo* in the Tehuacan Valley, Mexico. Can J Bot 77:203–208

Godínez-Álvarez H, Valverde T, Ortega P (2003) Demographic trends in the Cactaceae. Bot Rev 69:173–203

Harper JL (1977) Population biology of plants. Academic Press, London, UK

Haussmann NS, McGeoch MA, Boelhouwers JC (2010) Contrasting nurse plants and nurse rocks: the spatial distribution of seedlings of two sub-Antarctic species. Acta Oecol 36:299–305

Jiménez-Sierra C, Mandujano MC, Eguiarte LE (2007) Are populations of the candy barrel cactus (*Echinocactus platyacanthus*) in the desert of Tehuacan, Mexico at risk? Population projection matrix and life table response analysis. Biol Conserv 135:278–292

López González JJ, García Ponce G (2004) Distribución y evaluación de las poblaciones naturales del género *Ariocarpus* (Scheidweiler) en Coahuila, México. Cact Suc Mex 49:68–79

López-Flores D (2012) Demografía de *Astrophytum myriostigma* Lem. (Cactaceae). Tesis de Maestría. Facultad de Ciencias. Universidad Nacional Autónoma de México. México, DF, Mexico

Mancilla-Ramírez MR (2012) Dinámica poblacional de *Ariocarpus fissuratus* en poblaciones que difieren en su densidad. Tesis de Maestría. Instituto de Ecología. Universidad Nacional Autónoma de México. México DF, Mexico

Mandujano MC, Montaña C, Méndez I, Golubov J (1998) The relative contributions of sexual reproduction and clonal propagation in *Opuntia rastrera* in two habitats in the Chihuahuan Desert. J Ecol 86:911–921

Mandujano MC, Montaña C, Franco M, Flores A, Golubov J (2001) Integration of demographic annual variability in a clonal desert cactus. Ecology 82:344–359

Mandujano MC, Verhulst JAM, Carrillo-Ángeles IG, Golubov J (2007a) Population dynamics of *Ariocarpus scaphirostris* Bödecker (Cactaceae): evaluating the status of a threatened species. Int J Plant Sci 168:1035–1044

Mandujano MC, Golubov J, Huenneke LF (2007b) Effect of reproductive modes and environmental heterogeneity in the population dynamics of a geographically widespread clonal desert cactus. Popul Ecol 49:141–153

Mandujano MC, García Naranjo A, Rojas-Aréchiga M, Golubo J (2020). Conservation status, germination and establishment of the divine cactus, *Lophophora williamsii* (Lem, Ex Salm-Dyck). J. M. Coult. At Cuatrocienegas. In: Mandujano MC, Pisanty I, Eguiarte L (eds.) Plant diversity and Ecology in the Chihuahuan desert. Springer International, Cham, Switzerland

Martínez-Peralta C (2007) Biología floral de *Ariocarpus fissuratus* (Engelmann) Schumann (Cactaceae) en Cuatro Ciénegas, Coahuila, México. Tesis Profesional. Universidad Nacional Autónoma de México. México, DF, Mexico

Martínez-Peralta C, Mandujano MC (2011) Reproductive ecology on an endangered living rock cactus, *Ariocarpus fissuratus* (Cactaceae). J Botanical Torrey Soc 138:145–155

Martínez-Peralta C, Jiménez-Díaz J, Flores Vázquez JC, Mandujano MC (2020) Reproductive biology and conservation of living rock *Ariocarpus fissuratus*. In: Mandujano MC, Pisanty

I, Eguiarte LE (eds) Plant diversity and Ecology in the Chihuahuan desert. Springer International, Cham, Switzerland

Martorell C, Portilla-Alonso RM (2020). The ages of life: The changing forms and history of Coryphantha werdermanii throughout its development. In: Mandujano MC, Pisanty I, Eguiarte L (eds.) Plant diversity and Ecology in the Chihuahuan desert. Springer International, Cham, Switzerland

Nobel P (1989) Temperature, water availability and nutrient levels at various soil depths. Consequences for shallow-rooted desert succulents, including nurse plant effects. Am J Bot 76:1486–1492

Nobel PS, Geller GN, Kee SC, Zimmerman AD (1986) Temperatures and thermal tolerances for cacti exposed to high temperatures near the soil surface. Plant Cell Environ 9:279–287

Ortega-Baes FP (2001) Demografía de la cactácea columnar *Escontria chiotilla*. Tesis de Maestría. Instituto de Ecología. Universidad Nacional Autónoma de México. México, DF, Mexico

Peters EM, Martorell C, Ezcurra E (2008) Nurse rocks are more important than nurse plants in determining the distribution and establishment of globose cacti (Mammillaria) in the Tehuacán Valley, Mexico. J Arid Environ 72:593–601

Rosas-Barrera MD, Mandujano MC (2002) La diversidad de historias de vida de cactáceas, aproximación por el triángulo demográfico. Cact Suc Mex 47:33–41

SEMARNAT (2010) Secretaría de Medio Ambiente y Recursos Naturales. Norma Oficial Mexicana NOM-059-SEMARNAT-2010. Published on December 30, 2010 in Diario Oficial de la Federación, México, DF, Mexico

Silvertown J, Franco M, Pisanty I, Mendoza A (1993) Comparative plan demography—relative importance of life-cycle components to finite rate of increase in woody and herbaceous perennials. J Ecol 81:465–476

Stearns SC (1992) The evolution of life histories. Oxford University Press, Oxford, New York, NY, USA

Valencia CCM, Gutiérrez RE, Linaje TMS, Hernández TF, De la Fuente SN (2016) Ecología del falso peyote (*Ariocarpus fissuratus*) en la Reserva Ecológica Municipal Sierra y Cañón de Jimulco. Agrofaz 16:85–95

Valiente-Banuet A, Ezcurra E (1991) Shade as a cause of the association between the cactus *Neobuxbaumia tetetzo* and the nurse plant *Mimosa luisana* in the Tehuacan Valley, México. J Ecol 79:961–971

Valverde T, Quijas S, López-Villavicencio M, Castillo S (2004) Population dynamics of *Mammillaria magnimamma* Haworth. (Cactaceae) in a lava-field in Central Mexico. Plant Ecol 170:167–184

Villarreal-Quintanilla JA, Encina-Domínguez JA (2005) Plantas vasculares endémicas de Coahuila y algunas áreas adyacentes. Acta Bot Mexi 70:1–46

Villavicencio E, López J, Martínez O, Ponce G (2006) Distribución digitalizada y características ecológicas del género *Ariocarpus spp.* en Coahuila. SAGARPA-INIFAP. http://biblioteca.inifap.gob.mx:8080/jspui/bitstream/handle/123456789/440/203.pdf;jsessionid=6096390BA98533279F4CF7D9C29BD710?sequence=1

Chapter 12
Reproductive Biology and Conservation of the Living Rock *Ariocarpus fissuratus*

Concepción Martínez-Peralta, Jorge Jiménez-Díaz, Juan Carlos Flores-Vázquez, and María C. Mandujano

Abstract The globose cactus *Ariocarpus fissuratus*, known as "living rock," is renowned for the singular morphology of its extremely cryptic single stem/taproot, which results in a resemblance of their tubercles to rocky soil. Their beauty has placed *Ariocarpus* among the top rated illegally traded cactus. In the Cuatro Ciénegas Basin, important populations of *A. fissuratus* occur. This is the only *Ariocarpus* out of seven species in the genus, found outside Mexico, reaching southern Texas. In order to determine the influence of sexual reproduction to the conservation of healthy populations, we studied the reproductive ecology of *A. fissuratus* including floral morphology and phenology, interaction with insects, and mating and breeding systems. Results indicate that floral phenology is remarkably synchronic, and it occurs from middle October to early November through a few pulses that vary in onset and duration but lasting only 5 or less days. Genders are separated within the flowers in time and space, diminishing the probability of selfing. Numerous pollen grains and nectar are produced as rewards for pollinators, a feature that suggests a xenogamous system. *A. fissuratus* is an outcrosser, as indicated by forced pollination treatments. Floral visitors include bees, wasps, flies, butterflies, and beetles; only solitary bees behave as pollinators. Species assemblage of bee pollinators vary between years, suggesting that functional specialization of pollinators occurs in this species. In addition, we observed a high incidence of florivory by a tenebrionid beetle. We conclude that success of sexual reproduction is critical for the conservation of this cactus, but its short flowering phenology, its outcrossing mating system and dependence on pollinators, exotic floral visitors (the possible

C. Martínez-Peralta (✉)
Laboratorio de Ecología de la Polinización, Escuela de Estudios Superiores del Jicarero, Universidad Autónoma del Estado de Morelos, Jojutla, Mexico
e-mail: concepcion.martinez@uaem.mx

J. Jiménez-Díaz · J. C. Flores-Vázquez · M. C. Mandujano
Instituto de Ecología, Universidad Nacional Autónoma de México, UNAM, Mexico City, Mexico
e-mail: mcmandujano@iecologia.unam.mx

M. C. Mandujano et al. (eds.), *Plant Diversity and Ecology in the Chihuahuan Desert*, Cuatro Ciénegas Basin: An Endangered Hyperdiverse Oasis,
https://doi.org/10.1007/978-3-030-44963-6_12

negative effects of *A. mellifera*), and florivory may be caveats to the maintenance of healthy populations in this species.

Keywords Bee pollination · Native bees · Floral traits · Synchronous flowering · Cactaceae

Introduction

Despite the numerous recommendations for management and conservation of rare and threatened plants (e.g., Silva et al. 2015), knowledge of their floral and pollination ecology is still scarce. Even for iconic plants such as cacti, the paucity of information about phenology, biotic interactions, and reproductive output (Mandujano et al. 2010) certainly limits the design of adequate conservation strategies. Short phenologies, limited floral display, pollen limitation, scarcity of pollinators, and reduced fruit and seed set are frequent in rare and threatened plants (Byers and Meagher 1997). For example, reduced flower production, longer age to first reproduction, and less competitive performance make *Solidago shortii* narrowly endemic compared to its more common congener *S. altissima* (Walck et al. 2001). Hence, understanding how reproductive attributes interact with other components of biological rarity is fundamental for designing strategies in accordance with the biology of rare species.

Globose cacti are among the most illegally traded plants globally (Olmos-Lau and Mandujano 2016). The genus *Ariocarpus*, commonly called living rocks, are frequently extracted from natural populations, reducing effective population sizes and diminishing population recruitment (Mandujano et al. 2007). Even worse, complete *Ariocarpus* populations have disappeared because of change land use (Arroyo-Cosultchi et al. 2014). Although Mexican and international laws protect the seven species of *Ariocarpus*, only *A. fissuratus* (Fig. 12.1) is protected in part of its range by a natural protected area, the Cuatro Ciénegas Basin (CCB, formally called by the Mexican government "Area de Protección de Flora y Fauna de Cuatro Ciénegas," Souza et al. 2018; Souza and Eguiarte 2018).

In CCB, several populations of *A. fissuratus* occur, mainly in the transition areas between foothills of mountains and the valley. The population studied ranges in density from 0.14 to 1.64 ind/m^2 (Mancilla-Ramírez et al. 2020, this volume). We studied the reproductive biology of this cactus for several years in the locality Rancho Orozco (26°54′43.14″N, 102°07′21.54″W). We aimed to describe floral morphology and phenology, mating and breeding systems and its interaction with insect floral visitors in order to understand the influence of sexual reproduction to the conservation of healthy populations of this endangered species.

Fig. 12.1 Flowering *Ariocarpus fissuratus* in the Cuatro Ciénegas Basin. Due to high resemblance of its stems/taproot to the rocky soil, this cactus is commonly named "living rock"

Methods

Floral Phenology

To document the flowering period, we registered open flowers per day in a sample of plants from mid-October to early November, during 2005 and 2006. Individual anthesis was evaluated by measuring floral diameter (cm) in seven intervals from 7:00 to 19:00. In these intervals, the function of sexual organs was also evaluated: receptive stigmas have a wet appearance, and functional, dehiscent anthers have a granular surface because of pollen release. Percentage of functioning sexual organs was calculated for each time interval to determine whether sexual functions are simultaneous (homogamy) or separated in time (dichogamy; Cruden 1977, Kearns and Inouye 1993).

Floral Morphology and Breeding System

In 2005, we collected 34 flowers (each one from a different plant) in FAA (10:50:5:35 formalin, 95% ethanol, acetic acid, distilled water) to account for floral morphology (Martínez-Peralta et al. 2014). Flowers were longitudinally dissected to register a) flower length, b) flower width, c) pistil length, d) stamen length, e) number of ovules, f) number of stamens, and g) stigmatic lobes. To estimate pollen grains per flower, a smashed anther was mixed with 1 ml of distilled water; all pollen grains from a 100 μl aliquot were counted under a stereoscope. Pollen grains per flower was estimated by multiplying pollen grains per aliquot by the dilution factor (10X), and then by number of anthers per flower. Herkogamy, the spatial separation of sexual organs, was evaluated by means of a paired *t*-test between pistil length and stamen length (R Core Team 2018).

Breeding system was determined based on the proposal of Cruden (1977), as large corollas, dichogamy, and herkogamy are usually displayed by xenogamous species, while small corollas and coincident sexual functions are generally found in autogamous plants. During 2005, these traits were determined as described above, and the score for each trait was assigned following Cruden (1977). In addition, the ratio of male to female gametes (P/O ratio, see Cruden 1977) is higher in obligated xenogamous flowers, and follows a diminishing gradient that reaches a minimum in autogamous flowers. Mean pollen grains per flower and number of ovules were used to estimate mean P/O ratio, which was compared with the breeding system categories proposed by Cruden (1977).

Pollination Experiments

In order to determine mating system, individual flowers were assigned to one of the following pollination treatments (2005 season): (a) natural pollination, flowers were left available to floral visitors, (b) supplementary pollination, flowers were naturally pollinated, and an additional pollen load was applied, (c) outcrossing, flowers were emasculated and manually pollinated with a pollen load from 10 different donor plants located in a different patch, (d) manual selfing, flowers were manually self-pollinated and protected with mesh bags to prevent floral visitors, and (e) natural ("automatic") selfing, unmanipulated flowers that were covered with mesh bags (Kearns and Inouye 1993).

Fruits were collected 4 months after pollination, and seeds per fruit counted. Differences among pollination treatments were determined by means of GLM's: a binomial distribution and logit link function was applied for fruit set, and a Poisson distribution and log function for seed set (R Core Team 2018).

Pollinators and Nectar Production

Frequency, behavior, and species of insect visitors were registered by observing flowers for 3 years. During 2006, we observed 42 flowers (each flower from a different plant) at four intervals of 20 min each, from 10:00 to 16:00. During 2012 (38 flowers from 30 plants) and 2013 (43 flowers, each from a different plant), six 30-min intervals were established, from 10:00 to 15:00. During 2012 and 2013, revisits were also recorded, i.e., when an insect visited the same target flower several times consecutively.

In all 3 years of study, we collected a sample of floral visitors directly from the flowers and outside observation intervals; these insects were sacrificed in entomological jars with ethyl acetate and pinned for taxonomic identification. We obtained the list of species, and relative abundances per species per year. During 2006, a Tenebrionid beetle was observed eating flowers, so that we registered the percentage of damaged floral structures.

Finally, we measure nectar production in 2006. From a different plant we tagged 39 flowers that were covered with mesh bags and 35 flowers (each from a different plant) that were left uncovered, available to pollinators; we follow a repeated measurement design and registered nectar volume with a 2-μl microcapillary tube every 2 h, from 10:30 to 18:30. Cumulative nectar volume per flower was obtained and compared between treatments by means of a Wilcoxon rank test (R Core Team 2018).

Results

Floral Phenology

Floral longevity of living rock cactus flowers is between 1 and 3 days (Fig. 12.2, top). The flowers are diurnal, with a nyctinastic response (i.e., they close in darkness) (Fig. 12.2, middle). Flowering had an acute peak of 4–5 days in both years, and was slightly longer in 2005 (Fig. 12.2, bottom); although we did not register flower longevity during 2012 and 2013, flowering peak was apparently limited to only 2 days in middle October in both years (personal observation).

Floral Morphology and Breeding System

Flowers are herkogamous, as pistils are significantly longer than stamens ($t = 14.08$, d.f. = 33, $P = 1.67\text{e-}15$) (Fig. 12.3). Four hours after flower opening sexual functions overlapped, hence dichogamy is partial. The relatively wide flower diameter (Table 12.1), the herkogamy, as well as the partial dichogamy, should favor obligate xenogamy according to Cruden (1977). These floral traits, together with the high

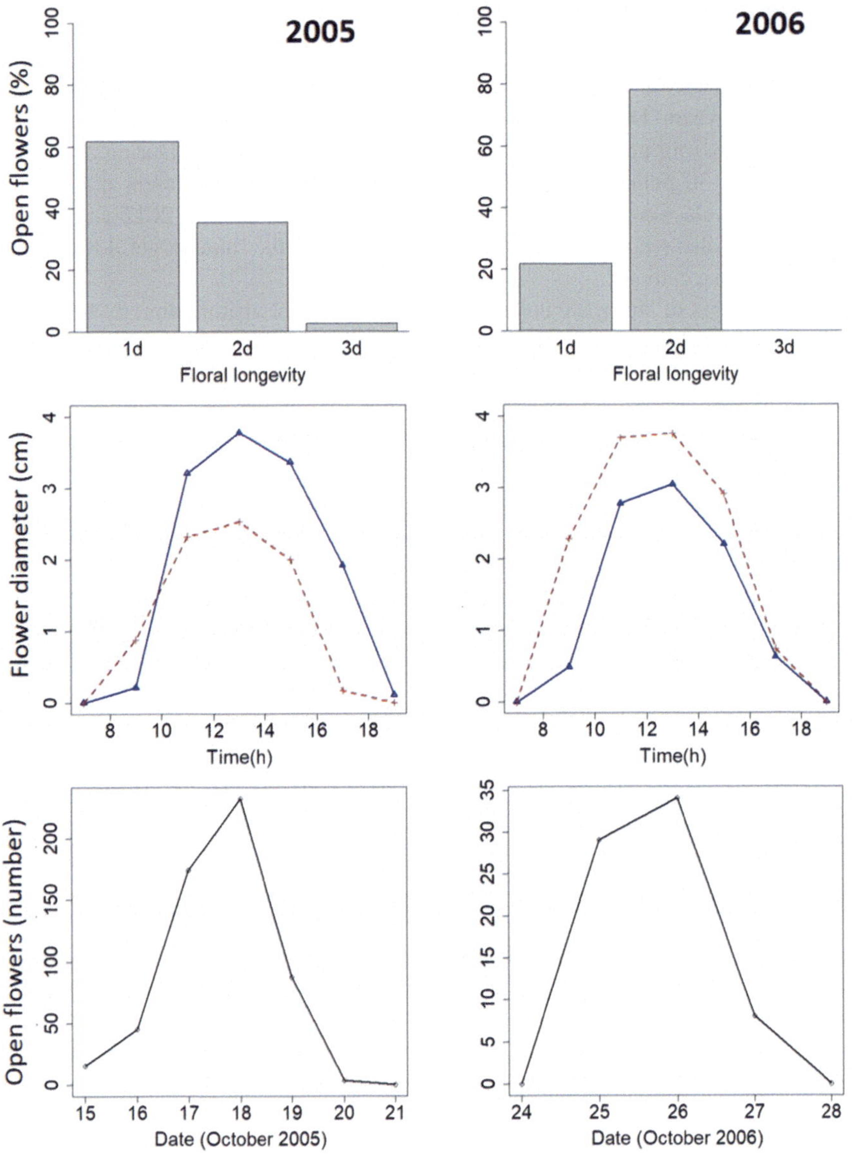

Fig. 12.2 Floral phenology of the living rock cactus in the Cuatro Ciénegas Basin during 2005 and 2006. Top: individual floral longevity. Middle: flower anthesis during 2 days of life (blue solid line with triangles for day one, red dotted line for day two). Bottom: population flowering peaks

Fig. 12.3 Flowering plant of *Ariocarpus fissuratus* from Cuatro Ciénegas Basin in 2006. Flowers are herkogamous because pistils (P) are significantly longer than stamens (St). Young buds (Bud) are covered with wool, and progressively emerge from it while maturing. Stem is constituted by tubercles (Tub) arranged in whorls with a high resemblance to the rocky soil

Table 12.1 Floral traits of the living rock cactus *Ariocarpus fissuratus* in the Cuatro Ciénegas Basin, central Coahuila, Mexico

Floral trait	Mean ± s.d.
Flower diameter	29.70 ± 5.42
Flower length	29.99 ± 3.97
Pistil length	18.64 ± 2.44
Stamen length	14.23 ± 1.71
Ovules	93.38 ± 30.78
Stamens	183.44 ± 59.35
Pollen grains	151,839 ± 49,126
P/O	1687 ± 422
Stigmatic lobes	6.58 ± 1.37

(26° 54′ 43.14″N, 102° 7′ 21.54″W) (*N* = 34 flowers each from a different plant), s.d. = standar deviation

P/O ratio (Table 12.1) (Cruden 1977) strongly suggest that *A. fissuratus* flowers are strictly xenogamous, adapted to cross-pollination (Martínez-Peralta et al. 2014).

Pollination Experiments

The natural ("automatic") selfing treatment set no fruits, and was excluded from further analyses, but indicated the necessity of a pollinator vector. GLM's showed that both, natural pollination and supplementary pollination were significantly more successful than manual outcrossing in both, fruit set and seed set ($\chi^2 = 87.89$, d.f = 3, P < 0.00001 and $\chi^2 = 92.99$, d.f. = 3, $P < 0.00001$, respectively) (Table 12.2). Manual selfing set only one fruit, with similar seed set to that from natural pollination.

Table 12.2 Fruit and seed set after pollination treatments (2005) in *Ariocarpus fissuratus* in the Cuatro Ciénegas Basin, central Coahuila, Mexico

Treatment	Flowers	Fruits	Fruit set (%) (Mean ± s.d.)	Seed set (Mean ± s.d.)
Natural pollination	96	51	53.1 ± 0.66 **a***	67.9 ± 36.8 **a***
Supplementary pollination	47	30	63.8 ± 0.65 **a***	64.4 ± 40.9 **a***
Outcrossing	40	9	22.5 ± 0.69 **b****	41.6 ± 27.5 **b***
Selfing (manual)	53	1	1.9 **c****	68 **a***
Selfing (natural)	0	0	0	–

(26° 54′ 43.14″N, 102° 7′ 21.54″W)
* $P < 0.0001$, ** $P < 0.001$, , s.d. = standar deviation

Table 12.3 Percentage of floral visits to *Ariocarpus fissuratus* per insect species during 3 years of observations

Species (family)	2006 (*N* = 42 flowers from 42 plants)	2012 (*N* = 38 flowers from 30 plants)	2013 (*N* = 43 flowers from 43 plants)
Apis mellifera (Apidae)	91.3	54.0 (73.6)	34.6
Lasioglossum sp. (Halicidae)	3		
Diadasia sp. (Apidae)	1.3		
Diadasia diminuta (Apidae)		9.9 (5.7)	21.2
Melissodes tristis (Apidae)		6.2 (1.9)	44.2
Ashmeadiella bigeloviae (Megachilidae)		Obs	
Agapostemon obliquus (Halictidae)		Obs	
Megachilidae sp. (Megachilidae)	Obs		
Wasps	4.4	6.6 (1.9)	
Others		23.3 (17)	

Numbers in parenthesis indicate percentage of revisits
Obs = observed and/or captured but not recorded during intervals

Pollinators and Nectar Production

Main floral visitors during all 3 years of observation were bees (Table 12.3; Figs. 12.4 and 12.5) and were the only insects that touched one or both sexual organs while visiting, thereby potentially transferring pollen among flowers. Other floral visitors were flies, orthopterans, lepidopterans, ants, and beetles. Bee assemblage was different across years; only the exotic *Apis mellifera* (Fig. 12.5) was found all 3 studied years, and during 2 seasons was the most frequent visitor. Some wild bee species were captured or observed foraging on *A. fissuratus* flowers, but not registered within observation intervals. During 2013 no revisits were registered; during 2012, *A. mellifera* recorded the highest proportion of revisits. The tenebrionid beetle registered during 2006 damaged 47% of the flowers ($n = 45$ flowers from 30 plants) (Martínez-Peralta and Mandujano 2011).

Variance of the cumulative néctar volume after 10 h of repeated measures per flower was higher than the mean in both treatments; for that reason, the Huber advanced mean and Huber standard deviation were obtained (R Core Team 2018). The advanced means and Huber SD values were 0.87 ± 0.58 μl for covered flowers and 1.01 ± 0.78 μl for uncovered flowers. The Wilcoxon rank test indicated that no

Fig. 12.4 The native bee *Melissodes tristis* (Apidae) was the most frequent floral visitor of *Ariocarpus fissuratus* during 2013 in the Cuatro Ciénegas Basin

Fig. 12.5 The honeybee *Apis mellifera* (Apidae) was the most frequent floral visitor of *Ariocarpus fissuratus* flowers during 2 years of study (2006 and 2012)

significant differences exist between covered flowers and those available to pollinators in the cumulative nectar volume ($W = 408.5$, $P = 0.81$). These results suggest that flowers produce more nectar than the requirements of the actual number of foragers.

Discussion

Flowering of the *A. fissuratus* population in CCB lasts only few days (5 days or less, Martínez-Peralta and Mandujano 2011). A similar highly synchronic phenological behavior has been described for species from tropical dry forests as *big bang* flowering (Gentry 1974), but it has been seldom reported in Cactaceae. Two species of cactus with highly synchronic flowering are the sister species *A. kotschoubeyanus*, from Northeastern Mexico, that has similar synchronic flowering pulses that last only between 4 and 7 days (Martínez-Peralta and Mandujano 2016), and *Cereus aethiops*, from North Argentina, where all the plants in a population flower in between 5 and 10 days (Eggli and Giorgetta 2015).

Outcrossing in mass flowering species is heavily dependent on pollinators that abandon their usual floral resources and foraging routes and change to visit the sudden and massive concentration of floral resources (Gentry 1974). In theory, an additional advantage of this flowering pattern is that predation of reproductive structures (in this case exerted by the Tenebrionid beetle), is mitigated (diluted) by the high floral density. Thus, a high floral display seems to be crucial for the successful reproduction of *A. fissuratus*, but less dense populations, or those individuals that flower outside the flowering peaks, could experiment less bee visitation and hence a possible failure in sexual reproduction.

Estimators of the breeding system, the OCI and the P/O ratio (Cruden 1977), indicate that *A. fissuratus* flowers are adapted to cross-pollination, as the species displays large flowers, herkogamy, a high pollen:ovule ratio, as well as pollen grains and nectar as rewards (Mandujano et al. 2010). The pollination experiments support these results, since there was almost no success of selfing, either natural or manual. By contrast, treatments involving cross-pollination were significantly more successful. Natural pollinated flowers set as many fruits and seeds as supplemental pollinated flowers, which indicates that sexual reproduction is not limited by the quantity of pollen deposited on stigmas (Larson and Barrett 2000, Martínez-Peralta et al. 2014).

Floral visitors of *A. fissuratus* encompass four orders (Coleoptera, Lepidoptera, Orthoptera, Hymenoptera), but the main frequency of visits comes from bees. Bee assemblage and relative frequency of each species varied across seasons, which suggests a functional specialization to bees in the pollination system of *A. fissuratus* (Ollerton et al. 2007). Functional specialization in pollination arises when a plant is pollinated by a group of related organisms; in this case, *A. fissuratus* is specialized to bees, even there are several species within the floral visitor assemblage. Wild bees are the visitors that probably co-evolved with *A. fissuratus* because they are regarded as native to the North American deserts, despite their low frequency during two seasons (they were dominant only during 2013). The behavior of wild bees in the flowers of *A. fissuratus*, touching anthers and stigmas, suggests that they have the higher potential as the more effective pollinators of the species (Schlindwein and Wittman 1995).

Apis mellifera was the most frequent visitor during two seasons, and the second most frequent in 2013. *Apis mellifera* is frequently recorded among floral visitors of cacti flowers (Martínez-Peralta et al. 2014, 2018), and is the most frequent species of the family Apidae across seasons and sites within CCB (Ávalos-Hernández et al. 2016). The role of *A. mellifera* in cactus pollination has been evaluated from null (Fagua and Ackerman 2011), or positive (Langley 2015) to even negative, as they might act mainly as pollen thieves (Ortega-Baes et al. 2011). However, many studies coincide that *A. mellifera* is an active pollen collector, and hence diminishes the available pollen for pollination. Our results indicate that *A. mellifera* elevates the risk of stigma clogging with self-pollen, because of the high percentage of revisits; stigma clogging limits outcrossing by avoiding pollen tube growth of cross pollen (Cesaro et al. 2004). The pollination experiment indicated that no pollen limitation exists; however, if *A. mellifera* promotes stigma clogging with self-pollen, flowers from both treatments, natural and supplemental pollination, were under the same effect of its visits, thereby masking negative effects. Finally, we observed that *A. mellifera* drives wild bees away from flowers; so, its presence could diminish wild bee visitation to *A. fissuratus* flowers.

Overall, massive and synchronous flowering period, the xenogamous breeding system, and the outcrossing mating system of *A. fissuratus* makes this species completely dependent on attracting enough pollinators to perform sexual reproduction. The high density of the CCB studied population permits a high floral display, assuring pollinator attraction and availability of mates for outcrossing. For conserving natural populations of *A. fissuratus*, it is fundamental to consider a threshold of minimum density. For that reason, any human activity (e.g., illegal extraction, change land use) that diminish the plant population density would have negative effects on its sexual reproduction.

Assemblage of floral visitors varied across seasons, but *A. fissuratus* flowers are clearly adapted to bee pollination, showing a functional specialization to this group (Ollerton et al. 2007). However, the role as pollinator of *A. mellifera* is difficult to define, as they can be important pollinators (because of its high frequency), but they can also remove large amounts of pollen without pollinating and hence negatively interfere with other perhaps better pollinators. Thus, for the conservation of healthy populations of the living rock cactus, to maintain both a large wild bee fauna and ecological bee availability is fundamental. Studying and evaluating wild bee abundance is still technically challenging but is undoubtedly urgent in the context of future pollinator decline (Vanbergen et al. 2013). Because its ubiquity, disentangling the performance of *A. mellifera* as pollinator is crucial for conserving pollination networks in the CCB and in similar arid and semiarid ecosystems.

References

Arroyo-Cosultchi G, Cárdenas-Ramos D et al (2014) Desaparición de una población de *Ariocarpus retusus* Scheidw. por la destrucción de su hábitat en Miquihuana, Tamaulipas, México. Cact Suc Mex 59:52–63

Ávalos-Hernández O, Cano-Santana Z et al (2016) Diversity and resource use patterns of Anthophile insects in Cuatro Ciénegas, Coahuila, México. Environ Entomol 45:1386–1397

Byers DL, Meagher TR (1997) A comparison of demographic characteristics in a rare and a common species of *Eupatorium*. Ecol Appl 7:519–530

Cesaro AC, Barrett SCH et al (2004) An experimental evaluation of self-interference in *Narcissus assoanus*: functional and evolutionary implications. J Evol Biol 17:1367–1376

Cruden RW (1977) Pollen-ovule ratios: a conservative indicator of breeding systems in flowering plants. Evolution 31:32–46

Eggli U, Giorgetta M (2015) Flowering phenology and observations of the pollination biology of South American Cacti. 2. *Cereus aethiops*. Haseltonia 21:19–40

Fagua JC, Ackerman JD (2011) Consequences of floral visits by ants and invasive honeybees to the hummingbird-pollinated, Caribbean cactus *Melocactus intortus*. Plant Spec Biol 26:193–204

Gentry AH (1974) Flowering phenology and diversity in tropical Bignoniaceae. Biotropica 6:64–68

Kearns CA, Inouye DW (1993) Techniques for pollination biologists. University Press of Colorado, Boulder, CO, USA

Langley LA (2015) The pollination ecology of *Sclerocactus brevihamatus* ssp. *tobuschii* (Cactaceae) in the Edwards Plateau Region of West-Central Texas. Master Thesis. Faculty of the College of Graduate Studies. Angelo State University, TX, USA

Larson BMH, Barrett SCH (2000) A comparative analysis of pollen limitation in flowering plants. Biol J Linn Soc 69:503–520

Mancilla-Ramírez MR, Flores-Vázquez JC, Martínez-Peralta C, Aguilar-Morales G, Mandujano MC (2020) Cuatro Ciénegas as a refuge for the living rock cactus, *Ariocarpus fissuratus*: demographic and conservation studies. In: Mandujano MC, Pisanty I, Eguiarte LE (eds.) Plant diversity and Ecology in the Chihuahuan desert. Springer International, Cham, Switzerland

Mandujano MC, Verhulst JAM et al (2007) Population dynamics of *Ariocarpus scaphirostris* Bödeker (Cactaceae): evaluating the status of a threatened species. Int J Plant Sci 168:1035–1044

Mandujano MC, Carrillo-Angeles IG et al (2010) Reproductive biology of Cactaceae. In: Ramawat KG (ed) Desert plants: biology and biotechnology. Springer-Verlag. Berlin, Heidelberg, Germany, pp 198–230

Martínez-Peralta C, Mandujano MC (2011) Reproductive ecology of the endangered living rock cactus, *Ariocarpus fissuratus* (Cactaceae). J Torrey Bot Soc 138:145–155

Martínez-Peralta C, Mandujano MC (2016) Pollen limitation in a rare cactus with synchronous mass flowering. Haseltonia 22:2–8

Martínez-Peralta C, Márquez-Guzmán J, Mandujano MC (2014) How common is self-incompatibility across species of the herkogamous genus *Ariocarpus*? Am J Bot 101(3):1–9

Martínez-Peralta C et al (2018) Polinizadores de cactáceas globosas en sitios con disturbio antropogénico. In: Ramírez-Bautista A, Pineda-López R (eds) Ecología y conservación de fauna en ambientes antropizados. REFAMA, UAQ, CONACyT, Queretaro city, Queretaro, Mexico, pp 347–367

Ollerton J, Killick A et al (2007) On the many ways to be a generalist flower. Taxon 56:717–728

Olmos-Lau VR, Mandujano MC (2016) An open door for illegal trade: online sale of *Strombocactus disciformis* (Cactaceae). Nature Conserv 15:1–9

Ortega-Baes P, Saravia M et al (2011) Reproductive biology of *Echinopsis terscheckii* (Cactaceae): the role of nocturnal and diurnal pollinators. Plant Biol 13:33–40

R Core Team (2018) R: a language and environment for statistical computing. R Foundation for Statistical Computing, Vienna, Austria. https://www.R-project.org/.

Schlindwein C, Wittman D (1995) Specialized solitary bees as effective pollinators of South Brazilian species of *Notocactus* and *Gymnocalycium* (Cactaceae). Bradleya 13:25–34

Silva L, Furtado-Dias E et al (2015) Towards a more holistic research approach to plant conservation: the case of rare plants on oceanic islands. Ann Bot-London 7:1–18

Souza V, Eguiarte LE (2018) In the beginning, there was fire: Cuatro Ciénegas Basin (CCB) and the long history of life on earth. In: Souza V et al (eds) Cuatro Ciénegas ecology, natural history and microbiology. Springer, Cham, Switzerland, pp 21–33

Souza V, Eguiarte LE et al (2018) A microbial saga: how to study an unexpected hot spot of microbial biodiversity from scratch? In: Souza V et al (eds) Cuatro Ciénegas ecology, natural history and microbiology. Springer, Cham, Switzerland, pp 1–20

Vanbergen AJ et al (2013) Threats to an ecosystem service: pressures on pollinators. Front Ecol Environ 11:251–259

Walck JL et al (2001) Why is *Solidago shortii* narrowly endemic and *S. altissima* geographically widespread? A comprehensive comparative study of biological traits. J Biogeogr 28:1221–1237

Chapter 13
Effect of Reproductive Modes on the Population Dynamics of an Endemic Cactus from Cuatro Ciénegas

María Dolores Rosas Barrera, Jordan Golubov, Irene Pisanty, and Maria C. Mandujano

Abstract Desert plants have several adaptations to cope with harsh environmental conditions; these adaptations range from the ability to store water to the diversification of reproductive modes, displaying sexual reproduction and clonality. *Grusonia bradtiana* is an endemic cactus from Coahuila which dominates vegetation at *bajada* landforms of Cuatro Ciénegas. The species reproduces sexually by seeds and clonally by fission of stems, allowing the parent plant to create genetically identical offspring. The type of recruitment influences the structure and dynamics of the population as well as the life history attributes that determine population growth. Negative environmental impacts can modify the reproductive success of plants; genotypic variation and sexual reproduction are reduced under stressful conditions while clonality is favored. Furthermore, the amount and quality of offspring via sexual reproduction decreases with high clonality, as inbreeding has negative genetic effects or if species are self-incompatible. However, clonality allows population growth or stability in harsh environments, although it may limit genotypic diversity. We studied the effect of sexual and clonal reproductive modes on the population structure and dynamics of *G. bradtiana* in two contrasting populations with different management history. The populations were located at Sierra de la Fragua [FRA-disturbed site], and at Sierra de San Marcos y Pinos [SMP-conserved site] in Cuatro Ciénegas Basin. Population structure of *G. bradtiana* is similar

M. D. Rosas Barrera (Deceased) · M. C. Mandujano (✉)
Instituto de Ecología, Universidad Nacional Autónoma de México, UNAM, Mexico City, Mexico
e-mail: mcmandujano@iecologia.unam.mx

J. Golubov
Department El hombre y su ambiente, Universidad Autónoma Metropolitana Xochimilco, Mexico City, Mexico

I. Pisanty
Department Ecología y Recursos Naturales, Facultad de Ciencias, Universidad Nacional Autónoma de México UNAM, Mexico City, Mexico

M. C. Mandujano et al. (eds.), *Plant Diversity and Ecology in the Chihuahuan Desert*, Cuatro Ciénegas Basin: An Endangered Hyperdiverse Oasis,
https://doi.org/10.1007/978-3-030-44963-6_13

between studied populations, despite their management history. In contrast, we found a higher density of individuals of *G. bradtiana* at SMP population, where clonal reproduction predominated, compared to FRA. According to our predictions, sexual fecundity was higher in individuals of all reproductive categories in SMP, where the sexual reproductive strategy predominates, while production of seeds was lower at FRA population due to strong inbreeding depression in the species. Interestingly, combination of both reproductive strategies allows populations to remain numerically stable.

Keywords Clonality · Disturbance · *Grusonia* · Population ecology · Sexual reproduction

Introduction

Plants display an enormous diversity of life history strategies that are partly the result of the variation in their demographic traits (Stearns 1992). Studying plant life histories through their demography allows calculation of different population parameters (Stearns 1992) and provides estimates of their conservation status and the role played by different environmental conditions on reproductive strategies and population success (Mandujano et al. 2015). Demographic data of organisms, structured in either states, sizes, or ages, is usually represented by life cycle graph models (transition matrix or life tables; Stearns 1992; Caswell 2001; Fig. 13.1), which are used to estimate population growth rate, compare between life history attributes under different environmental scenarios, or establish guidelines for species management (Silvertown et al. 1993; Caswell 2001; Mandujano et al. 2001).

Desert plants inhabit under extreme environmental conditions of temperature and unpredictable rainfall regimes that strongly influence the species' birth and death rates along their lifetime, generating a widely varying demographic behavior in populations (Polis 1991; Mandujano et al. 2001, 2007). In these environments there are a number of succulent plant species that have developed adaptations such as diversification of their reproductive modes, because recruitment through sexual means is usually the main demographic bottleneck along the life cycle (Turner et al. 1966; Bravo-Hollis 1978; Mandujano et al. 2001), and, therefore, populations are usually maintained by clonal recruitment (Mandujano et al. 2007; Carrillo-Ángeles et al. 2011). Production of sexual and clonal offspring results in a widely scheme of demographic strategies that modify survival and growth (Mandujano et al. 2001). In addition, how the vital rates (survival, growth, and reproduction) are modified by environmental clues can point towards the role

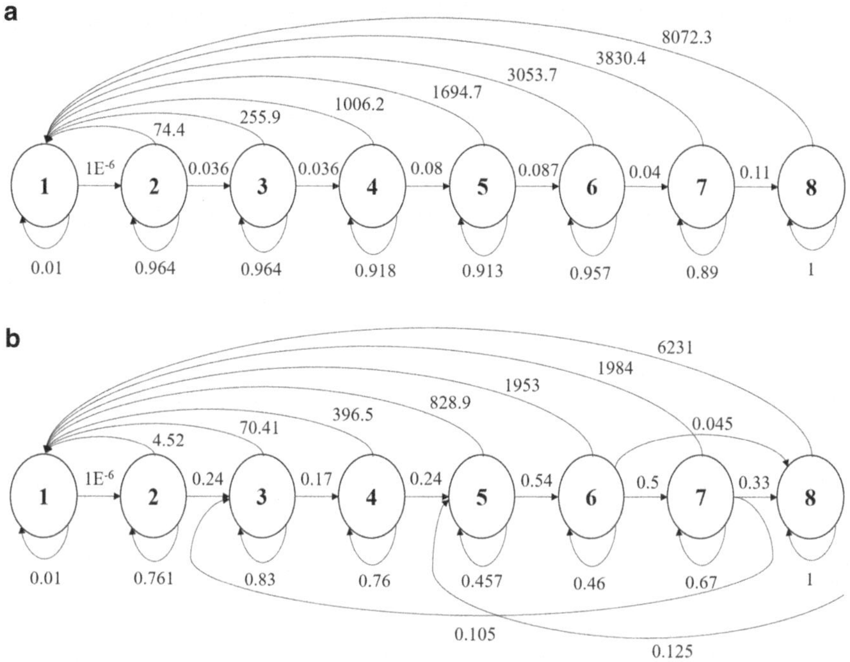

Fig. 13.1 Life cycle of *Grusonia bradtiana*. Demographic data were obtained at Cuatro Ciénegas Basin in the Chihuahuan desert, Coahuila, Mexico. The parameters shown correspond to the matrix at (**a**) SMP-conserved site and (**b**) FRA-disturbed site (2000–2001). Nodes represent size classes (1 seed class, 2 to 8 in number of branches: 1–30, 31–100, 101–200, 201–300, 301–450, 451–600, >600); the lower figure is the probability of stasis; the upper figure (bold) is average individual fecundity (number of seeds·individual^{-1}·year^{-1}). Shorter arrows between nodes contain probabilities of growth to the next size class. For (b) Lower arrows connecting nodes contain average individual contributions of clonality (clonal recruits·individual^{-1}·year^{-1})

played by environmental heterogeneity on population dynamics (Stearns 1992; Caswell 2001).

Dominance of some type of recruitment markedly affects populations. For example, clonal offspring have higher rates of recruitment than sexual offspring, but clonal propagules do not form banks. On the other hand, sexual seedlings are recruited with less frequency, but seed may form seed banks which can wait opportunity windows for establishment when conditions are favorable (Mandujano et al. 2001, 2007; García-Morales et al. 2018). Clonality allows for fast growth, which increases survival and reproductive potential, reduces generation time, maintains genetic combinations (Cook 1985; Mandujano et al. 2007), and does not incur in the costs of sex (Crow 1986). These attributes make clonality advantageous in stressful or highly variable environments (Cook 1979). However, clonality dilutes genetic structure by generating identical progeny, as found by

García-Morales et al. (2018) for *Opuntia microdasys* (Cactaceae), a species in which juveniles of several cohorts belong to a single or a few genotypes. Clonality does not increase genetic variation unless mutation accumulates within individual clones, and as such may lead to the complete loss of sexual reproduction if mutations that compromise the latter arise (Eriksson 1992; Eckert et al. 1999) or in species that have strong inbreeding depression or incompatibility which will reduce sexual reproductive success (Mandujano et al. 1996; Charpentier 2002; Carrillo-Ángeles et al. 2011). Genetic homogeneity of a clonal population has the disadvantage of being susceptible to disease (Cook 1979, 1985; Silvertown and Lovett-Doust 1993) as exemplified in elms (Burdon and Shattock 1980) and *Agave tequilana* (Eguiarte et al. 2015).

However, most species in natural populations usually combine sexual and clonal reproduction, which gives rise to mixed reproductive strategies that greatly vary among populations (Carrillo-Ángeles et al. 2011). Population dynamics reflects the predominant type of recruitment, influenced by resources and biotic interactions (Mandujano et al. 2001). In addition, there are several human induced factors that determine establishment, survival and, therefore, the future of desert plants. Human disturbance such as habitat destruction, cattle rising, agriculture, and resource extraction in desert lands are the major threats for many species, cacti included (Saunders et al. 1991; Novoa et al. 2014; Goettsch et al. 2015). Unfortunately, desert ecosystems have low resilience and slow or null means to recover after strong disturbance (Macmahon 1988; Thomey et al. 2014).

Grusonia bradtiana (old man cactus) is a unique clonal cactus, endemic to the Chihuahuan desert, locally abundant in the lower slopes of calcareous soils in Coahuila Mexico (Mandujano and Golubov 2000). Columnar stems of the species are covered with trichomes and spines, and it is commonly distributed as a dominant component of the flora in *bajadas* (Fig. 13.2, Photo, Flores-Vázquez et al. 2020, this volume). This species has horizontal branches formed of long chains of stems, so plants prostrate, creeping along the ground and forming huge masses of stems with the tips that can reach up to 1 m height pointing upward. Stem segments are light green and 4 to 7 cm in diameter, with 8 to 10 longitudinal ribs, 15 to 20 spines per areole (Bravo-Hollis 1978). Only stems with tips pointing upward produce yellow actinomorphic flowers, and ellipsoidal fruits (Fig. 13.2a–c). The species has a sexual system of partial self-incompatibility that determines fruit production (Plasencia-López et al. 2020, this volume), and the plant is non-palatable for most herbivores, including cattle. In CCB, the populations of *G. bradtiana* are found in areas that have a long history of management, including extraction of candelilla wax from *Euphorbia antisyphilitica* (Fig. 13.3), cattle ranching, agriculture, and human settlements (Martínez-Ballesté and Mandujano 2013; Martínez-Ballesté et al. 2020, this volume). In this chapter, we estimate the impact of disturbance regimes on seed production, population structure, dynamics, and life history attributes of *G. bradtiana* using life table models.

Fig. 13.2 (**a**) *Grusonia bradtiana* grows in the landform *bajada*, gentle slopes at foot of hills and mountains, on gravel soils, and form thick stands where the crassicaule scrubland vegetation type dominates. (**b**) The cactus is a prickly plant with cylindrical or short columnar chain of stems; it has spiny floral buds that are produced in branches with the tips pointing upward. (**c**) Flower of *Grusonia bradtiana* with bright yellow perianth (corolla) and lobulated stigma, stamens with red filaments which move inward when pollinators enter a flower. (Photos: a by Helga Ochoterena, and b and c by María C. Mandujano)

Methods

Demography and Life History

Populations of *G. bradtiana* were studied with a demographic approach at two representative *bajadas* that had contrasting management history and degree of chronic disturbance, Sierra de San Marcos y Pinos (hereafter SMP-conserved site) and

Fig. 13.3 *Euphorbia antisyphilitica* in bloom, a native species from the Chihuahuan desert, abundant at Cuatro Ciénegas. The species is used to extract high quality wax for cosmetic, alimentary, pharmaceutic and industrial products. (Photo by Eder Ortiz-Martínez)

Sierra de la Fragua (hereafter FRA-disturbed site). The study sites were located at the protected area known as Area de Protección de Flora y Fauna de Cuatro Ciénegas, in the state of Coahuila, northern Mexico (26° 45′ to 27° 05′ N and 102° 05′ to 102° 20′ W), with a mean annual precipitation of 211 mm year^{-1} and a mean temperature of 21.9 °C (Montiel-González et al. 2018). At each population three permanent plots, each measuring 20 × 50 m, were established and all individuals of *G. bradtiana* were mapped and tagged. For each individual, plant cover and height (cm), total number of stems and branches, new stems produced, flowers and fruits were recorded through 2000–2001.

SMP-conserved site had a single owner who keeps livestock at low intensity (15 heads of cattle/ha) until the land was bought by a conservation organization in 2000, at this time, all cattle was removed. The population at FRA-disturbed site has been in use for over two centuries, and the most recent human settlements date from 1940. Here, the extraction of candelilla wax with dirt access roads used by pack animals, cattle and human activities has impacted vegetation and soil (Martínez-Ballesté and Mandujano 2013). Even though *G. bradtiana* is the dominant species in both sites, other common species include: *Larrea tridentata, Agave lechuguilla, Acacia* sp., *Cylindropuntia kleiniae, Jatropha dioica, Hechtia scariosa* and *Cylindropuntia leptocaulis* in the SMP site, and *Fouquieria splendens* in the FRA site (Pinkava 1984; Flores-Vázquez et al. 2020, this volume; Martínez-Ávalos et al. 2020, this volume).

Martorell and Peters (2005) propose that anthropogenic chronic disturbance can be estimated by several indicators such as land use, soil compaction, presence of animal feces, and proximity to human settlements, among others. We estimate disturbance by applying a principal components analysis (Kachigan 1986) on several disturbance agents and found that all plots from SMP-conserved site (left side of Fig. 13.4) are similar among each other and had less chronic disturbance than plots from FRA-disturbed site (right side, Fig. 13.4). The most important parameters to define chronic disturbance at SMP are livestock trails and there was also an effect on adjacency to human settlements. Disturbance at FRA is determined by the presence of vegetation islands, and number and cover of human roads, erosion, and number of livestock trails, and there was also an effect on adjacency to human settlements, density of cattle feces, and browsing (Fig. 13.4).

Populations of *G. bradtiana* were structured into one stage class (seeds) and seven size categories in terms of total number of branches (1–30, 31–100, 101–200, 201–300, 301–450, 451–600, >600, Fig. 13.1). Size structures between sites were similar ($\chi^2 = 0.0012$, df = 6, $P > 0.05$) (Fig. 13.5). Fecundity was calculated as the mean number of fruits per capita × mean number of seeds per fruit (calculated from 50 collected fruits at each site and eliminating aborted seeds). The population growth rate (λ) of *G. bradtiana* at each site was calculated using matrix population models. The contribution of each demographic process was determined through prospective analysis of sensitivity and elasticity (Caswell 2001) in R with popbio

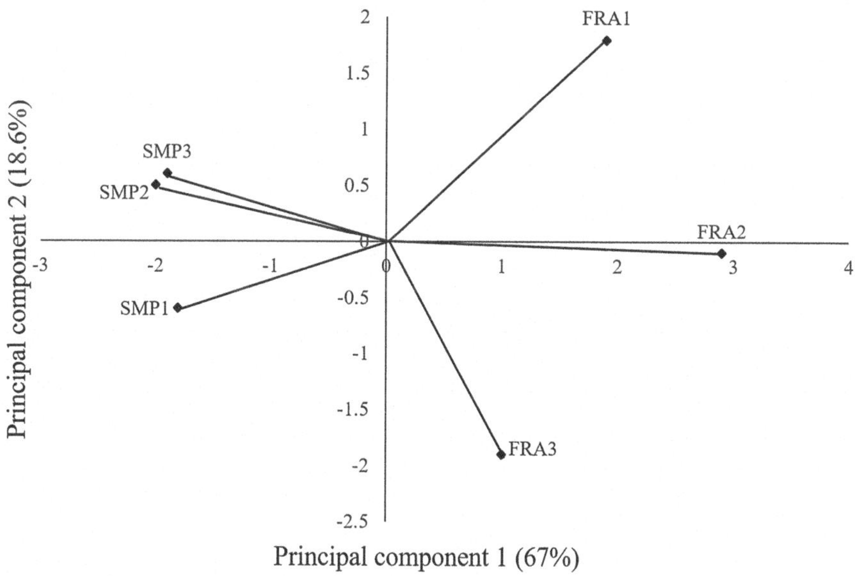

Fig. 13.4 Principal component analysis of different variables to estimate chronic disturbance on 6 permanent plots at Cuatro Ciénegas Basin. First principal component explains 67.7% and second component explains 18.6%, overall 86.3% of the total variance. The most important parameters to define chronic disturbance are number and cover of human roads, erosion, and number of livestock trails, and there was also an effect on adjacency to human settlements, density of cattle feces, and browsing. Plots at Sierra de San Marcos y Pinos (SMP 1–3) are conserved sites and plots at Sierra de la Fragua (FRA 1–3) are the disturbed sites. The demographic data of *Grusonia bradtiana* were gathered at both sites from 2000 to 2001

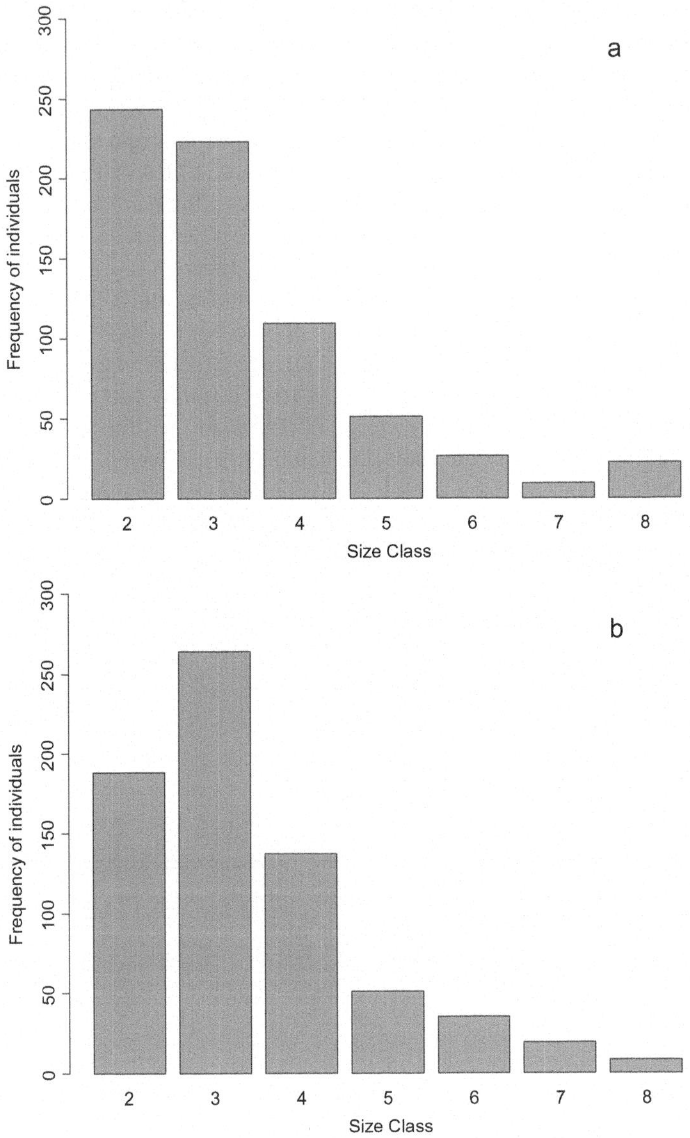

Fig. 13.5 Population structure of *Grusonia bradtiana* at two sites in Cuatro Cienégas Basin with different anthropogenic chronic disturbance. (**a**) Sierra de San Marcos y Pinos (SMP) is a conserved site and (**b**) Sierra de la Fragua (FRA) disturbed site. (data from demographic census from 2000 and 2001)

(Stubben and Milligan 2007), which are common demographic tools to estimate population trends and contribution of demographic process and types of recruitment to population dynamics.

Results and Discussion

The density of *G. bradtiana* individuals is amazingly high (SMP 2280 ± s.e. 9 ind/ha, FRA 2350 ± s.e. 65 ind/ha), and abundance of plants decreased with plant size (Fig. 13.5). The biggest plants were tagged at FRA with an average of 1503 stems per plant in comparison to 1219 stems at SMP (last size class, Figs. 13.1 and 13.5). The high density of these populations is frequent in prickly plants that dominate scrublands in the Chihuahuan Desert, such as *Opuntia rastrera* (Mandujano et al. 2001). Other prickly pear cacti have low population densities as the purple prickly pear, from 140 ind/ha in the population where recruitment relays upon seeds, and up to 600 ind/ha where it depends on clonality, that is, populations of *Opuntia macrocentra* with clonal recruitment have 4.2 times more density than populations that lack or have few clonal recruits (Mandujano et al. 2007). Fecundity of *G. bradtiana* increased with plant size and seeds were produced by all size classes (Fig. 13.1), but differed between sites, with a marked increase in all size categories in the SMP site ($P < 0.001$, Fig. 13.1). Of all seed samples ($N = 50$ fruits per site) 93.8% had a developed embryo in the SMP site and the FRA site had almost 20% less viable seeds (73.6%). Fruits have an average of 92.4 ± 10.1 seeds (Plasencia-López et al. 2020, this volume). The species fruit set and seeds/fruit ratio can vary among populations, but it can be difficult to find any seeds at all (Hamilton 1970); for example, up to 83% fruits can be empty and the rest can show a reduced seed count (26 ± 4.6 seeds/fruit), as found in another population (Mandujano and Golubov 2000). In addition, newly formed seeds do not germinate because they have an imposed dormancy period of around one year. Seeds 3.5 years old that were stored in laboratory conditions were sown in 1% bacteriological agar, 12 h photoperiod, at 25 °C germinated less than 5% (Rosas Barrera obs. pers.). Previous reports found that germination is slow and attains low percentages (from 1% to 10%) under different treatments (Hamilton 1970). The lack or reduced germination and the low seed numbers are due to a combined effect of inbreeding depression, pollen limitation, and florivory (Hamilton 1970; Mandujano and Golubov 2000; Plasencia-López et al. 2020, this volume).

The population models allowed to determine that both populations were at numerical equilibrium. We estimated a rate of population growth of $\lambda = 1.0051$ (95% confidence interval, 1.0789–0.9315) at SMP-conserved site, and $\lambda = 1.0848$ (1.1994–0.9702) at FRA-disturbed site [lambda, $\lambda = n_{t+1}/n_t$, where n_t is the population's size at time t or $t + 1$]. If $\lambda = 1$ the population is stable, $\lambda > 1$ indicates population increases and $\lambda < 1$ population decrease, (Caswell 2001)]. Fecundity had low contributions to λ, but there were important differences among the studied populations. Life cycles and prospective analyses showed that clonal reproduction predominates at FRA-disturbed site and sexual reproduction at SMP-conserved site. In both populations, survival was the demographic process that had the highest contribution towards lambda, followed by growth which was higher at the FRA-disturbed site (Table 13.1). Mortality was low at both populations and sexual fecundity (seed production) was higher at the SMP-conserved site. Bigger plants had the highest reproductive values for both populations, but FRA reproductive value is twice the reproductive value of individuals from SMP (Fig. 13.6). Reproduction occurs in early life cycle stages for both populations of *G. bradtiana* (Fig. 13.1), but more reproductive individuals were recorded at SMP (56.87% reproductive plants) than at FRA (36.46% reproductive plants). Plants at FRA-disturbed site are under stressful conditions, and higher clonality results in less fruit and seed production. Clonality hinders sexual reproduction, probably because geitonogamy and inbreeding depression are very high in this species (Carrillo-Ángeles et al. 2011; Plasencia-López et al. 2020, this volume).

The differences in chronic disturbance between SMP-conserved site and FRA-disturbed site are mainly due to the type and intensity of the disturbance agent. Main disturbance factors in SMP-conserved site were livestock trails and proximity to settlements, but as the site has limited human access, the latter play a minor role. This population has a better condition of vegetation and for *G. bradtiana* there is an increase in sexual reproduction and some seedling recruitment, favored by nurse plant availability, that would potentially generate higher genetic diversity. At FRA-disturbed site, disturbance factors were fragmentation of the vegetation, proximity to human settlements, and manmade paths all associated to the extraction of resources (mainly *Euphorbia antisyphilitica*) and in which *G. bradtiana* has a predominant clonal recruitment and redundancy of genotypes should be expected, as found in other predominantly clonal cacti species (García-Morales et al. 2018). The *bajadas* where *G. bradtiana* is distributed are diverse in endangered cacti species and are therefore targets for conservation (Pinkava 1984; Flores-Vázquez et al. 2020, this volume; Martínez-Ávalos et al. 2020, this volume). Both populations of *G. bradtiana* show a log normal population structure that have been interpreted as coming from self-perpetuating populations found in stable environments that have continuous recruitment (Whipple and Dix 1979). Undoubtedly the populations of *G. bradtiana* at both sites are in equilibrium with minimum recruitment occurring, a phenomenon that is common for most cacti species (Turner et al. 1966; Martorell and Peters 2005; Mandujano et al. 2015).

Table 13.1 Elasticity or contribution of different demographic process to rate of population increase (λ). Studied populations of *Grusonia bradtiana* are SMP (conserved population located at Sierra de San Marcos y Pinos) and FRA (disturbed population located at Sierra de la Fragua) at Cuatro Ciénegas Basin, Coahuila, Mexico

Population	Fecundity	Survival	Growth	Clonality
SMP-conserved stie	0.00365192	0.9718677	0.02451674	0
FRA-disturbed site	9.74×10^{-4}	7.78×10^{-1}	1.67×10^{-1}	5.36×10^{-2}

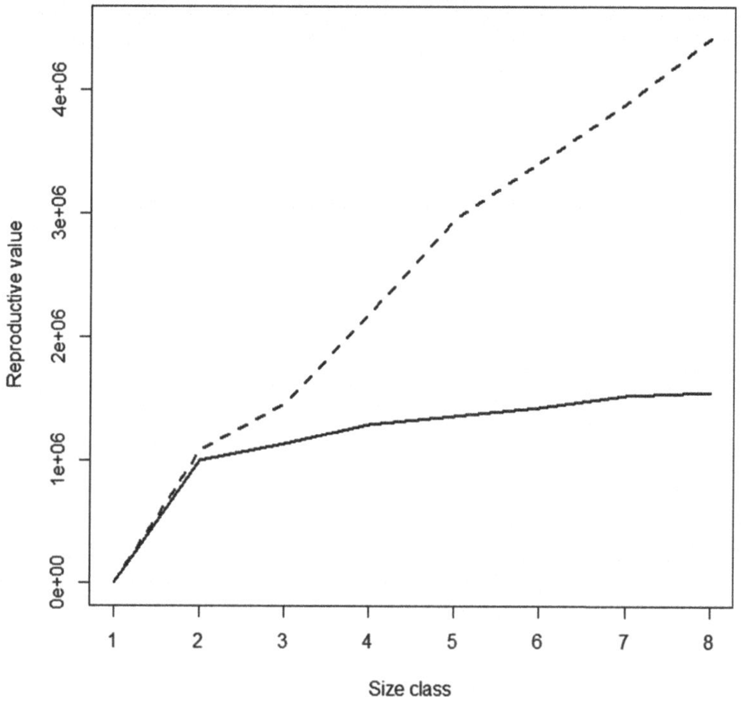

Fig. 13.6 Reproductive value of *Grusonia bradtiana* at two sites in Cuatro Cienégas Basin with different anthropogenic chronic disturbance. Sierra de San Marcos y Pinos (SMP-conserved site, solid line) and Sierra de la Fragua (FRA-disturbed site, dotted line)

Even though the environment is heterogeneous and determined most population trends, the chronic anthropogenic disturbance affected traits related with reproductive performance and population density. The populations of *G. bradtiana* show a common population structure; however, the means by which this structure is achieved differs by the reproductive strategy (sexual or clonal) that predominates at each site. Offspring are commonly clonal in FRA-disturbed site while sexual is in SMP-conserved site. The number of small individuals reflected in the population structure indicates either some recruitment of new individuals, or propagule survival or permanence on the same stage for prolonged time periods (Mandujano et al. 2001). Interestingly, despite the difference in population density, there is a common population structure across disturbance regimes. This is unexpected, for previous studies normally report differing population structures at each habitat or disturbance regime for example, Parker (1987) in *Stenocereus thurberi*, Bullock et al. (1994) in *Cirsium vulgare*, Mandujano et al. (2001) in *Opuntia rastrera*, Mandujano et al. (2007) in *Opuntia macrocentra*, and Silva (1996) in *Pachycereus pringlei*. The disturbance regime seems to affect other components of the life history of *G. bradtiana*, like an increase in fecundity at SMP-conserved site, as well as an increase in density at FRA-disturbed site and a reduction of fruit and seed set in the latter.

Recruitment through sexual means brings long-term fitness advantages with just the ability of purging deleterious alleles rapidly (Silvertown and Lovett-Doust 1993), accelerated evolutionary rates (Maynard Smith 1978), spread of beneficial mutations, de novo genetic combinations (Crow 1986), and long distance dispersal (Cook 1985; Stearns 1987). There is, however, a cost in producing the structures for sexual reproduction (Crow 1986; Silvertown and Lovett-Doust 1993; Silvertown et al. 1992), the seedling stage is usually linked to high mortality rates (Charlesworth 1980), and the genetic combinations might not be adaptive.

In clonal species, there is a short-term trade-off between prevalence of sexual reproduction and clonality (Stearns 1992; Silvertown et al. 1993). Parent plants can assign resources towards clonality or sexual reproduction reaching possible optimum combinations (Harada and Iwasa 1996). When sexual reproduction is absent or low, clonality is often enough to ensure the stability and, often, the growth of a population (Cook 1979, 1985; Silvertown et al. 1992, 1993). This trade-off represents an important limitation towards the evolution of life histories because how individuals solve the sexual-clonal conflict will ultimately determine fitness (Silvertown and Lovett-Doust 1993).

Natural selection will ultimately favor genotypes capable of optimizing the balance between the costs and benefits of sexual and clonal reproduction (Silvertown and Lovett-Doust 1993; Mandujano et al. 2001, 2007). The diversity of life history strategies is a consequence of the temporal and spatial variation in ecological conditions (Patridge and Harvey 1988). The optimal phenotypic expression of reproduction in *G. bradtiana* differs between disturbance regimes with an increase of sexual reproduction in SMP-conserved site and a more vigorous clonality in FRA-disturbed site, a strategy that has allowed *G. bradtiana* to be dominant species of the Cuatro Ciénegas Basin (Flores-Vázquez et al. 2020, this volume, Martínez-Ávalos et al. 2020, this volume). The high density of *G. bradtiana* in FRA probably indicates a species that can show a ruderal life history and stress tolerant one in the less disturbed sites (Silvertown et al. 1992; Martorell and Peters 2005). Anthropogenic chronic disturbance evidently tends to favor clonal propagation rather than sexual reproduction in *G. bradtiana*.

Acknowledgements Financial support was provided by projects IN205500 PAPIIT-DGPA UNAM and National Council of Science and Technology (CONACyT) project 221362 to MC Mandujano. Luis E. Eguiarte for comments on previous versions of the chapter and Rosa Mancilla for doing life cycle figures. PRONATURA provided free stay at La Becerra, Cuatro Ciénegas, during our field trips. Personnel from the Area de Protección de Flora y Fauna Cuatrociénegas provided logistic support. We appreciate the help during field work to Manuel Rosas, Luci Plasencia, Nicolas Palleiro, and Gisela Aguilar Morales. In memoriam of our dearest friend and student Dolores Rosas Barrera (1972–2018).

References

Bravo-Hollis H (1978) Las Cactáceas de México, vol 1. Universidad Nacional Autónoma de México, México, DF, Mexico

Bullock JM, Clear Hill B, Silvertown J (1994) Demography of *Cirsium vulgare* in a grazing experiment. J Ecol 82:101–111

Burdon JJ, Shattock RC (1980) Disease in plant communities. Appl Biol 5:145–220

Carrillo-Ángeles IG, Mandujano MC, Golubov J (2011) Influences of the genetic neighborhood on ramet reproductive success in a clonal desert cactus. Popul Ecol 53:449–458

Caswell H (2001) Matrix population models. Construction, analysis, and interpretation. Sinauer Associates, Inc, Sunderland, MA, USA

Charlesworth B (1980) The cost of sex in relation to mating system. J Theoretical Biol 84:655–671

Charpentier A (2002) Consequences of clonal growth for plant mating. Evol Ecol 15:521–530

Cook RE (1979) Asexual reproduction: a further consideration. Am Nat 113:769–772

Cook RE (1985) Growth and development in clonal plant populations. In: Jackson JBC, Buss LW, Cook RE (eds) Population biology and evolution of clonal organisms. Yale University Press, New Haven, CT, USA, pp 259–296

Crow JF (1986) Basic concepts in population, quantitative, and evolutionary genetics. W.H. Freeeman, New York, NY, USA

Eckert CG, Dorken ME, Mitchell SA (1999) Loss of sex in clonal populations of a flowering plant, *Decodon verticillatus* (Lythraceae). Evolution 53:1079–1092

Eguiarte LE, Sheinvar E, Golubov J, Mandujano MC (2015) Ecología y genética del maguey en México. Parte III La naturaleza. In: Vera JL, Fernández R (eds) Agua de las verdes matas. Tequila y Mezcal. Artes de México y del Mundo y el Instituto Nacional de Antropología e Historia, México, DF, Mexico, pp 183–212

Eriksson O (1992) Evolution of seed dispersal and recruitment in clonal plants. Oikos 63:439–448

Flores-Vázquez JC, Rosas-Barrera MD, Sánchez-Gallen I, Mandujano MC (2020) Ecological importance of Bajadas in the Chihuahuan desert. In: Mandujano MC, Pisanty I, Eguiarte LE (eds) Plants diversity and ecology in the Chihuahuan desert. Cuatro Ciénegas Basin, Chihuahuan desert, Mexico. an endangered hyperdiverse oasis. Springer International Publishing AG, part of Springer Nature, Cham, Switzerland

García-Morales E, Carrillo-Ángeles IG, Golubov J, Piñero D, Mandujano MC (2018) Influence of fruit dispersal on genotypic diversity and migration rates of a clonal cactus from the Chihuahuan desert. Ecol Evol 8:12559–12575

Goettsch B, Hilton-Taylor C, Cruz-Piñon G et al (2015) High proportion of cactus species threatened with extinction. Nature Plants 1:15142. https://doi.org/10.1038/NPLANTS.2015.142

Hamilton MW (1970) Seedling development of *Grusonia bradtiana* (Cactaceae). Am J Bot 57:599–603

Harada Y, Iwasa Y (1996) Analyses of spatial patterns and population processes of clonal plants. Res Popul Ecol 38:153–164

Kachigan SK (1986) Statistical analysis: an interdisciplinary introduction to univariate and multivariate methods, 2nd edn. Radius Press, New York, NY, USA

MacMahon JA (1988) Warm deserts. In: Barbour MG, Billings WD (eds) North American terrestrial vegetation. Cambridge University Press, New York, NY, USA, pp 231–264

Mandujano MC, Golubov J (2000) *Grusonia bradtiana* en la zona calcárea del Bolsón de Mapimí. Cact Suc Mex 45:66–68

Mandujano MC, Montaña C, Eguiarte LE (1996) Reproductive ecology and inbreeding depression in *Opuntia rastrera* (Cactaceae) in the Chihuahuan desert: why are sexually derived recruitments so rare? Am J Bot 83:63–70

Mandujano MC, Montaña C, Franco M, Golubov J, Flores-Martinez A (2001) Integration of demographic annual variability in a clonal desert cactus. Ecology 82:344–359

Mandujano MC, Golubov J, Huenneke L (2007) Effect of reproductive modes and environmental heterogeneity in the population dynamics of a geographically widespread clonal desert cactus. Popul Ecol 49:141–153

Mandujano MC, Bravo Y, Verhulst J, Carrillo-Angeles I, Golubov J (2015) The population dynamics of an endemic collectible cactus. Acta Oecol 63:1–7

Martínez-Ávalos JG, Martínez-Gallegos R, Guerra-Pérez A et al (2020) Diversity and distribution of cacti species in the Cuatro Ciénegas Basin. In: Mandujano MC, Pisanty I, Eguiarte LE (eds) Plants diversity and ecology in the Chihuahuan desert. Cuatro Cienégas Basin, Chihuahuan desert, Mexico. an endangered hyperdiverse oasis. Springer International Publishing AG, part of Springer Nature, Cham, Switzerland

Martínez-Ballesté A, Mandujano MC (2013) The consequences of harvesting on regeneration of a non-timber wax producing species (*Euphorbia antisyphilitica* Zucc.) of the Chihuahuan desert. Econ Bot 67:121–136. https://doi.org/10.1007/s12231-013-9229-4

Martínez-Ballesté A, Iglesias Chacón T, Mandujano MC (2020) Between the arid and the opulent; plant resources of the Mexican desert. In: Mandujano MC, Pisanty I, Eguiarte LE (eds) Plants diversity and ecology in the Chihuahuan desert. Cuatro Ciénegas Basin, Chihuahuan desert, Mexico. an endangered hyperdiverse oasis. Springer International Publishing AG, part of Springer Nature, Cham, Switzerland

Martorell C, Peters E (2005) The measurement of chronic disturbance and its effects on the threatened cactus *Mammillaria pectinifera*. Biol Conserv 124:199–207

Maynard Smith J (1978) The evolution of sex. Cambridge University Press, Cambridge, UK

Montiel-González G, Bautista F, Delgado C, García-Oliva F (2018) The climate of Cuatro Ciénegas Basin: drivers and temporal patterns. In: Souza V, Olmedo-Álvarez G, Eguiarte LE (eds) Cuatro Ciénegas ecology, natural history and microbiology, Cuatro Ciénegas Basin: an endangered hyperdiverse oasis. Springer International Publishing AG, part of Springer Nature, Cham, Switzerland, pp 35–42

Novoa A, Le Roux JJ, Robertson MP, Wilson JRU, Richardson DM (2014) Introduced and invasive cactus species: a global review. AoB Plants 7:1–14

Parker KC (1987) Seedcrop characteristics and minimum reproductive size of organ pipe cactus (*Stenocereus thurberi*) in southern Arizona. Madroño 34:294–303

Patridge L, Harvey PH (1988) The ecological context of life-history evolution. Science 241:1449–1455

Pinkava DJ (1984) Vegetation and Flora of the Bolsón of Cuatro Ciénegas region, Coahuila, México: IV. Summary, endemism and corrected catalogue. J Arizona-Nevada Acad Sci 19:23–47

Plasencia-López L, Rojas-Aréchiga M, Mandujano MC (2020) Reproductive biology of *Grusonia bradtiana* (Cactaceae): a dominant species and endemic clonal Cactus from Cuatro Ciénegas Basin and contiguous areas in the Chihuahuan desert. In: Mandujano MC, Pisanty I, Eguiarte LE (eds) Plants diversity and ecology in the Chihuahuan desert, Cuatro Ciénagas Basin, Chihuahuan desert, Mexico. an endangered hyperdiverse oasis. Springer International Publishing AG, part of Springer Nature, Cham, Switzerland

Polis GA (1991) Desert communities: an overview of patterns and processes. In: Polis GA (ed) The ecology of desert communities. The University of Arizona Press, Tucson, AZ, USA

Saunders DA, Hobbs RJ, Margules CR (1991) Biological consequences of ecosystem fragmentation. Conserv Biol 5:18–32

Silva PC (1996) Demografía comparativa de *Pachycereus pringlei* en dos unidades geomorficas contrastantes del paisaje en Baja California Sur, México. Masters's thesis, Universidad Nacional Autónoma de México, México, DF, Mexico

Silvertown J, Lovett-Doust J (1993) Introduction to plant population biology. Blackwell Scientific Publications, Oxford, UK

Silvertown J, Franco M, McConway K (1992) A demographic interpretation of Grime's triangle. Funct Ecol 6:130–136

Silvertown J, Franco M, Pisanty I, Mendoza A (1993) Comparative plant demography-relative importance of life-cycle components to the finite rate of increase in woody and herbaceous perennials. J Ecol 81:465–476

Stearns SC (1987) The evolution of sex and its consequences. Birkhäuser Verlag, Boston, MA, USA

Stearns SC (1992) The evolution of life histories. Oxford University Press, London, UK

Stubben C, Milligan B (2007) Estimating and analyzing demographic models. J Stat Soft 22:1–23

Thomey ML, Ford PL, Reeves MC, Finch DM, Litvak ME, Collins SL (2014) Review of climate change impacts on future carbon stores and management of Warm deserts of the United States. General Technical Report RMRS-GTR-316. U.S. Department of Agriculture, Forest Service, Rocky Mountain Research Station, Fort Collins, CO, USA

Turner RM, Alcon SM, Booth JA (1966) The influence of shade, soil and water on saguaro seedling establishment. Bot Gaz 127:95–102

Whipple SA, Dix RL (1979) Age structure and successional dynamics of a Colorado subalpine forest. Am Midl Nat 101:142–158

Chapter 14
Conservation Status, Germination, and Establishment of the Divine Cactus, *Lophophora williamsii* (Lem. Ex Salm-Dyck) J. M. Coult., at Cuatro Ciénegas

Maria C. Mandujano, Alejandra García Naranjo, Mariana Rojas-Aréchiga, and Jordan Golubov

Abstract *Lophophora williamsii,* commonly known as "peyote" or "divine cactus", has a wide distribution in the Chihuahuan Desert, but is under threat due to over-harvesting for religious ceremonies or psychedelic tourism, and many populations no longer exist. Natural populations can be rehabilitated or aided by sowing seeds or reintroducing specimens, which requires information on germination requirements and establishment conditions. We assessed seed germination and establishment to identify ecological factors that may determine the formation of seed banks, the effect of solar radiation on establishment, and seed aging after ex situ storage. This information can be used for conservation programs of peyote populations at Cuatro Ciénegas. Seeds collected from the wild and a sample of seeds kept for several years in a botanical garden's germplasm collection were used for experimental trials. Germination experiments included seeds of different ages (1–9 years old) and three different light conditions. Germination decreased with seed age and was higher under shade mesh conditions. Peyote seeds remained viable for several years, which suggest they may form a seed bank as well as survive storage at room temperature (20 ± 2 °C), with a germination decay of up to 25% indicating a loss of viability process. Although Cuatro Ciénegas Basin holds well preserved populations of the divine cactus, they may be threatened by biological and anthropogenic factors, so it is urgent to drive efforts towards the future conservation of this outstanding cactus.

M. C. Mandujano (✉) · M. Rojas-Aréchiga
Departamento de Ecología de la biodiversidad, Instituto de Ecología, Universidad Nacional Autónoma de México, UNAM, Mexico City, Mexico
e-mail: mcmandujano@iecologia.unam.mx

A. García Naranjo
Colectividad Razonatura A.C, Mexico City, Mexico

J. Golubov
Lab. Ecología, Sistemática y Fisiología Vegetal, Depto. El Hombre y Su Ambiente, Universidad Autónoma Metropolitana Xochimilco, Mexico City, Mexico

M. C. Mandujano et al. (eds.), *Plant Diversity and Ecology in the Chihuahuan Desert*, Cuatro Ciénegas Basin: An Endangered Hyperdiverse Oasis,
https://doi.org/10.1007/978-3-030-44963-6_14

Keywords Cactaceae · Establishment · Nurse plant · Peyote · Seed bank · Seed longevity

The Search of Peyote

The peyote, *Lophophora williamsii*, (Cactaceae), is a widespread species from the Chihuahuan Desert. It grows on both banks of the Rio Grande and in scattered places in Aguascalientes, Chihuahua, Durango, Coahuila, Hidalgo, Jalisco, Nuevo Léon, Querétaro, San Luis Potosí, Tamaulipas, and Zacatecas along the Chihuahuan Desert in Mexico (Anderson 2001). This species receives several common names according to different ethnic groups: peyote (Nahuatl), kamaba (Tepehuan), hicore, hikuli or jiculi (Huichol), huaname (Tarahumara), wokow (comanche), señi (kiowa), mescalito and "mescal buttons" (commercial name) (Schultes and Hoffman 1982).

When the crowns (stems) of peyote are cut off and dried, they form the so-called *mescal buttons* which are eaten in religious ceremonies. Several Indians of Mexico like Tarahumaras and Huicholes, and others from the southern plains make annual "pilgrimages" to gather it (Anderson 1980). Tribes too distant to visit the peyote-fields procure their supplies by mail from merchants in lower Texas who deal exclusively with mescal buttons (Schultes 1938; Trout and Terry 2016). The peyote is a North American cactus threatened to extinction due to over collection for its hallucinogenic properties, and its collection and consumption are allowed only for native groups which use it in magical-religious rituals; otherwise possession is a federal offense (Anderson 1980). Peyote was noticed by the Spaniards since 1560, as Jesuit missionaries of the XVII testify that the Mexican indigenous people used peyote for ceremonial and medicinal purposes. The Spaniards found that the consumption of peyote was firmly established within the native religions, and efforts to exterminate them displaced the cult towards the mountains, where it still exists nowadays (Schultes and Hoffman 1982) as part of a medicinal-religious cult. Through various hallucinations, the healer communicates with the malicious spirits which cause disease and death. In each ceremony 4 to 30 cacti are consumed per person. Currently the plant is prized by Tarahumaras, Huicholes, and other Mexican cultures, as well as by the members of the Native American Church in the USA and Western Canada (Schultes and Hoffman 1982; Bátis and Rojas-Aréchiga 2002). Peyote has been used for more than 7000 years, and evidence for the peyote cult has been found in Cuatro Ciénegas Basin (CCB), dating back to 810 to 1070 A.C. It was used in Monte Alban (Oaxaca) in 200 B.C. and in Colima in 100 B.C. (Schultes and Hoffman 1982). The narcotic properties of peyote have attracted wide attention for generations. There is a native culture with magic and religious practices which has a special law regime that regulates and allows its consumption by native tribes like Huicholes, Coras, and Tarahumaras in Mexico, and Navajos, Comanches and Athabascans in North America (Benciolini and Gutierrez del Angel 2016). Additionally, there is an increasing phsychedelic tourism looking for experiences with peyote. Schultes (1938) describes that chemical intoxication with peyote is divided into two general phases: "a period of contentment and over-sensitivity, and a period of nervous calm and muscular sluggishness, often accompanied by hypo-cerebrality, colored visual hallucinations, and abnormal synaesthesia. Alterations in tactile sensation, very slight muscular incoordination, disturbances in space and time perception, and auditory hallucinations may accompany severe peyote-intoxication. The most striking characteristic, however, is the occasionally induced peyote vision which is often fantastically colored." For consumption, the collection of the whole peyote plant or of some of its parts takes place in natural populations, so the knowledge on requirements and adequate conditions for germination and establishment can be the basis for recovery programs in diminished populations.

Peyote, *Lophophora williamsii*

Lophophora belongs to the Cactaceae family, subfamily Cactoideae, and the tribe Cacteae (Guzmán et al. 2003) and has two described species: *L. williamsii* and *L. diffusa*, both inhabiting in desert areas and generally thriving on calcareous soils. They are morphologically and chemically distinct (Rojas-Aréchiga and Flores 2016).

Lophophora williamsii is a globose plant, frequently the apex is flattened, 2 to 6 cm high and 4 to 11 cm in diameter. Stems are bluish-green, in some occasions yellowish-green, sometimes with a red tint. It has 4 to 14 ribs which are almost always present and well defined, with variable height, and sometimes they can form high tubercles. Rounded areoles have 0.9 to 1.5 cm between each other with a diameter of 2 to 4 mm.

Individuals of *Lophophora williamsii* have variable shapes throughout their life cycle: at the beginning they have three tubercles, then five, and so on, but while they increase in number, their size is reduced, and the initial spiral changes toward a rib arrangement (Bravo-Hollis and Sánchez-Mejorada 1978). The growth rate of the plants of *L. williamsii* is very slow and they require more than 5 years to reach a diameter of 15 mm. The age and size of the plants are two factors which apparently determine the number of ribs in some populations; young plants usually present 5 ribs, while adults present 5 to 14 (Anderson 1980).

Peyote flowers, varying from red to pink or white, are borne on the apical areolae at the top of the crown during June and July (Fig. 14.1). Flowers measure between 1 and 2.4 cm in length and 1 to 2.2 cm in diameter. The perianth of the flower is pink, sometimes white, but rarely yellow. The plants found at CCB have pink flowers. Flowering begins when plants are still small (Bravo-Hollis and Sánchez-Mejorada 1978), usually between March and September (Anderson 1969). The fruit is a claviform berry of 15 to 20 mm length and 2 to 3.4 mm diameter. It is a naked fruit of red or pinkish color which matures and emerges quickly from the wooly apex of the plant and remains on the plant, partially contained within the hairs of the apical depression (Bravo-Hollis and Sánchez-Mejorada 1978). In San Luis Potosí, fructification has been registered from July to September (Trujillo Hernández 2002).

Fig. 14.1 (**a**) Flowering plant of *Lophophora williamsii* (peyote) with a stem of 8 ribs (center) and a small plant of 6 ribs (left). Plants are growing associated with *Jatropha dioica*, a nurse plant species for this cactus. (**b**) Adult plants of peyote at dry season. (Photographs: (**a**) Juan Carlos Flores Vázquez, (**b**) Mariana Rojas-Aréchiga)

Most seeds are broadly oval, medium-sized, black-brown, matt (Barthlott and Hunt 2000), and weigh 1.19 ± 0.15 mg ($N = 50$) (Rojas-Aréchiga et al. 2013). *Lophophora williamsii* has sexual and vegetative reproduction. The vegetative propagation results in a tussock like growth, given that adventitious shoots emerge from the base of the stem. This growth is considered a response to damage caused by herbivores or by cutting the little heads during harvest. The shoots produced this way can be separated and develop adventitious roots, allowing the formation of new colonies (Anderson 1969). Peyote plants regularly form colonies, although solitary individuals may also be found. Peyote plants tend to grow in the shade of shrubs in a facilitation interaction known as nurse plant effect. Some nurse plants are *Prosopis* spp., *Acacia farnesiana, Mimosa biuncifera, Agave lechuguilla, Jatropha dioica,* and *Euphorbia antisyphilitica* (Anderson 1969; Lumbreras and Barrón 1976; Bravo-Hollis and Sánchez-Mejorada 1991; García Naranjo and Mandujano 2010, Fig. 14.1). They coexist with other succulent plants such as *Ariocarpus retusus, Coryphantha* sp., *Echinocactus horizonthalonius, E. platyacanthus, Ferocactus* sp., *Hechtia glomerata*, and *Mammillaria* spp. (Islas Huitrón 1999).

There are several studies on the taxonomy of this species, anthropological cult, biosynthesis, and chemistry of its alkaloids, as well as its medicinal effects (Bruhn and Holmstedt 1974). Nevertheless, studies on its ecology are rare and very little is known about its natural populations (Rojas-Aréchiga and Flores 2016). Human activities and constant collection of specimens have considerably reduced the number of individuals or altered the life form of peyote's populations in several localities in Texas and Mexico (Anderson 1969; Islas Huitrón 1999). Peyote is a cherished cactus, appreciated by both professional and amateurs collectors. It is always taken from the wild, so the species is now depleted from several areas, and many populations along its distribution range no longer exist (Trujillo Hernández 2002; Mandujano and Briseño-Sánchez 2019).

Germination and Establishment

The life cycle of plants includes two distinct phases according to their mobility: a sessile and a dispersal phase. The recruitment of new individuals in a population represents the interphase between the two phases (Eriksson and Ehrlen 1992). The establishment phase involves the germination of the seed and the emergence of the seedling (Harper 1977), which are critical in the life cycle of a plant, given their high vulnerability to the influence of unfavorable factors (Martorell and Portilla-Alonso 2020, this volume). Mortality in this stage is characteristically high. The events which occur in these early stages significantly influence the dynamics of their populations. Increments in a population through time are mainly the product of the recruitment of new individuals which already have passed through this crucial stage of their life cycle (Ruedas et al. 2000).

Water availability in the soil, temperature, light, and biotic interactions such as herbivore-plant and nurse plant facilitation generally controls seed germination, establishment and growth of plants that inhabit arid environments. For succulent desert plants, soil humidity near the surface is very important, given that the roots sometimes only extend to a depth of 0.1 m (Jordan and Nobel 1982; Mandujano et al. 1998; Flores Rivas 2001). Water scarcity and predation are factors which cause the highest mortality in cacti seedlings (Mandujano et al. 1998). Seedling survival is often the main determinant for the distribution of adult plants (De Villiers et al. 2001).

Most cacti produce fruits with a high number of quiescent seeds which reach high germination rates under different environmental controlled conditions (Rojas-Aréchiga and Vázquez-Yanes 2000; Rojas-Aréchiga et al. 2013). Nevertheless, the seedlings grow extremely slow and the patterns of biomass assignment are relatively rigid. The combination of these features with their long-life cycles and their specific preferences for habitat result in a low and variable recruitment through time (Ruedas et al. 2000; Mandujano et al. 2001; Martínez Berdeja and Valverde 2008; Martorell and Portilla-Alonso 2020, this volume).

Light may be a stimulatory signal to promote the germination of several cacti. Some species germinate well under darkness and light conditions, while others are strictly positive photoblastic (i.e., they require light to germinate; Rojas-Aréchiga et al. 1997). This is known for several globose and cylindrical cacti (Rojas-Aréchiga and Vázquez-Yanes 2000; Rojas-Aréchiga et al. 2013), and light requirement for germination is phylogenetically constrained in tribes Cacteae and Pachycereeae (Rojas-Aréchiga et al. 2013; Rojas-Aréchiga 2014).

Extreme temperatures do not favor the germination of cacti; for example, temperatures below 12 °C or above 28 °C produce low germination values. The temperature range at which germination occurs depend sometimes on the age of the seed, but species may have different requirements (Trujillo Hernández 2002). Under experimental conditions cacti generally germinate between 10 °C and 35 °C (Nobel 1998). Nevertheless, there are no significant differences in seed germination for several globose and columnar cacti species incubated under constant temperatures above 20 °C (Rojas-Aréchiga and Vázquez-Yanes 2000).

In some species, a great proportion of the seeds germinate soon after their dispersal, but a small fraction remains in the soil, creating a seed reserve which can germinate later. This reserve is known as a seed bank and is important because it spreads out germination over time (Wulff 1995). Species that form seed banks show certain morphological and ecological features which allow them to persist in the soil for some time, including light requirement for germination, small seed size, some dormancy mechanisms, a period of post maturation for germination, and high ecological longevity (Bowers 2000; Rojas-Aréchiga and Batis 2001). Positive photoblastism prevents germination if seeds are deeply burrowed and dormancy mechanisms may allow the maintenance of seed viability after dispersal, although it is not a requisite for persistence in the soil (Thompson 2000). Also, the avoidance of germination under burial conditions increases the probabilities for establishment (Fenner 1995; Ruedas et al. 2000).

Not all species are able to form seed banks, a reserve of viable seeds in the soil or attached to the parent plant (aerial seed banks) (Harper 1977). Representative species of plant communities exposed to unpredictable disturbances, such as those caused by crop farming, fire, fluctuations in water level, or when colonization opportunities arise randomly are more prone to form seed banks (Fenner 1995). In such situations, in which reproductive adults die frequently or stop reproducing, a seed bank may ensure the persistence of the population through sexual propagules, when germination and establishment occur at the right place and moment (Wulff 1995, Peralta-García et al. 2020, this volume). In this way, risk of germination under not adequate conditions is spread over time.

In this study we searched for some ecological factors which partially determine the population viability and establishment of peyote plants. We compared the germination of seeds of *L. williamsii* of different ages (Fig. 14.2), in order to determine the loss of viability through time and the probability of forming a persistent seed bank in the soil. We also evaluated the effect of different light radiations on the germination and seedling establishment of *L. williamsii* individuals under controlled conditions. We expect to find differences in the germination of seeds of different ages, due to loss of viability which could influence their persistence in the soil seed bank and that differential Photosynthetically active radiation (PAR) significantly affects germination and establishment, given that in natural conditions individuals establish under nurse plants where seedlings are exposed to lower radiation.

Seed Germination of Different Ages

Germination was assessed with a sample of 500 seeds of different ages. Peyote seeds were collected in different periods (1994, 1995, 1996, 1998, 1999, 2000, and 2002) and from populations from the Chihuahuan Desert including sites from San Luis Potosi, Durango, Chihuahua, and Sierra de San Marcos y Pinos, and Sierra La Madera at CCB, Coahuila, Mexico. Seeds were kept at room temperature (20 ± 2 °C)

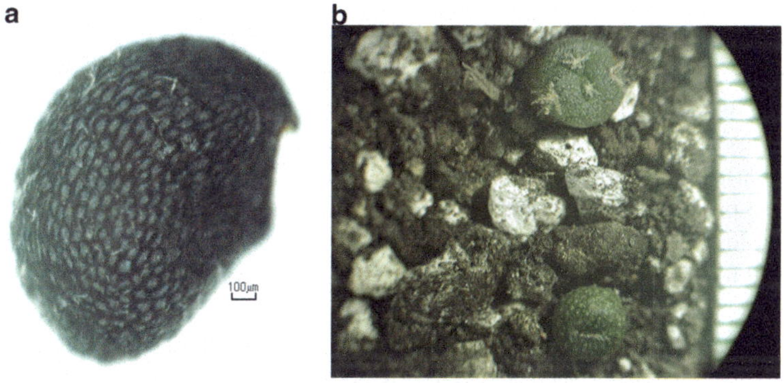

Fig. 14.2 (**a**) Seed of *Lophophora williamsii* under optical microscope (4×). (**b**) Seven month old peyote seedlings under stereoscopic microscope (7.5×), scale in mm. (Photographs: Mariana Rojas-Aréchiga)

since they were harvested. With the available seeds two different experiments were performed.

For germination experiments, we used 200 seeds from different harvest periods. According to their availability, we used 1 to 2 seeds from 1994 and 5 to 6 seeds of each of the other six collection dates. The seeds were placed in glass jars with a 1:1 black soil and pumice especially prepared substrate and watered at soil capacity. Each jar contained 32 seeds of different ages (seeds collected in 1995, 1996, 1998, 1999, 2000, and 2002) and 2 to 3 seeds from 1994 selected randomly (2 jars contained 34 and 4 jars 33 seeds). At the bottom of the jar a grid was drawn with a label, identifying each seed. A seed was considered germinated once the radicle appeared. The jars were placed in a germination chamber at 25 °C, with a photoperiod of 12:12 h light-darkness (Lab-Line 844L). Germination was counted every day during two months until no germination was observed. A generalized linear model, with binary response variable and binomial error was fitted, using the statistical package GLIM 4 (1992 Royal Statistical Society, London), in order to evaluate if there were significant differences in the germination of seeds of different ages. We estimated the annual loss of viability ($LG = P_t - P_{t+1}$, where P is the mean percentage of germination of seeds of age t and $t + 1$, respectively) due to seed aging as a difference between total germination percentages between pairs of seed ages and with regression analysis. Seeds that do not germinate were assumed dead, and mortality rate was estimated as $q_x = 1 - lx$, were $lx = lx + 1/lx$, the standardized number of surviving at the start of age interval x.

First date of germination (FDG) was the third day after initiating the experiment. This FDG value indicates a faster initiation of germination (Kader 1998). The latest germination took place 45 days after sowing the seeds, indicating that the seeds can take a significant long time to germinate (Kader 1998), in comparison to seeds that germinate in high percentages in a single week, as some globose cacti and columnar species like *Neobuxbaumia* spp., *Myrtillocactus geometrizans, Pachycereus pectenaboriginum* (Rojas-Aréchiga 2013, 2014). Overall percentage of germination was 64%. In general, older seeds showed lower germination percentage. One-year old seeds germinated with the highest percentage (78.13%) and nine-years old showed the lowest (12.5%). The low percentage of germination observed in the nine years old seeds must be considered with caution, given that the number of seeds available was very low ($n = 8$). The results of the experiment show that there are significant differences in the germination of seeds of different age ($X^2 = 13.68$, df = 6, $P = 0.033$, Fig. 14.3a), which indicated that the seeds loose germination capacity with age. From the predicted germination value we obtained the percentage of germination capacity loss. Three year old seeds lost 3% germination capacity per year, and increased to 4% after 8 years (Fig. 14.3b). Seed mortality also increased with seed aging (Fig. 14.3c). Seeds lost their viability when exposed to long-term storage at room temperature, by seed aging processes (Priestley 1986). Regression model showed that increasing seed age reduced germination 3.2% yearly ($P < 0.001$, $R^2 = 0.996$), however other factors affecting germination should be considered, besides seed age. Nevertheless, it is difficult to make conclusions about the effect of

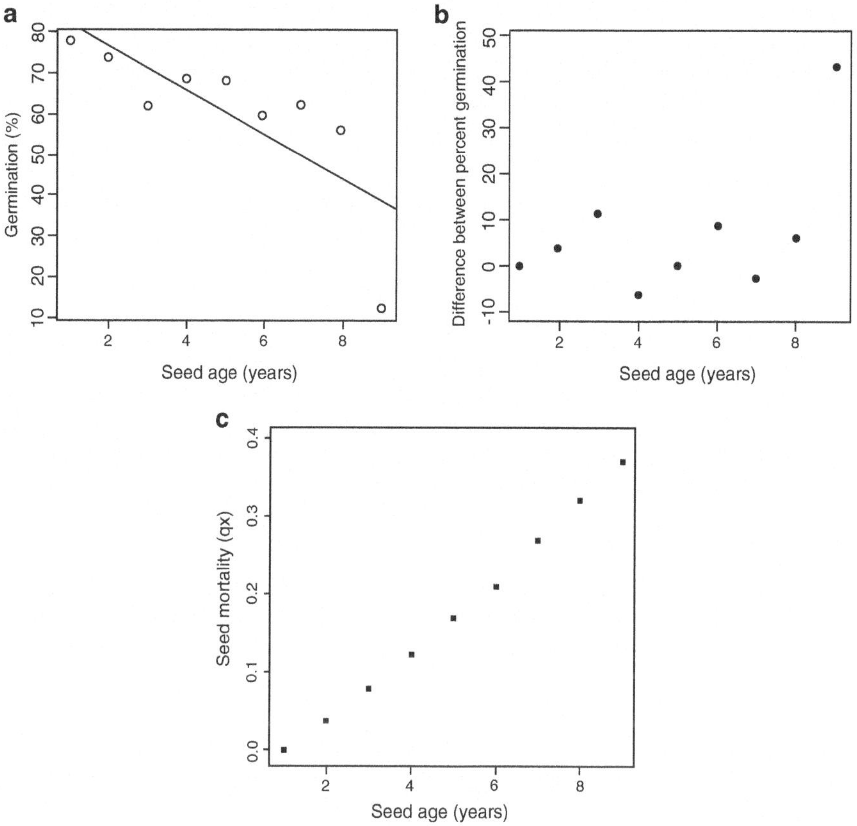

Fig. 14.3 (**a**) Germination percentages and regression line, (**b**) loss of germination-viability (*LG*), and (**c**) mortality (q_x) of *Lophophora williamsii* seeds of different ages

age of seeds on germination in this case, because the seeds originated from different harvest periods.

Germination of Seeds Under Different Light Conditions

This experiment was a completely random factorial design which consisted of three conditions of solar radiation under which germination was evaluated. Seeds originated from a mix of 300 seeds of different ages (1994, 1995, 1996, 1998, 1999, 2000, 2002), collected in CCB. Five replicates of 20 seeds each were used for each treatment ($N = 100$ seeds per treatment combination). Treatments consisted of three solar radiation conditions, that were established using greenhouse mesh fixed with rubber bands to the pots (replicates) where seeds were sown. The treatments were double mesh, single mesh, and no mesh. The Photosynthetically Active Radiation

(PAR, in μMol/sec/m^2) was measured with a quantum sensor LI-COR model LI-189. For each treatment we took five light measurements at different hours of the day (8:15, 11:15, 14:15, and 17:15 h) at the site where the pots were placed on a table. The average PAR for the treatment without shade mesh was 393.8 μMol/sec/m^2 and was considered 100% solar incidence; for the simple shade mesh a PAR of 264.7 μMol/sec/m^2 was obtained, allowing 67.1% of light to pass; for the double shade mesh PAR value was 181.4 μMol/sec/m^2, corresponding to 45.3% of the solar radiation. The pots contained a moistened substrate (1:1 black soil and pumice) on which the seeds were sown and kept inside a greenhouse at the Ecology Institute, Universidad Nacional Autónoma de Mexico (UNAM) at Mexico City.

The necessary amount of water for each irrigation was determined with a pilot experiment. Pots with dry substrate were weighed with a precision scale, then watered until they reached field capacity and then weighed again. The pots were placed under solar radiation in the same place where the experiment would be carried out inside the greenhouse, and each day at the same time the pots were weighed. The amount of water evaporated with the different treatments was determined by weight difference. In this pilot experiment, which lasted two weeks, three replicates per light treatment were used.

Starting on day the seeds were sown, the pots were watered daily, and their position was changed randomly on the experimental table. The maximum and minimum temperature was registered daily within the greenhouse. The maximum average temperature was of 38.3 °C and the minimum average temperature of 11.8 °C. The number of seeds that germinated was registered daily, until we ceased to observe germination (48 days).

The germination percentage was higher and the germination time shorter when the seeds of *L. williamsii* were under reduced solar radiation, in comparison with the treatment without mesh (Fig. 14.4). Mean percentage of germination was 46% under the double shade mesh (45.3% PAR); 30% germinated under the single shade mesh (67.1% PAR) and without mesh (100% PAR) only 15% germinated. The univariate analysis of variance used to assess the differences on germination response under the different PAR treatments showed that there were differences in the germination of the seeds between treatments ($F = 7.58$, $P = 0.007$). Contrast *t*-tests revealed that the treatments differed between each other, especially the treatment of extreme shade versus the treatment without shade mesh ($t = 3.89$, $P = 0.002$). The treatment of intermediate shade showed no differences with the no mesh treatment ($t = 2.06$, $P = 0.06$) nor with the double mesh treatment ($t = 1.83$, $P = 0.09$, Fig. 14.4).

Discussion

There was a slow loss in the germination capacity of peyote seeds due to seed age, as seeds may remain viable for several years under controlled conditions. Seeds of different ages started to germinate early or within a 40 day lapse, consistent with the variability found for other cacti species (Ruedas et al. 2000). The seeds sown under

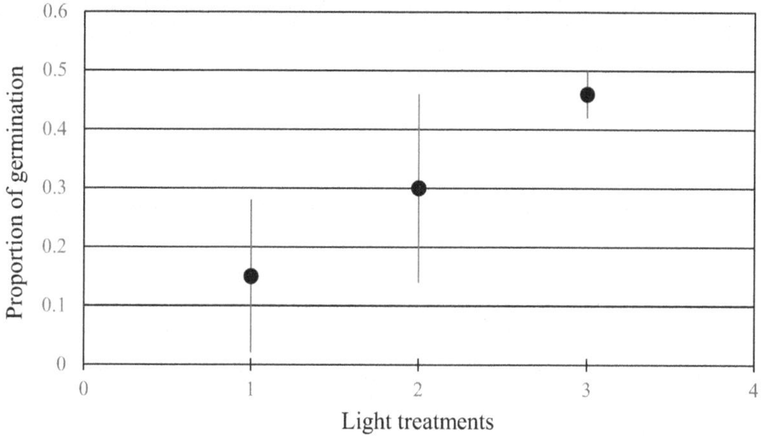

Fig. 14.4 Germination ((proportion ± s.d.) of *Lophophora williamsii* seeds under three conditions of light; where light treatments 1 = without shade mesh (100% exposition to PAR), 2 = one shade mesh (67.1% exposition to PAR), and 3 = double shade mesh (45.3% exposition to PAR)

different light conditions started to germinate after 12 days and continued during 48 days, which suggests that under harsh conditions, germination velocity is lower, which was also found in *Mammillaria crucigera* (Contreras and Valverde 2002). Germination percentage was high for all ages (64% in total) with the exception of nine-year old seeds, but this result needs further confirmation due to small sample size. This germination percentage coincides with germination reported of *L. williamsii* for fresh seeds (60%) (Rojas-Aréchiga et al. 2013) and aged seeds (≈ 65%) (Trujillo Hernández 2002) at 25 °C. The seeds of this species are able to remain viable, while buried in the soil until environmental conditions are optimal for germination and establishment (Fenner 1995).

Despite the differences that were found in the seed germination percentages of seeds from different ages, results showed that seed age explains some of the variation, suggesting that there are other factors which determine seed germination, such as temperature. For example, seeds of the genus *Lophophora* may develop secondary dormancy triggered by low temperatures (Trujillo Hernández 2002).

Lophophora williamsii has small seeds which may have higher dispersal probabilities and may be easily incorporated into the soil substrate. Furthermore, their ability for burial due to its small size is an effective strategy against predation (Fenner 1995), and is characteristic of unpredictable ecosystems, like deserts (Wulff 1995). Despite that *L. williamsii* seeds have some features necessary to form soil seed banks, under natural conditions the predation pressure can be so high that it may prevent the formation of a seed bank, even a transient one (Rojas-Aréchiga and Batis 2001). Peyote seeds are positive photoblastic (Rojas-Aréchiga et al. 2013) in accordance with several species of cylindrical and globose cacti. This light requirement prevents germination if seeds are deeply buried in the soil and increases the chances for seedling establishment under optimal conditions (Rojas-Aréchiga et al. 1997). Restricting germination to superficial dry soil layers and the need to fulfill light requirement ensures that germination will only take place after rains moisten the soil for a long period of time (Kigel 1995).

The establishment of individuals under nurse plants (García Naranjo and Mandujano 2010) is consistent with the higher germination observed under treatment with some form of shade, but germination can occur under a wide range of environmental conditions. However, the most successful germination was obtained in the treatment with PAR reduction, coinciding with what has been reported for *Stenocereus thurberi* (Nolasco et al. 1997) and *Mammillaria magnimamma* (Ruedas et al. 2000). Probably light is not the unique determining factor for establishment, because under shade, humidity may remain for a longer period of time. This is consistent with the results obtained for other species of cacti, in which once water availability is enough, germination can reach high percentages shortly after seeds were sown (Rojas-Aréchiga et al. 1997; Ruedas et al. 2000).

We have shown the importance of shade for germination and establishment of *L. williamsii* seedlings. A seed from a desert environment needs to handle several obstacles to reach the germination stage. Afterwards, during establishment, the aridity of the environment constantly prevents the development of the following stages. Humidity, radiation protection, and temperature reduction generated by nurse plants are important factors which determine the survival and establishment of several succulent plants, like cacti (Steenbergh and Lowe 1969; Jordan and Nobel 1979, 1981).

The establishment of individuals under nurse plants coincides with the high germination success under shade conditions. The environmental severity decreases if the seedling is located at a favorable site where it is sheltered. The nurse plant provides the seed with a microclimate which, when it coincides with the adequate amount of rain, allowing the establishment of new individuals (Jordand and Nobel, 1979, 1981).

Cultivating peyote in nurseries to satisfy collectors can aid conservation. However, actual legal regulation precludes such practices as it is a felony to possess a peyote, this contradiction between legislation and the need for conservation strategies, due to illegal collection, clearly indicates that regulations need urgent revision. It is important to allow peyote cultivation to decrease the pressure on wild populations. In addition, collected seeds can be stored for several years under laboratory conditions and our results suggest that peyote seeds have the potential to maintain a persistent seed bank which may dampen over collection in natural populations.

Acknowledgements We thank the Botanical Garden of UNAM for the donation of seeds. This paper is in memory of Dr. Juan José López, "El Centauro del Norte", a cactus enthusiast and researcher that led us to visit the peyote localities in Cuatro Ciénegas.

References

Anderson EF (1969) The biogeography, ecology, and taxonomy of *Lophophora* (Cactaceae). Brittonia 21:299–310

Anderson EF (1980) Peyote, the divine cactus. The University of Arizona Press, Tucson, Arizona, USA

Anderson EF (2001) The cactus family. Timber Press, Portland, OR, USA
Barthlott W, Hunt D (2000) Seed diversity in the Cactaceae. David Hunt, Milborne Port, UK
Bátis AI, Rojas-Aréchiga M (2002) El peyote y otros cactos alucinógenos de México. Biodiversitas 40:12–17
Benciolini M, Gutierrez del Angel A (2016) From solid to frothy: use of peyote in the Cora and Huichol Easter in Western Mexico. In: Labate BC, Cavnar C (eds) Peyote-history, tradition, politics and conservation. ABC-CLIO, Santa Barbara, CA. USA, pp 171–190
Bowers JE (2000) Does *Ferocactus wislizeni* (Cactaceae) have a between-year seed bank? J Arid Environ 45:197–205
Bravo-Hollis H, Sánchez-Mejorada H (1978) Las cactáceas de México, vol I. Universidad Nacional Autónoma de México, México, DF, Mexico
Bravo-Hollis H, Sánchez-Mejorada H (1991) Las cactáceas de México, vol II. Universidad Nacional Autónoma de México, México, DF, Mexico
Bruhn JG, Holmstedt B (1974) Early peyote research. An interdisciplinary study. Econ Bot 28:353–390
Contreras C, Valverde T (2002) Evaluation of the conservation status of a rare cactus (*Mammillaria crucigera*) through the analysis of its population dynamics. J Arid Environ 51:89–102
De Villiers AJ, Van Rooyen MW, Theron GK (2001) The role of facilitation in seedling recruitment and survival patterns, in the Strandveld Succulent Karoo, South Africa. Journal of Arid Environments 49 (4):809–821
Eriksson O, Ehrlen J (1992) Seed and microsite limitation of recruitment in plant populations. Oecol 91:360–364
Fenner M (1995) Ecology of seed banks. In: Kigel J, Galili G (eds) Seed development and germination. M. Dekker, New York, NY, USA, pp 507–528
Flores Rivas JD (2001) Dinámica del establecimiento de plantas de diferentes formas de vida del desierto de Tehuacán, México. PhD dissertation. Instituto de Ecología, A.C. Xalapa, Veracruz, Mexico
García Naranjo A, Mandujano MC (2010) Patrón de distribución espacial y nodricismo del peyote (*Lophophora williamsii*) en Cuatrociénegas, México. Cact Suc Mex 55:51–60
Guzmán U, Arias S, Dávila P (2003) Catálogo de Cactáceas Mexicanas. Universidad Nacional Autónoma de México y Comisión Nacional para el conocimiento y uso de la Biodiversidad, México, DF, Mexico
Harper JE (1977) Population biology of plants. Academic Press, London, UK
Islas Huitrón H (1999) Estudio ecológico de *Lophophora williamsii* (Lem.) Coulter en una comunidad vegetal perturbada del desierto de San Luis Potosí. Bachelors dissertation. Universidad Nacional Autónoma de México, México, DF, Mexico
Jordan PW, Nobel PS (1979) Infrequent establishment of seedlings of *Agave deserti* (Agavaceae) in the Northwestern Sonoran Desert. Am J Bot 66:1079–1084
Jordan PW, Nobel PS (1981) Seedling establishment of *Ferocactus acanthodes* in relation to drought. Ecology 62:901–906
Jordan PW, Nobel PS (1982) Height distributions of two species of cacti in relation to rainfall, seedling establishment, and growth. Bot Gaz 143:511–517
Kader M (1998) Notes on various parameters recording the speed of seed germination. J Agric Tropics Subtropics 99:147–154
Kigel J (1995) Seed germination in arid and semiarid regions. In: Kigel J, Galili G (eds) Seed development and germination. Marcel Dekker, New York, NY, USA, pp 645–700
Lumbreras, Barrón U (1976) Contribución al conocimiento del "peyote" *Lophophora williamsii* Lem. Coulter. Bachelor dissertation. Universidad Nacional Autónoma de México México, DF, Mexico
Mandujano MC, Briseño-Sánchez I (2019). Retos para la conservación de la biodiversidad: El peyote, *Lophophora diffusa* (Cactaceae). La Biodiversidad en Querétaro: Estudio de Estado. 2008. Comisión Nacional para el Conocimiento y Uso de la Biodiversidad (CONABIO), Universidad Autónoma de Querétaro (UAQ). Querétaro city, Querétaro, Mexico

Mandujano MC, Montaña C, Golubov J (1998) The relative contributions of sexual reproduction and clonal propagation in *Opuntia rastrera* from two habitats in the Chihuahuan Desert. J Ecol 86:911–921

Mandujano MC, Montaña C, Franco M, Golubov J, Flores-Martínez A (2001) Integration of demographic annual variability in a clonal desert cactus. Ecology 82:344–359

Martínez Berdeja A, Valverde T (2008) Growth response of three globose cactus cacti to radiation and soil moisture: an experimental test of the mechanism behind the nurse effect. J Arid Environ 72:1766–1774

Martorell C, Portilla-Alonso RM (2020). The ages of life: The changing forms and history of Coryphantha werdermanii throughout its development. In: Mandujano MC, Pisanty I, Eguiarte L (eds.) Plant diversity and Ecology in the Chihuahuan desert. Springer International, Cham, Switzerland

Nobel PS (1998) The environmental biology of agaves and cacti. Cambridge University Press, Cambridge , UK

Nolasco H, Vega-Villasante F, Diaz-Rondero A (1997) Seed germination of *Stenocereus thurberi* (Cactaceae) under different solar irradiation levels. J Arid Environ 36:123–132

Peralta-García C, Pisanty I, Orozco-Segovia A, Sánchez-Coronado ME, Rodríguez-Sánchez M (2020) Germination of riparian species in natural and experimental conditions. In: Mandujano MC, Pisanty I, Eguiarte LE (eds) Plant diversity and Ecology in the Chihuahuan desert. Springer International, Cham, Switzerland

Priestley DA (1986) Seed aging: implications of seed storage and persistence in the soil. Cornell University Press, Ithaca, NY, USA

Rojas-Aréchiga M (2014) Patrones de respuesta fotoblástica en semillas de cactáceas: un enfoque ecológico y filogenético. PhD dissertation. Universidad Autónoma Metropolitana, México, DF, Mexico

Rojas-Aréchiga M, Batis AI (2001) Las semillas de cactáceas, ¿forman bancos en el suelo? Cact Suc Mex 46:76–82

Rojas-Aréchiga M, Flores J (2016) An overview of cacti and the controversial peyote. In: Labate BC, Cavnar C (eds) Peyote-history, tradition, politics and conservation. ABC-CLIO, Santa Barbara, CA, USA, pp 21–42

Rojas-Aréchiga M, Vázquez-Yanes C (2000) Cactus seed germination: a review. J Arid Environ 44:85–104

Rojas-Aréchiga M, Orozco-Segovia A, Vázquez-Yanes C (1997) Effect of light on germination of seven species of cacti from Zapotitlan Valley in Puebla, Mexico. J Arid Environ 36:571–578

Rojas-Aréchiga M, Mandujano MC, Golubov J (2013) Seed size and photoblastism in species belonging to tribe Cacteae (Cactaceae). J Plant Res 126:373–386

Ruedas M, Valverde T, Castillo Arguero S (2000) Respuesta germinativa y crecimiento de plántulas de *Mammillaria magnimamma* (Cactaceae) bajo diferentes condiciones ambientales. Bol Soc Bot Mex 66:25–35

Schultes RE (1938) The appeal of peyote (*Lophophora williamsii*) as a medicine. American Anthropologist. October–December 1938, New Series 40(4/1):698–715

Schultes RE, Hoffman A (1982) Plantas de los Dioses: orígenes del uso de los alucinógenos. Fondo de Cultura Económica, México, DF, Mexico

Steenbergh WF, Lowe CH (1969) Critical factors during the first years of life of the saguaro (*Cereus giganteus*) at saguaro National Monument, Arizona. Ecology 50:825–834

Thompson K (2000) The functional ecology of soil seed banks. In: Fenner M (ed) Seeds-the ecology of regeneration in plant communities. CABI Publishing, London, UK, pp 215–235

Trout K, Terry M (2016) Decline of the genus *Lophophora* in Texas. In: Labate BC, Cavnar C (eds) Peyote-history, tradition, politics and conservation. ABC-CLIO, Santa Barbara, pp 1–20

Trujillo Hernández A (2002) Ecología fisiológica de la germinación de las cactáceas del género *Lophophora*. MSc dissertation, Universidad Nacional Autónoma de México, México, DF, Mexico

Wulff RD (1995) Environmental maternal effects on seed quality and germination. In: Kigel J, Galili G (eds) Seed development and germination. Marcel Dekker, New York, NY, USA, pp 491–506

Chapter 15
Genetic and Ecological Characterization of the Invasive Wetland Grasses *Arundo donax* and *Phragmites australis* in the Cuatro Ciénegas Basin

Ricardo Colin and Luis E. Eguiarte

Abstract *Arundo donax* and *Phragmites australis* are two of the most aggressive invasive grasses worldwide, both are associated with wetlands and can be very abundant, becoming dominant in these ecosystems. These two species are common in northern Mexico. Genetic and ecological characterization of *A. donax* in two populations from the state of Coahuila (North of Mexico) indicate that they are less clonal and more variable, as well as with a higher genetic diversity compared to populations in other parts of the world and suggest that their genotypes are adapted to different environmental conditions and may represent independent introductions. On the other hand, genealogical analyses show that two independent lineages of *P. australis* are present in Mexico, the Gulf Coast subspecies, *P. australis* ssp. *berlandieri*, found across Mexico, including the state of Coahuila, and the endemic native subspecies, *P. australis* ssp. *americanus*, found in a population from Cuatro Ciénegas Basin (CCB) (Coahuila, Mexico). Here, we conduct a review of the genetic and ecological characteristics of both species in the Chihuahuan Desert, mainly focusing in CCB. The aim is to provide a better understanding in the evolutionary ecology of these two closely related and ecologically similar species and determine if these species of grasses represent a risk for the ecosystem and the valley's biota.

Keywords Asexual propagation · Chloroplast haplotypes · Clonal diversity Invasive species · ISSRs

R. Colin (✉) · L. E. Eguiarte
Departamento de Ecología Evolutiva, Instituto de Ecología, Universidad Nacional Autónoma de México, UNAM, Mexico City, Mexico
e-mail: fruns@unam.mx

M. C. Mandujano et al. (eds.), *Plant Diversity and Ecology in the Chihuahuan Desert*, Cuatro Ciénegas Basin: An Endangered Hyperdiverse Oasis,
https://doi.org/10.1007/978-3-030-44963-6_15

The Biology of Invasive Plants

The classification of plants as either native or introduced is a fundamental biological distinction, as it determines virtually all aspects of how the organisms are treated in scientific and applied contexts. For example, in ecological studies, native species are treated as vital participants in natural services and processes, whereas ecological studies of invasive plants focus on their harmful effects (D'Antonio and Vitousek 1992; Kolar and Lodge 2001; Sakai et al. 2001; Lambrinos 2004; Mack et al. 2000, and references therein). If the species is considered native, then extraction efforts will inevitably require management with a view toward conservation. On the other hand, if it is known to be invasive, then intensive use and even eradication are usually viewed as permissible and even desirable (Rejmánek 2000; D'Antonio and Meyerson 2002; Vilà et al. 2011; Simberloff et al. 2013; Blackburn et al. 2014; Latombe et al. 2016, and references therein).

It is well known that the genetic characteristics of populations will determine their capacity of establishment and range expansion (Mooney and Cleland 2001; Sakai et al. 2001; Lee 2002; Allendorf and Lundquist 2003; Dlugosch and Parker 2008; Suarez and Tsutsui 2008). Since species can evolve during their initial establishment and during their range expansion (Sakai et al. 2001), knowledge of the levels of diversity and of the genetic structure in the native and introduced range can help us to better understand the underlying demographic processes and the adaptive evolution that lead to the invasion (Mooney and Cleland 2001; Sakai et al. 2001; Lee 2002; Allendorf and Lundquist 2003; McCauley et al. 2003; Trewick et al. 2004; Kang et al. 2007; Prentis et al. 2008). That is, the genetic reshuffle at founding will determine the outcome and future impact for non-native populations (Mooney and Cleland 2001; Lee 2002; Novak and Mack 2005; Prentis et al. 2008).

Molecular markers (genetic data from the nuclear and organellar genomes) are critical to identify and study invasive species, pinpointing areas of origin of invaders, distinguishing among species in groups that are difficult to differentiate morphologically (i.e., identification of cryptic species), understanding types and distance of migration, tracking dispersal and spread, determining whether or not hybridization is occurring, and detecting introgression (Hufbauer 2004; Booth et al. 2007; Ward et al. 2008), which are crucial aspects of the basic biology of invaders that we need to know to develop an appropriate management and control strategy of these species.

Among the great diversity of flowering plants, grasses constitute a major group of invasive plants and their negative impact (e.g., they can dramatically alter native plant community structure and ecosystem processes such as fire frequency, nutrient cycling, and water circulation) has long been under great scrutiny by ecologists and conservation scientists (D'Antonio and Vitousek 1992; Strauss et al. 2006).

The Protected Area of Cuatro Ciénegas and Two Potentially Invasive Grasses

Protected natural areas constitute a valuable instrument of environmental policy to safeguard biological richness and to carry out actions for the preservation of biodiversity (SEMARNAP 1997; INE 1999; CONANP 2018). In Mexico there are 182 protected natural areas that are subject to special protection, conservation, restoration, and development regimes (CONANP 2016). However, the management of wild habitats depends almost entirely on whether dominant species are regarded as important native keystones to be protected or harmful exotics to be eradicated (Dudley and Collins 1995; SEMARNAP 1997; INE 1999; Mack et al. 2000; D'Antonio and Meyerson 2002; Hendrickson and McGaugh 2005; Vilà et al. 2011; Simberloff et al. 2013; CONABIO 2018; CONANP 2018). Despite the importance of the native versus naturalized distinction for science and applied situations, the status of many species remains ambiguous, even for some species that dominate vast areas of habitat.

The Cuatro Ciénegas Basin (CCB) (Coahuila, Mexico) is considered the most important wetland in the Chihuahuan Desert, and one of the most important wetlands in Mexico (INE 1999). At the international level, it is classified as a RAMSAR site and is considered a priority wetland for the world (INE 1999), an area of protection of flora and fauna under Mexican government, and a priority site for Conservation of Nature by the World Wildlife Fund for Nature and UNESCO (INE 1999; Souza et al. 2012). CCB is a remarkable area, having the highest endemism than any other place in North America (Stein et al. 2000), and much of the valley's biota is classified either as endangered, threatened, or in special protection status by the Mexican government and the Convention on International Trade of Endangered Species (CITES) (Minckley 1992; INE 1999; Souza et al. 2004; Souza et al. 2006; Instituto Nacional de Ecología y Cambio Climático 2016). Current threats to its biodiversity include water exploitation, species invasions, industrial development, rapidly increasing tourism and population growth. Therefore, actions have been implemented to promote the conservation of terrestrial and aquatic ecosystems in the region. One of these actions is the control and eradication of invasive species (INE 1999; Souza et al. 2004; Hendrickson and McGaugh 2005; Instituto Nacional de Ecología y Cambio Climático 2016).

The giant reed, *Arundo donax* and the common reed, *Phragmites australis* are considered two of the most aggressive invasive grasses worldwide and the negative ecological impact of these species is well known (Perdue 1958; Dudley and Collins 1995; Bell 1997; Dudley 2000; Saltonstall 2002; Saltonstall and Hauber 2007; Saltonstall and Stevenson 2007; Lambert et al. 2010; Meyerson et al. 2010). These two species of grasses are associated with wetlands and can be very abundant and become dominant in these ecosystems. Both species are common in northern Mexico; however, the more ecological approach has left out many of the genetic or evolutionary aspects of their introduction in the region.

In this chapter we carry out a review of the genetic and ecological characteristics of *A. donax* and *P. australis* in general and in particular for the ecological region of the Chihuahuan Desert, with special focus in the CCB. The aim is to provide a better understanding of the evolutionary ecology of these two related and ecologically similar species, and determine if these grass species represent a risk for the ecosystem and the valley's biotic diversity.

The Complex Case of the Invasive *Arundo donax*

Systematics and distribution of *Arundo donax*. *Arundo* L. (Poaceae, tribe Arundineae) is a cosmopolitan genus that includes three to five taxa distributed from tropical Asia to the Mediterranean Basin (Conert 1961; Bell 1997; Grass Phylogeny Working Group 2001; Danin 2004). The species *Arundo donax* (Fig. 15.1 a, b, and c) is the largest species in the genus and is one of the tallest herbaceous grasses (up to 10 m tall). Since giant reed has been cultivated in Asia, southern Europe, North Africa, and the Middle East for thousands of years (Perdue 1958; Zohary 1962; Bell 1997), the native range is a matter of speculations because the biogeographic and evolutionary origin of this species has been obscured through

Fig. 15.1 Pictures showing inflorescences, stems, and stands of *Arundo donax* (a-c) and *Phragmites australis* (d and e). Photographs of *Arundo* by Ricardo Colin and photographs of *Phragmites* by Luis E. Eguiarte

ancient and widespread cultivation, and has been variously reported as southern Asian (Bell 1997; Dudley 2000), eastern Asian (Polunin and Huxley 1987), and from countries surrounding the Mediterranean Sea, where it occurs along the other *Arundo* species, *A. plinii* Turra, *A. collina* Tenore and *A. mediterranea* Danin (Perdue 1958). Nevertheless, a more recent study has provided evidence that *A. donax* likely originated in Asia and subsequently spread into the Mediterranean region (Mariani et al. 2010). The ancient spread of *A. donax* has been accounted mainly due to its multiple uses, including the making of baskets, mats, fishing rods, walking-sticks, fences, roof thatching, plant stakes, musical instruments such as the reeds for clarinets and saxophones, shading or as ornamental, and more recently for erosion control, paper, pulp, and rayon (viscose) (Perdue 1958; Zohary 1962). Therefore, it has become widely dispersed by humans and it is currently found growing into all of the tropical-subtropical and warm-temperate areas of the world (Dudley 2000; Lewandowski et al. 2003; Khudamrongsawat et al. 2004; Ahmad et al. 2008; Mariani et al. 2010; Tarin et al. 2013).

In the USA, giant reed is believed to have been initially introduced into southern California from the Mediterranean in the early 1800s for erosion control, with later introductions being made in Texas and Florida as late as the 1940s (Bell 1997; Perdue 1958). It was also used for roof thatching and widely cultivated for the production of reeds for musical instruments (Perdue 1958; Bell 1997). Since its introduction, *A. donax* has escaped cultivation and become a major invasive weed of riparian habitats, such as in Southern California, Florida, and along the Rio Grande in the border between Texas and Mexico (Bell 1997; Dudley and Collins 1995; Dudley 2000), where it not only displaces native species but also dramatically modifies ecological and successional processes (Bell 1997; Dudley 2000).

In Mexico, *A. donax* (Fig. 15.1 a, b, and c) is a common introduced species, growing in a variety of climates and habitats, including disturbed marshes, wetlands, rivers, lakes, riparian zones, and along roads. In the CCB, *A. donax* has been recently identified and efforts to document its invasion began in 2004 (Hendrickson and McGaugh 2005). Although it is not clear when *Arundo* arrived to the area, at present it has been documented to occur in a variety of mostly small to relatively large stands widely scattered throughout the CCB and surrounding areas (Fig. 15.2) (Hendrickson and McGaugh 2005). Nowadays, *Arundo* is regarded as one of the major threats to native riparian and aquatic ecosystems in the CCB (Hendrickson and McGaugh 2005; CONABIO 2018), and control measures have been implemented in the valley, including control using chemical herbicides, cutting and removing biomass, as well as prescribed fire (Hendrickson and McGaugh 2005; Contreras-Arquieta and Cruz-Nieto 2007).

Genetic characterization. In Mexico, Colin et al. (unpublished) used Inter Simple Sequence Repeats (ISSRs) to estimate the current geographic distribution of genetic diversity of *A. donax* in 20 natural populations across different geographic regions, and found a total of 77 different genotypes (clones) along the Mexican territory evidencing that all the Mexican populations are multiclonal (including from 3 to 9 different genotypes). In the case of northern Mexico, they analyzed 60 individuals from two populations of the state of Coahuila (CCB, $N = 32$ and Valle Cruya,

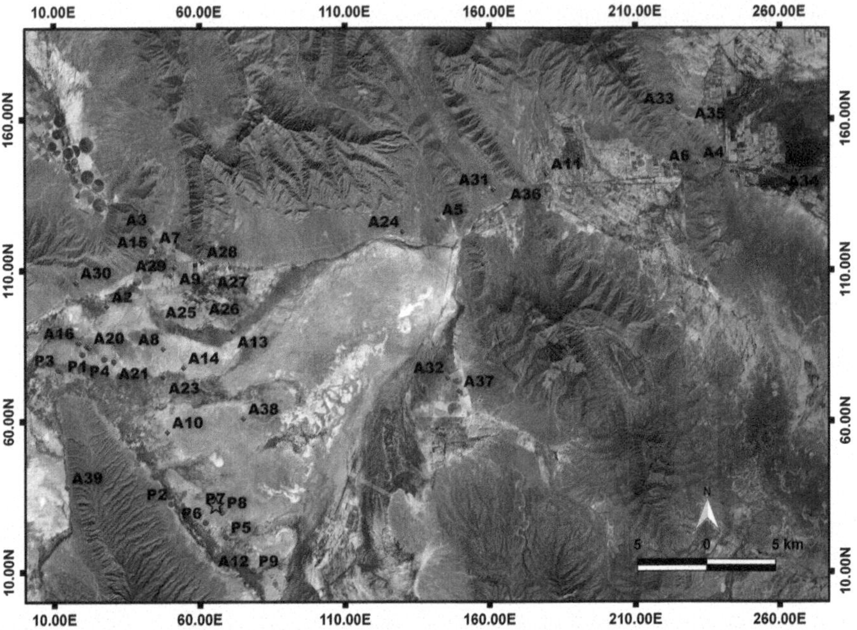

Fig. 15.2 Known localities of *Arundo donax* and *Phragmites australis* stands. Diamonds in red indicate *Arundo* locations in the CCB and surrounding areas. Circles in blue show the *Phragmites australis* ssp. *berlandieri* localities and the green stars indicate the locations of *Phragmites australis* ssp. *americanus*

N = 28), detecting 13 unique genotypes not found in any other populations (five and eight genotypes in the CCB and Valle Cruya, respectively; Fig. 15.3). In addition, they found high values of clonal diversity (Table 15.1), as suggested by the proportion of distinguishable genotypes (G/N, where G= unique genotypes, and N= number of analyzed individuals (Ellstrand and Roose 1987)). We can conclude that Mexican populations are less clonal and genetically more variable than many populations in other parts of the world (Table 15.1).

Reproduction. *Arundo donax* has been observed flowering annually between August and October, and although it produces abundant flowers, viable seeds have not been observed in most areas where it has been introduced (Perdue 1958), including North America, Europe, and Australia (Dudley 2000; Di Tomaso and Healey 2003; Lewandowski et al. 2003; Williams et al. 2008). Thus, asexual propagation through stem layering and rhizome proliferation is believed to be the primary mode of reproduction (Dudley 2000; Boland 2006; Johnson et al. 2006; Williams et al. 2008; Mariani et al. 2010; Haddadchi et al. 2013; Lewandowski et al. 2003).

In Mexico, genetic analyses of random mating by evaluating linkage disequilibrium among loci showed evidence of linkage disequilibrium for the population of CCB, suggesting an asexual mode of propagation (Colin et al. unpublished), in accordance to findings in previous studies (Khudamrongsawat et al. 2004; Ahmad et al. 2008; Mariani et al. 2010; Haddadchi et al. 2013). In contrast, no evidence of

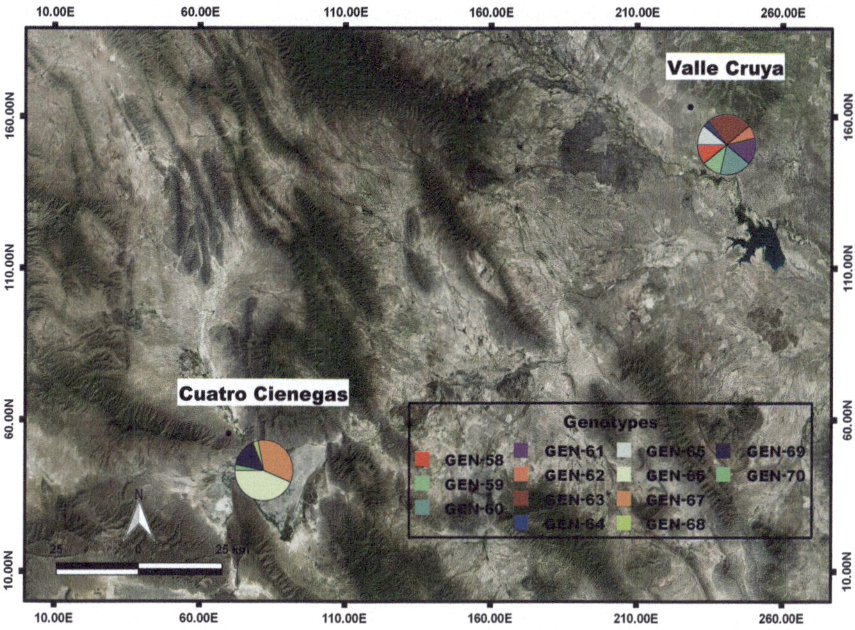

Fig. 15.3 Populations of *Arundo donax* depicting the frequency of genotypes found in northern Mexico. Data were obtained from Colin et al. (unpublished). Colors indicate different genotypes

linkage disequilibrium among loci was found in Valle Cruya, being consistent with (at least partially) sexual recombination (Colin et al. unpublished). Although the frequency at which rare sexual reproductive events may occur in *A. donax* remains unclear, viable seeds (Perdue 1958) have been reported in Asian populations (Afghanistan, South Western Pakistan, and Iran), and more recently, Johnson et al. (2006) found a low frequency of ovules that may have been viable in florets collected in the USA (California, Nevada, Colorado, New Mexico, Texas, Georgia, and Washington D. C.), as well as in northern Mexico, particularly in the state of Nuevo Leon. Thus, the evidence for sexual reproduction that was detected in Valle Cruya is not surprising, but it should be taken with caution (Colin et al. unpublished), since there is no knowledge about the biological reproduction in the Mexican *A. donax* populations.

Eventually, it will be important to conduct studies on the fruit production, seed set rates, viability, and germination of seeds as well as on their ability to spread in the different Mexican populations of *A. donax*. These analyses of reproductive biology along with molecular markers should allow us to better disentangle the roles of seed and asexual propagation, and to understand the importance of the reproductive systems in governing the organization and levels of genetic diversity within the species.

Genetic structure and genetic relationships. It is important to describe the genetic differentiation (also called genetic structure) of populations because it

Table 15.1 Genetic diversity in northern Mexican populations and other regions for *Arundo donax*

Country	Genetic markers	N	G	G/N	References
United States	Isozyme	87	8	0.092	Khudamrongsawat et al. (2004)
United States	RAPD	87	40	0.460	Khudamrongsawat et al. (2004)
United States	SRAP	185	2	0.011	Ahmad et al. (2008)
United States	TE-based	185	3	0.016	Ahmad et al. (2008)
Southern France	SRAP	20	1	0.050	Ahmad et al. (2008)
Southern France	TE-based	20	1	0.050	Ahmad et al. (2008)
Italy	ISSRs	12	1	0.083	Mariani et al. (2010)
Mediterranean Basin	AFLP	16	1	0.062	Hardion et al. (2012)
Australia	ISSRs	58	38	0.655	Haddadchi et al. (2013)
New World[a]	microsatellites	159	6	0.038	Tarin et al. (2013)
Old World[b]	microsatellites	203	129	0.635	Tarin et al. (2013)
Italy	microsatellites	86	8	0.093	Pilu et al. (2014)
Mexico					
Cuatro Ciénegas, CCB	ISSRs	32	5	0.156	Colin et al. (unpublished)
Valle Cruya	ISSRs	28	8	0.285	Colin et al. (unpublished)
North Region[c]	ISSRs	60	13	0.216	Colin et al. (unpublished)

[a] List of the countries sampled and number of analyzed plants in New World: Texas/ Rio Grande Basin = 105, Southeast U.S = 12, California/Nevada = 8, Mexico = 29, Argentina = 5 (Tarin et al. 2013). [b]List of the countries sampled in Old World and number of analyzed plants: Spain = 132, Turkey = 6, Israel = 3, Greece = 6, Italy = 24, Portugal = 11, Morocco = 3, France = 6, Algeria = 12 (Tarin et al. 2013). [c]North region is constituted by the Cuatro Ciénegas (CCB) and Valle Cruya populations (Colin et al. unpublished).

N = Sample size, *G*= Number of detected genotypes, *G/N* = Proportion of distinguishable genotypes (Ellstrand and Roose 1987)

reflects the biological processes of the past that modeled the evolution of the species (Templeton 2006; Pleines et al. 2009; Zhao et al. 2013). In addition, analyzing how variation is partitioned within and between populations in introduced species is critical to understand their potential to establish and spread in a novel range (Sakai et al. 2001; Facon et al. 2006; Marrs et al. 2008).

Genetic structure analyses of *A. donax* in Mexico, carried out by means of several methods (i.e., principal coordinate analysis, agglomerative hierarchical clustering analysis, and AMOVA), suggest the existence of four genetic groups showing a clear genetic differentiation (coancestry coefficient $\theta = 0.830$), and low levels of gene flow between clusters (Colin et al. unpublished). In the particular case of the populations from the state of Coahuila, the analysis indicates that the Valle Cruya population is more similar to populations near the coasts of the Pacific and of the Gulf of Mexico, while all the plants from CCB remained separated, forming a distinct group (Colin et al. unpublished). In addition, these authors also identify that these two populations are genetically very different, as they do not share genotypes between them (Fig. 15.3) (Colin et al. unpublished).

Ecological and genetic characterizations. To evaluate the role of environmental conditions and to detect climate differences and/or similarities between the two populations located in the state of Coahuila, a canonical correspondence analysis was carried out using the genotype distribution data and the environmental variables (Hijmans et al. 2005) (Colin et al. unpublished). The ecological characterization indicates that the distribution and abundance of genotypes are influenced by environmental factors. Temperature seasonality appeared as the strongest environmental variable correlated with the North region, and it is directly associated with genotypes distributed in the area, suggesting that the Northern genotypes are adapted to stronger seasonality with scarce rainfall.

On the other hand, the comparisons in the global distribution of the species indicate that the Valle Cruya population is associated with the bioclimatic space occupied by the invasive occurrence records from the USA and some areas from Europe, while the population of Cuatro Ciénegas was more similar to the environmental space occupied by native records from Asia and some areas from Australia and Africa (Colin et al. unpublished).

General observations and perspectives on *Arundo donax*. The apparently obligate asexual propagation of *A. donax* in introduced ranges of Mediterranean region, Europe, and the USA may keep their genetic diversity low (Ahmad et al. 2008). On the other hand, in the case of Mexican *A. donax* populations, the higher genetic diversity could be likely maintained by different modes of dispersal (i.e., asexual propagation by means of broken stems or rhizome fragments, and some levels of sexual reproduction), suggesting that the frequency at which rare sexual reproductive events may occur in *A. donax* may be variable among populations (Colin et al. unpublished).

Another alternative to sexual reproduction that can explain the level of variation observed in *A. donax* could be multiple introductions from different source regions, as suggested by Khudamrongsawat et al. (2004) and evidenced by Tarin et al. (2013) in the USA. Somatic mutations could also be contributing to the genetic variation in *A. donax* (Khudamrongsawat et al. 2004) and to determine the importance of this type of mutations on the genetic variation found in the Mexican populations of *Arundo*, it would be necessary to compare the genetic composition of plants with known rhizome connections (Khudamrongsawat et al. 2004; Haddadchi et al. 2013).

However, the differences in the values in a given geographic area may be simply due to different sampling schemes and the use of different genetic markers (Mohammadi and Prasanna 2003; Zhao et al. 2006). It will be important to use a standardized method to study the species worldwide (Colin et al. unpublished). It is also important to include in the future large samples from more areas in Asia, where the species apparently originated (Mariani et al. 2010) and that have been poorly studied.

As mentioned above, multiple introductions or a single introduction of multiple genotypes from diverse source populations can result in enhanced genetic diversity in the introduced range (e.g., Simberloff 2009; Lavergne and Molofsky 2007; Kelager et al. 2013), and this may result in the development of locally adapted genotypes (or ecotypes) through natural selection (Sexton et al. 2002; Lavergne and Molofsky 2007; Prentis et al. 2008). In relation to this, the ecological characterization of genotypes of *A. donax* in the North region of Mexico suggests that the genotypes in these populations are influenced by fluctuations in temperature and altitude ranges (Colin et al. unpublished), and point out that *A. donax* may have been introduced different times from disjunct regions (perhaps from the USA and Asia).

The use of genetic data to reconstruct invasion histories, and reveal how a non-native plant adapts and expands into new territory, can also lead to more effective management strategies (Sakai et al. 2001; Allendorf and Lundquist 2003; Prentis et al. 2008). That is, populations having high genetic variation may be more difficult to control because of naturally variable genotypes within the introduced population or the possibility of newly emerged resistant plants, as a result of ongoing natural selection (Sexton et al. 2002; Sterling et al. 2004; Prentis et al. 2008).

In recent years efforts have been made to identify specific phytophagous insects in giant reed as biocontrol agents in the USA (Goolsby and Moran 2009; Goolsby et al. 2009; Goolsby et al. 2013; Goolsby et al. 2015). However, the high regional differentiation of giant reed in Mexico and the multiple genotypes detected in the North region of Mexico imply that these diverse populations may show different levels of susceptibility or resistance to pathogen or other biocontrol agents (Burdon et al. 1981; Sexton et al. 2002; Bruckart et al. 2004; Sterling et al. 2004; Prentis et al. 2008). Therefore, the genetic diversity among populations may be sufficiently great to warrant different control strategies. Differential responses to the same management method have been observed in genetically diverse populations (Sexton et al. 2002; Sterling et al. 2004; Goolsby et al. 2006), so knowledge of existing genetic diversity in the populations provides further insight into the responses of populations to specific management strategies, including the use of chemical control and biocontrol programs (Burdon and Marshall 1981; Chapman et al. 2004; Gaskin et al. 2005; Ward et al. 2008).

Further research of biogeographic relationships in *A. donax* should include populations across a broader range from North America (USA and Mexico), Europe (mainly from Mediterranean basin), and Asia in order to address the question of multiple introductions. Other molecular tools, for instance, chloroplast DNA, could be used in conjunction with ISSR or microsatellite data (Tarin et al. 2013) coupled with ecological analysis to identify if there are different lineages through the global distribution and to determine the possible divergences between those lineages. Additionally, potential source populations of the invader can also be identified, and these populations may help to further locate associated natural enemies, which may be useful later as biological control agents (Ellstrand and Schierenbeck 2000; Clark et al. 2013; Kelager et al. 2013; Ndlovu et al. 2013).

Phragmites australis, an Important Native Species with Invasive Potential

Systematics and distribution of *Phragmites australis* The genus *Phragmites* Adans. (Poaceae, tribe Arundineae) has a worldwide distribution, from the Arctic regions to the tropics (Den Hartog et al. 1989). Historically a number of species, subspecies, and varieties in the genus have been described; currently four species are recognized: *P. australis* (Cav.) Trin. ex Steud., *P. karka* (Retz.) Trin. ex Steud., *P. mauritianus* Kunth and *P. japonicus* Steud (Clevering and Lissner 1999; Saltonstall et al. 2004; Lambertini et al. 2006).

Within the genus, the species *Phragmites australis* (Fig. 15.1 d and e) is the most widely distributed, as it is found in Europe, Asia, Africa, America, and even Australia (Conert 1961; Bjôrk 1967; Clevering and Lissner 1999; Saltonstall 2003a). *Phragmites australis* has been present for at least 40,000 years in southwestern United States (Hansen 1978) and chloroplast genetic data from (cpDNA) indicate that three distinct lineages are found in North America. *P. australis* ssp. *americanus* (comprising thirteen native North American haplotypes: A–H, AA, AB, AC, S, and Z) found throughout the USA and called the "Native North America" lineage (Saltonstall et al. 2004). *P. australis* ssp. *berlandieri*, with only one haplotype I, that grows in Southern United States (from Florida to California), Mexico, Central America, and Asia, named the "Gulf Coast" lineage, (Saltonstall 2002; Saltonstall and Hauber 2007; Saltonstall and Stevenson 2007); this lineage is usually not considered invasive (Meyerson et al. 2009; Meyerson et al. 2010). The third lineage is the haplotype M of *P. australis*, considered an invasive lineage that was probably introduced from Eurasia to North America since ca. 1600 and now is the most common *P. australi*s found from the south of Canada and most USA (Saltonstall 2002, 2003a, 2003b; Saltonstall et al. 2004; Saltonstall and Hauber 2007; Saltonstall and Stevenson 2007; Meyerson et al. 2010).

*Phragmite*s *australis* in Mexico (Fig. 15.1 d and e) is a common species, growing in different environments from disturbed to well-preserved, along roads rivers, marshes, wetlands, and lakes. In the CCB *P. australis* (Fig. 15.1 d and e) is a very common species, however, as in *Arundo*, it is not clear when *Phragmites* arrived to the region. Presently, it is found in different stands widely scattered throughout the CCB (Fig. 15.2), and it is considered a native species of the valley (Colin and Eguiarte 2016).

Genetic characterization. Two lineages of *Phragmites* are present along Mexico: The Gulf Coast lineage, *P. australis* subsp. *berlandieri* found across of Mexico, and the Native North America lineage *P. australis* subsp. *americanus,* found in north of Mexico, in the state of Coahuila (Colin and Eguiarte 2016).

Phragmites australis subsp. *berlandieri* has been characterized by a single cpDNA haplotype (haplotype I) and low genetic diversity in the USA, Central America, and northern South America (Saltonstall 2002, 2003a). Nevertheless, 13 new haplotypes (BH, BM- BX) were found in Mexico, strongly suggesting expansion and diversification in this country (Colin and Eguiarte 2016). From this subspecies four populations were analyzed in the CCB, (Mojarral, Vereda, Desviación, and Mezquite) for a total of 58 individuals (Colin and Eguiarte 2016). Chloroplast haplotype diversity (H_d) in these populations ranged from zero to 0.5333, and nucleotide diversity (π) among these populations ranged from zero to 0.0003 (Table 15.2).

Phragmites australis subsp. *berlandieri* populations in CCB are characterized by a lower genetic variation when compared with central and southern regions of Mexico (Table 15.2), and by the presence of two haplotypes (BH and BM). The

Table 15.2 Genetic variation of *Phragmites australis* in the Cuatro Ciénegas Basin (CCB), Central, and Southern regions of Mexico

Lineages	Population	N	H_d	π	S	h
Gulf Coast	Mojarral	14	0	0	0	1
Gulf Coast	Vereda	15	0.34286	0.00019	1	2
Gulf Coast	Desviacion	15	0.5333	0.00029	1	2
Gulf Coast	Mezquite	14	0.49451	0.00027	1	2
Native North America	Poza X	13	0.65385	0.00042	3	4
Regions						
Cuatro Ciénegas Basin CCB[a]	4	58	0.44828	0.00025	1	2
Central[b]	8	125	0.64503	0.00051	8	8
Southern[c]	14	215	0.66712	0.00051	5	9

[a] Cuatro Ciénegas Basin (CCB): 4 populations in de valley (Mojarral, Vereda, Desviacion, and Mezquite). [b]Central: 8 populations from the states of Guanajuato, Michoacan, and Jalisco. [c]Southern: 14 populations belonging to the states of Veracruz, Tabasco, Campeche, Yucatan, Quintana Roo, and Oaxaca (Colin and Eguiarte 2016)
N = sample size, H_d = haplotype diversity, π = nucleotide diversity, S = segregating sites, and h = number of haplotypes. Data were obtained from Colin and Eguiarte (2016)

Mojarral population presented only one haplotype (BH; Fig. 15.4), whereas the remaining populations had two haplotypes, with greater predominance of haplotype BM in the populations Desviación and Mezquite, while the haplotype BH was more frequent in the Vereda population (Fig. 15.4). It is interesting to mention that the haplotype I has been reported in the Texas side of the Rio Grande and even in the Cuatro Ciénegas Valley (K. Saltonstall, Smithsonian Tropical Research Institute, personal communication). Nevertheless, Colin and Eguiarte (2016) sampling did not detect this haplotype in the CCB (Fig. 15.4), and it was detected only in populations from southern and central Mexico, being more common in the southern areas.

On the other hand, *P. australis* subsp. *americanus* includes thirteen native North American cpDNA haplotypes (A–H, AA, AB, AC, S, and Z), showing high genetic diversity (Saltonstall 2002, 2003a, 2003b). In Mexico, four new haplotypes (BI–BL) were found in the Poza X population from CCB (Fig. 15.4); these haplotypes are related to the native lineage of North America, particularly to haplotype B (Saltonstall 2002, 2003a), increasing the range of *P. australis* subsp. *americanus* (Colin and Eguiarte 2016). As in the USA, this Mexican population (Poza X) showed high genetic variation compared with the populations of *P. australis* subsp. *berlandieri* (Table 15.2). The finding of these haplotypes in the CCB may be related to the high diversity and levels of endemism of the basin (Souza et al. 2012; Carson et al. 2015).

Reproduction. Previous studies in *P. australis* (Rice et al. 2000; Ishii and Kadono 2002; Saltonstall 2002, 2003a; Brisson et al. 2008; Howard et al. 2008; Fer and Hroudova 2009; Baldwin et al. 2010; Belzile et al. 2010; McCormick et al. 2010; Kirk et al. 2011) have documented both sexual and asexual (e.g., lateral rhizome extension, rhizome fragments, and tillering) propagation. Flowers are

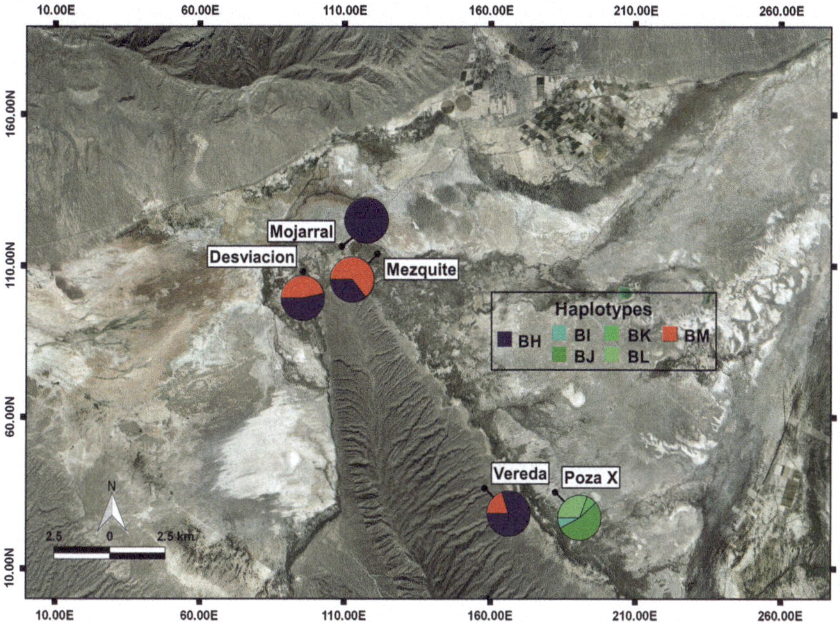

Fig. 15.4 Populations of *Phragmites australis* depicting the frequency of haplotypes found in Cuatro Ciénegas Basin (CCB). Haplotypes BH and BM (violet and red colors, respectively) belong to *Phragmites australis* ssp. *berlandieri*. Haplotypes BI, BJ, BK, and BL (green color gradient) depicting haplotypes that were more closely related to the native North American lineage *Phragmites australis* ssp. *americanus*. Data were obtained from Colin and Eguiarte (2016)

monoecious and wind-pollinated. Each infructescence can contain thousands of seeds, which have long silky hairs on the rachilla, facilitating wind dispersal. It was originally suggested that the species is self-incompatible (Clevering and Lissner 1999), but more recent evidence suggests that self-pollination is possible (Lambert and Casagrande 2007). In addition, environmental conditions can affect production and seed viability; seeds require high exposure to light, as well as diurnal temperature fluctuations (range 10° to 30° C) to germinate (Marks et al. 1994; Campbell 2007; Meyerson et al. 2010).

Phragmites australis subsp. *berlandieri* flowers annually from late October to November, and although it produces abundant flowers, viable seeds have not been observed in most areas from the USA, Mexico, Central America, and northern South America, and vegetative propagation through stem layering and rhizome proliferation is believed to be the primary mode of reproduction (Pellegrin and Hauber 1999; Saltonstall 2002, 2003a; Meyerson et al. 2009: Meyerson et al. 2010). Colin and Eguiarte (2016), working with populations across the species' geographical range in Mexico, found that some populations contain only a single haplotype (such as the population Mojarral; see Table 15.2 and Fig. 15.4), suggesting that in some cases the populations may derive from a single founder, and subsequently spread by asexual propagation. In contrast, in the majority of the populations—as in

Desviación, Mezquite, and Vereda in CCB (see Table 15.2 and Fig. 15.4)—different haplotypes were detected, indicating that different individuals originated these populations (Colin and Eguiarte 2016).

Phragmites australis subsp. *americanus* flowers earlier, between June and October (Meyerson et al. 2009; Meyerson et al. 2010). It propagates both by seed, mainly dispersed by wind, and by vegetative propagules likely dispersed by water and maybe by animals. However, apparently this lineage has lower seed germination rates (Meyerson et al. 2010). In Mexico, genetic diversity for this subspecies was high (see Poza X population in Table 15.2 and Fig. 15.4). Again, as in the other subspecies, this indicates either that dispersal events involved multiple founders by vegetative propagules or that dispersal by seeds is important for the establishment of new populations (Colin and Eguiarte 2016).

These findings suggest that in Mexico the genetic composition of the populations from both lineages is likely the result of different modes of spread and dispersal, including at least some sexual reproduction. However, as discussed by Colin and Eguiarte (2016), it is important to conduct future studies including aspects of the reproductive ecology, seed germination, and dispersal that shall help us to determine the importance of reproductive systems in the life history of the species. In addition, it will be relevant to use nuclear DNA (e.g., microsatellites or SNPs) to understand the roles of seed and asexual propagation and to better describe the evolutionary genetics of *P. australis* in the CCB and Mexico (Colin and Eguiarte 2016).

Genetic structure and genetic relationships. Using a representative set of samples from the Mexican range, Colin and Eguiarte (2016) investigated the genetic structure patterns within and between the introduced *P. australis* subsp. *berlandieri* populations by conducting a spatial analysis of molecular variance (SAMOVA). Colin and Eguiarte (2016) found seven genetic groups ($K = 7$) in Mexico. The CCB was characterized by the presence of two genetic groups ($K1$ and $K2$). The first cluster ($K1$) included the Mojarral and Vereda populations that were grouped together with populations belonging to the states of Michoacán, Veracruz, Tabasco, Jalisco, Campeche, Yucatán, and Quintana Roo. The second cluster ($K2$) comprises only the Desviacion and Mezquite populations.

The historical demography of these groups was inferred by examining hypotheses of demographic expansion (Colin and Eguiarte 2016). For the first cluster, $K1$ a recent range expansion was inferred and the estimate of the age of population expansion was older (ca. 0.73 Myr before the present) for $K1$. In contrast, the analysis for the $K2$ cluster suggests that these populations have maintained a constant population size (Colin and Eguiarte 2016).

Ecological and genetic characterizations. At present, there is no data available on whether or not environmental factors affect the distribution and abundance of haplotypes for both lineages in the CCB. However, the patterns of genetic diversity found across the species' geographical range in Mexico suggest that dispersal was from the South towards the North, with a reduction of genetic diversity in the process (Table 15.2) (Colin and Eguiarte 2016). This gradient of genetic diversity suggests that the Pleistocene climatic changes were relevant for the historical dynamics of *P. australis* in Mexico. In consequence, the populations in the CCB may be relatively recent, and this may also explain why the *P. australis* populations from Mexico—even in the northern areas— are so different from the populations in the USA (Colin and Eguiarte 2016).

General observations and perspectives on *P. australis*. As found by Colin and Eguiarte (2016) in the CCB we can find two independent and native lineages of *Phragmites*: *P. australis* subsp. *berlandieri* and *P. australis* subsp. *americanus*, we provided insights into the dynamic of the populations of *P. australis* subsp. *berlandieri* in the CCB, and Colin and Eguiarte (2016) also demonstrated that the invasive lineage of *P. australis* (haplotype M) is absent in the CCB.

The CCB is characterized by the presence of two (Fig. 15.4) of the thirteen haplotypes belonging to *P. australis* subsp. *berlandieri* found in Mexico, and by a low genetic variation (Table 15.2). The detection of new haplotypes in the Poza X population from *P. australis* subsp. *americanus* increased the proposed range of this lineage (Fig. 15.4) (Colin and Eguiarte 2016).

Populations in the CCB have significantly lower genetic variation when compared with populations from the central and southern Mexican regions (Colin and Eguiarte, personal observation). Genetic diversity can be reduced in expanding populations, as only a few individuals contribute with genetic variation to the newly colonized populations (Buckley et al. 2012; Hallatschek and Nelson 2008). Previous studies have shown that there is a correlation between genetic diversity and environmental heterogeneity *in P. australis* (Curn et al. 2007; Hansen et al. 2007; Engloner 2009), but whether or not climatic factors affect the distribution and abundance of haplotypes in *P. australis* subsp. *berlandieri* in Mexico has yet to be determined (Colin and Eguiarte in preparation).

The genetic structure of populations is not always reflected in the geographical proximity of individuals, and individuals with different geographical locations are not necessarily genetically differentiated (Evanno et al. 2005). Among multiple-population clustering of the Mexican population, group *K1* included different populations separated by long distances, and this cluster included the Mojarral and Vereda CCBs populations, that were grouped together with populations from Michoacán, Veracruz, Tabasco, Jalisco, Campeche, Yucatán, and Quintana Roo states. In contrast, the cluster *K2* that included Desviación and Mezquite was restricted to a small area within CCB (Colin and Eguiarte, personal observation). These results suggest that genetic clustering in *P. australis* in Mexico is best explained as a function of genetic divergence rather than of geographical distance, in contrast to what happens in the USA (Saltonstall 2003a) and, perhaps in China (An et al. 2012).

Finally, results suggest that haplotype U distributed in Australia and Asia (Saltonstall 2002, 2003a) is the closest relative to haplotypes of *P. australis* subsp. *berlandieri* found in Mexico (Colin and Eguiarte 2016); however, their history is complex. Genetically, it has been shown that the haplotype I has affinity to different geographic regions (Saltonstall 2002, 2003a; Lambertini et al. 2006) and has been reported to have a closer relationship with *Phragmites* in Asia (Saltonstall 2002, 2003a), as well as with the African species *P. mauritianus* (Lambertini et al. 2006); also morphological analyses indicate that haplotype I may correspond to *P. karka*, from tropical Africa (Saltonstall and Hauber 2007; Ward 2010). Future genetic and morphological analyses of plants from Africa and South America will be needed to understand the origin and evolution of *P. australis* in Mexico and of the *Phragmites* genus in general.

Concluding Remarks: *A. donax* Is an Aggressive Invasive Plant, But *P. australis* Is Native to CCB

Here we presented a review of the evolutionary biology and invasive potential of two related and ecologically similar grasses (*A. donax* and *P. australis*) in the Chihuahuan Desert, with special emphasis in the CCB. In general, in the state of Coahuila there are 13 genotypes of *A. donax* and two independent native lineages of *Phragmites* (*P. australis* subsp. *berlandieri* and *P. australis* subsp. *americanus*), although the invasive lineage (M) was not found. We hope that this review will shed new light on the understanding of ecological, evolutionary, and biogeographic aspects of the complex studied species and can also assist in the development of conservation, management, and policy strategies of these species in the region, and in particular in the CCB.

The genetic data available for *A. donax* and *P. australis* reflect that both sexual and asexual propagation seem to be relevant but the role of each reproductive mode varies in each species. Gaps in our knowledge are particularly apparent in the dispersal and establishment phases of the life cycle of these two species, for example rates of seed set and germination have not yet been determined in Mexico. Indeed, it will be important to conduct studies on the viability and germination of seeds as well as on their ability to spread in native and introduced environments to assess the fecundity of these species and of their different genetic lineages. At the same time, as we mentioned above, it will be relevant to use nuclear DNA (in particular microsatellites or SNPs) markers to better understand the roles of seed and asexual propagation and evaluate their importance to determine genetic diversity within and among populations.

Recent studies have shown that when species with a broad climatic and geographical range colonize or invade a new environment or spread over large geographical ranges, they may come upon suitable habitats to colonize and can establish in contrasting climatic regions, displaying environmental tolerances and physiological plasticities that promote their potential to adapt and invade these environments (Lee 2002; Sax et al. 2007; Vellend et al. 2007; Prentis et al. 2008). Thus, the novel environmental conditions found in the range of introduction may act as strong selection forces and can lead to rapid evolution (in centuries, decades, or even just in years) during the establishment and spread of invasive species (García-Ramos and Rodríguez 2002; Colautti and Lau 2015).

In *A. donax* the data shown here suggest that the genotypes distributed in the Northern region are adapted to stronger seasonality with scarce rainfall, and point out that *A. donax* may have been introduced in different occasions from disjunct regions. On the other hand, for *P. australis*, we cannot say at this time whether the haplotypes are influenced by fluctuations in temperature, rainfall, and/or altitude ranges; however, evidence suggests a post-glacial colonization or recolonization of *P. australis* in the CCB. Thus, more research is needed to evaluate whether or not the environmental factors play an important factor in the distribution of haplotypes in the region.

Besides natural selection, lack of gene flow between the native and introduced ranges can generate genetic differentiation and divergence between them and may eventually lead to allopatric speciation (García-Ramos and Rodríguez 2002; Lee 2002; Sax et al. 2007; Vellend et al. 2007; Prentis et al. 2008; Futuyma 2013; Colautti and Lau 2015). In Mexico, for *A. donax* 77 different genotypes were detected by Colin et al. (unpublished), among which 13 unique genotypes were found in northern Mexico (5 and 8 genotypes in the CCB and Valle Cruya, respectively (Fig. 15.3), but we still do not know if they are the same or different haplotypes from those found in the Mediterranean region, Europe, Australia, and/or the USA. Thus, we need to use a standardized method (e. g., chloroplast DNA, in conjunction with microsatellite markers or SNPs) to evaluate if there are different lineages and to determine their biogeographical ranges in the species worldwide distribution.

Phragmites australis in Mexico seems to still be in the process of differentiation and diversification (Colin and Eguiarte personal observation). Of the thirteen (BH, BM- BX) new haplotypes found now in Mexico, two of them (BH and BM) were found in the CCB. Furthermore, 4 new haplotypes belonging to the native North America Lineage were also found in the CCB (Fig. 15.4). Experimental and ecological data including viability and germination of seeds information are needed to determine the importance of the propagation mode (sexual vs. asexual) in the life history of the different lineages of *P. australis* in Mexico.

Further research in morphology, ecology, and genetics of more samples from *A. donax*, especially from the USA, Europe, Asia, and the Mediterranean region and from *P. australis* in particular from South America, and Africa, are still needed to link the observed patterns in genetic variation with the environmental factors in order to determine more clearly the evolutionary processes in both species and address in detail their origin, ecology, and invasive potential in Mexico and in particular in the CCB.

Acknowledgments The manuscript was written during a sabbatical leave of LEE in the University of Minnesota, Department of Plant and Microbial Biology, in the laboratory of Peter Tiffin, with support of the program PASPA- DGAPA, UNAM. We deeply acknowledge the detailed reviews of the manuscript by Irene Pisanty and by Erika Aguirre Planter. Ricardo Colin Nuñez was a doctoral student from Programa de Doctorado en Ciencias Biomédicas, Universidad Nacional Autónoma de México (UNAM) and received fellowship 215751 from CONACyT. We thank the Programa de Posgrado de Doctorado en Ciencias Biomédicas, Universidad Nacional Autónoma de México (UNAM) and the Laboratorio de Evolución Molecular y Experimental, Instituto de Ecología, UNAM.

References

Ahmad R, Liow P, Spencer D et al (2008) Molecular evidence for a single genetic clone of invasive *Arundo donax* in the USA States. Aquatic Botany 88:113–120

Allendorf FW, Lundquist LL (2003) Introduction: population biology, evolution, and control of invasive species. Conservation Biology 17:24–30

An JX, Wang Q, Yang J et al (2012) Phylogeographic analyses of *Phragmites australis* in China: Native distribution and habitat preference of the haplotype that invaded North America. Journal of Systematics and Evolution 50:334–340

Baldwin AH, Kettenring KM, Whigham DF (2010) Seed banks of *Phragmites australis*-dominated brackish wetlands: Relationships to seed viability, inundation, and land cover. Aquatic Botany 93:63–169

Bell GP (1997) Ecology and management of *Arundo donax*, and approaches to riparian habitat restoration in southern California. In: Brock JH, Wade M, Pysek P, Green D (eds) Plant Invasions: Studies from North America and Europe. Backhuys publishers, Leiden, The Netherlands, pp 103–113

Belzile F, Labbe J, Leblanc M et al (2010) Seeds contribute strongly to the spread of the invasive genotype of the common reed (*Phragmites australis*). Biological Invasions 12:2243–2250

Bjôrk S (1967) Ecologic investigation of *Phragmites communis*. Studies in the theoretic and applied limnology. Folia Limnologica Scandinavica 14:1–248

Blackburn TM, Essl F, Evans T (2014) A unified classification of alien species based on the magnitude of their environmental impacts. PLoS Biology 12:e1001850

Boland JC (2006) The importance of layering in the rapid spread of *Arundo donax* (giant reed). Madrono 53:303–312

Booth D, Provan J, Maggs CA (2007) Molecular approaches to the study of invasive seaweeds. Botanica Marina 50:385–396

Brisson J, Paradis E, Bellavance M (2008) Evidence of sexual reproduction in the invasive common reed (*Phragmites australis* subsp. *australis*; Poaceae) in eastern Canada: A possible consequence of global warming. Rhodora 110:225–230

Bruckart W, Cavin C, Vajna L et al (2004) Differential susceptibility of Russian thistle accessions to *Colletotrichum gloeosporioides*. Biol Control 30:306–311

Buckley J, Butlin RK, Bridle JR (2012) Evidence for evolutionary change associated with the recent range expansion of the British butterfly, *Aricia agestis*, in response to climate change. Molecular Ecology 21:267–280

Burdon JJ, Groves RH, Cullen JM (1981) The impact of biological control on the distribution and abundance of *Chondrilla juncea* in south-eastern Australia. Journal of Applied Ecology 18:957–966

Burdon JJ, Marshall DR (1981) Biological control and the reproductive mode of weeds. Journal of Applied Ecology 18:649–658

Campbell AL (2007) Sexual Reproduction in Non-Native Common Reed, *Phragmites australis*. M.S. thesis. Ohio State University. Columbus, OH, USA

Carson EW, Souza V, Espinosa-Perez H et al (2015) Mitochondrial DNA diversity and phylogeography of *Lucania interioris* inform biodiversity conservation in the Cuatro Ciénegas basin, Mexico. Western North Naturalist 75:200–208

Chapman H, Robson B, Pearson ML (2004) Population genetic structure of a colonising, triploid weed, *Hieracium lepidulum*. Heredity 92:182–188

Clark LV, Evans KJ, Jasieniuk M (2013) Origins and distribution of invasive *Rubus fruticosus* L. agg. (Rosaceae) clones in the western United States. Biol Invasions 15:1331–1342

Clevering OA, Lissner J (1999) Taxonomy, chromosome numbers, clonal diversity and population dynamics of *Phragmites australis*. Aquatic Botany 64:185–208

Colautti RI, Lau JA (2015) Contemporary evolution during invasion: evidence for differentiation, natural selection, and local adaptation. Molecular Ecology 24:1999–2017

Colin R, Eguiarte LE (2016) Phylogeographic analyses and genetic structure illustrate the complex evolutionary history of *Phragmites australis* in Mexico. American Journal of Botany 103:876–887
CONABIO (2018) Sistema de información sobre especies invasoras en México. Comisión Nacional para el Conocimiento y Uso de la Biodiversidad. http://www.biodiversidad.gob.mx/especies/Invasoras/invasoras.html. Accessed Sep 2018
CONANP (2018) Programa Nacional de Áreas Naturales Protegidas, 2014–2018. Comisión Nacional de Áreas Naturales Protegidas. Mexico City, Mexico
CONANP (2016). https://www.gob.mx/conanp/acciones-y-programas/areas-naturales-protegidas-decretadas Accessed Sep 2018
Conert HJ (1961) Die Systematik und Anatomie der Arundinae. Cramer, Weinheim, West Germany
Contreras-Arquieta A, Cruz-Nieto M (2007) Control de la planta invasora *Arundo donax* en Cuatro Ciénegas, un centro de gran biodiversidad de especies endémicas de flora y fauna de Norteamérica. Informe técnico final de Pronatura Noreste, AC, US Fish & Wildlife Service. Monterrey, Nuevo León, Mexico
Curn V, Kubatova B, Vavrova P et al (2007) Phenotypic and genotypic variation of *Phragmites australis*: comparison of populations in two human-made lakes of different age and history. Aquatic Botany 86:321–330
D'Antonio CM, Meyerson LA (2002) Exotic plant species as problems and solutions in ecological restoration: a synthesis. Restoration Ecology 10:703–713
D'Antonio CM, Vitousek PM (1992) Biological invasion by exotic grasses, the grass-fire cycle and global change. Annual Review of Ecology and Systematics 23:63–88
Danin A (2004) *Arundo* (Gramineae) in the Mediterranean reconsidered. Willdenowia 34:361–369
Den Hartog C, Kvet J, Sukopp H (1989) Reed: a common species in decline. Aquatic Botany 35:1–4
Di Tomaso JM, Healey EA (2003) Aquatic and riparian weeds of the west. University of California—Agriculture and Natural Resources Publication 3421:254–262
Dlugosch KM, Parker IM (2008) Founding events in species invasions: genetic variation, adaptive evolution, and the role of multiple introductions. Molecular Ecology 17:431–449
Dudley TL (2000) *A. donax L.* In: Bossard CC, Randall JM, Hoshovsky MC (eds) Invasive Plants of California's Wildlands. University of California Press, Berkeley, CA, USA, pp 53–58
Dudley TL, Collins B (1995) Biological Invasions in California Wetlands: The Impacts and Control of Non-indigenous Species in Natural Areas. Pacific Institute for SIDES, Oakland, CA, USA
Ellstrand NC, Roose ML (1987) Patterns of genotypic diversity in clonal plant species. American Journal of Botany 74:123–131
Ellstrand NC, Schierenbeck KA (2000) Hybridization as a stimulus for the evolution of invasiveness in plants? Proc Natl Acad Sci USA 97:7043–7050
Engloner AI (2009) Structure, growth dynamics and biomass of reed (*Phragmites australis*) A review. Flora 204:331–346
Evanno G, Regnaut S, Goudet J (2005) Detecting the number of clusters of individuals using the software STRUCTURE: a simulation study. Molecular Ecology 14:2611–2620
Facon BB, Genton J, Shykoff J (2006) A general eco-evolutionary framework for understanding bioinvasions. Trends in Ecology and Evolution 21:130–135
Fer T, Hroudova Z (2009) Genetic diversity and dispersal of *Phragmites australis* in a small river system. Aquatic Botany 90:165–171
Futuyma DJ (2013) Evolution, 3rd edn. Sinauer, Sunderland, MA, USA
García-Ramos G, Rodríguez D (2002) Evolutionary speed of species invasions. Evolution 56:661–668
Gaskin JF, Dao-Yuan Z, Bon M-C (2005) Invasion of *Lepidium draba* (Brassicaceae) in the western United States: distributions and origins of chloroplast DNA haplotypes. Molecular Ecology 14:2331–2341
Goolsby JA, De Barro PJ, Makinson JR et al (2006) Matching the origin of an invasive weed for selection of a herbivore haplotype for a biological control programme. Molecular Ecology 15:287–297

Goolsby JA, Moran PJ (2009) Host range of *Tetramesa romana* Walker (Hymenoptera: Eurytomidae), a potential biological control of giant reed, *Arundo donax* L. in North America. Biological Control 49:160–168

Goolsby JA, Moran PJ, Adamczyk JA et al (2009) Host range of the European, rhizome-stem feeding scale *Rhizaspidiotus donacis* (Leonardi) (Hemiptera: Diaspididae), a candidate biological control agent for giant reed, *Arundo donax* L. (Poales: Poaceae) in North America. Biocontrol Science and Technology 19:899–918

Goolsby JA, Moran PJ, Racelis AE et al (2015) Impact of the biological control agent Tetramesa romana (Hymenoptera: *Eurytomidae*) on *Arundo donax* (Poaceae: Arundinoideae) along the Rio Grande River in Texas. Biocontrol Science and Technology 26:47–60

Goolsby JA, Racelis AE, Goolsby JB et al (2013) Evaluation of biogeographical factors in the native range to improve the success of biological control agents in the introduced range. Biocontrol Science and Technology 23:1213–1230

Grass Phylogeny Working Group (2001) Phylogeny and subfamilial classification of the grasses (Poaceae). Annals of Missouri Botanical Garden 88:373–457

Haddadchi A, Gross CL, Fatemi M (2013) The expansion of sterile *Arundo donax* (Poaceae) in southeastern Australia is accompanied by genotypic variation. Aquatic Botany 104:153–161

Hallatschek O, Nelson DR (2008) Gene surfing in expanding populations. Theor Popul Biol 73:158–170

Hansen RM (1978) Shasta ground sloth food habits, Rampart Cave, Arizona. Paleobiology 3:302–319

Hansen DL, Lambertini C, Jampeetong A et al (2007) Clone-specific differences in *Phragmites australis*: effects of ploidy level and geographic origin. Aquatic Botany 86:269–279

Hardion L, Verlaque R, Baumel A et al (2012) Revised systematics of Mediterranean Arundo (Poaceae) based on AFLP fingerprints and morphology. Taxon 61:1217–1226

Hendrickson DA, Mcgaugh S (2005) *Arundo donax* (Carrizo Grande / Giant Cane) in Cuatro Ciénegas. http://desertfishes.org/cuatroc/organisms/non-native/arundo/Arundo.html. Accessed September 2018

Hijmans RJ, Cameron SE, Parra JL et al (2005) Very high resolution interpolated climate surfaces for global land areas. International Journal of Climatology 25:1965–1978. http://www.worldclim.org Accessed September 2018

Howard R, Travis SE, Sikes BA (2008) Rapid growth of a Eurasian haplotype of *Phragmites australis* in a restored brackish marsh in Louisiana, USA. Biological Invasions 10:369–379

Hufbauer RA (2004) Population Genetics of Invasions: Can We Link Neutral Markers to Management? Weed Technology 18:1522–1527

INE (1999) Programa de Manejo del Área de Protección de Flora y Fauna Cuatrociénegas. Instituto Nacional de Ecología, Tlacopac, DF, Mexico

Instituto Nacional de Ecología y Cambio Climático (2016). https://www.gob.mx/inecc/acciones-y-programas/analisis-de-la-variacion-del-nivel-de-los-principales-cuerpos-de-agua-de-cuatrocienegas Accessed Sep 2018

Ishii J, Kadono Y (2002) Factors influencing seed production of *Phragmites australis*. Aquatic Botany 72:129–141

Johnson M, Dudley T, Burns C (2006) Seed production in *Arundo donax*? California Invasive Plant Council 14:12–13

Kang M, Buckley YM, Lowe AJ (2007) Testing the role of genetic factors across multiple independent invasions of the shrub Scotch broom (*Cytisus scoparius*). Molecular Ecology 16:4662–4673

Kelager A, Pedersen JS, Bruun HH (2013) Multiple introductions and no loss of genetic diversity: invasion history of Japanese Rose, *Rosa rugosa*, in Europe. Biol Invasions 15:1125–1141

Khudamrongsawat J, Tayyar R, Holt J (2004) Genetic diversity of giant reed (*Arundo donax*) in the Santa Ana River, California. Weed Science 52:395–405

Kirk H, Paul J, Straka J et al (2011) Long-distance dispersal and high genetic diversity are implicated in the invasive spread of the common reed, *Phragmites australis* (Poaceae), in northeastern North America. American Journal of Botany 98:1180–1190
Kolar CS, Lodge DM (2001) Progress in invasion biology: predicting invaders. Trends in Ecology and Evolution 16:199–204
Lambert AM, Casagrande RA (2007) Characteristics of a successful estuarine invader: evidence of self-compatibility in native and non-native lineages of *Phragmites australis*. Mar. Ecol. Prog. Ser. 337:299–301
Lambert AM, Dudley TL, Saltonstall K (2010) Ecology and Impacts of the Large-Statured Invasive Grasses *Arundo donax* and *Phragmites australis* in North America. Invasive Plant Science and Management 3:489–494
Lambertini C, Gustafsson MHG, Frydenberg J et al (2006) A phylogeographic study of the cosmopolitan genus *Phragmites* (Poaceae) based on AFLPs. Plant Systematics and Evolution 258:161–182
Lambrinos JG (2004) How interactions between ecology and evolution influence contemporary invasión dynamics. Ecology 85:2061–2070
Latombe G, Pyšek P, Jeschke JM et al (2016) A vision for global monitoring of biological invasions. Biological Conservation 213:295–308
Lavergne S, Molofsky J (2007) Increased genetic variation and evolutionary potential drive the success of an invasive grass. Proc Natl Acad Sci USA 104:3883–3888
Lee CE (2002) Evolutionary genetics of invasive species. Trends in Ecology and Evolution 17:386–391
Lewandowski I, Scurlock JMO, Lindvall E et al (2003) The development and current status of perennial rhizomatous grasses as energy crops in the US and Europe. Biomass Bioenergy 25:335–361
Mack RN, Simberloff D, Lonsdale WM (2000) Biotic invasions: causes, epidemiology, global consequences, and control. Ecological Applications 10:689–710
Mariani C, Cabrini R, Danin A et al (2010) Origin, diffusion and reproduction of the giant reed (*Arundo donax* L.): a promising weedy energy crop. Annals of Applied Biology 57:191–202
Marks M, Lapin B, Randall J (1994) *Phragmites australis* (*P. communis*): threats, management, and monitoring. Nat. Areas J. 14:285–294
Marrs RA, Sforza R, Hufbauer RA (2008) When invasion increases population genetic structure: a study with *Centaurea diffusa*. Biological Invasions 10:561–572
McCauley DE, Smith RA, Lisenby JD et al (2003) The hierarchical spatial distribution of chloroplast DNA polymorphism across the introduced range of *Silene vulgaris*. Molecular Ecology 12:3227–3235
McCormick MK, Kettenring KM, Baron HM et al (2010) Extent and reproductive mechanisms of *Phragmites australis* spread in brackish wetlands in Chesapeake Bay, Maryland (USA). Wetlands 30:67–74
Meyerson LA, Lambert AM, Saltonstall K (2010) A Tale of Three Lineages: Expansion of Common Reed (*Phragmites australis*) in the U.S. Southwest and Gulf Coast. Invasive Plant Science and Management 3:515–520
Meyerson LA, Saltonstall K, Chambers RM (2009) *Phragmites australis* in eastern North America: A historical and ecological perspective. In: Silliman BR, Grosholz E, Bertness MD (eds) Salt Marshes Under Global Siege. University of California Press, California, USA, pp 57–82
Meyerson LA, Viola DV, Brown RN (2010) Hybridization of invasive *Phragmites australis* with a native subspecies in North America. Biological Invasions 12:103–111
Minckley WL (1992) Three decades near Cuatro Ciénegas. Mexico: photographic documentation and a plea for area conservation. J Ariz-Nev Acad Sci 26:89–118
Mohammadi SA, Prasanna BM (2003) Analysis of genetic diversity in crop plants—salient statistical tools and considerations. Crop Science 43:1235–1248
Mooney HA, Cleland EE (2001) The evolutionary impact of invasive species. Proceedings of the National Academy of Sciences of the United States of America 98:5446–5451

Ndlovu J, Richardson DM, Wilson JRU et al (2013) Elucidating the native sources of an invasive tree species, *Acacia pycnantha*, reveals unexpected native range diversity and structure. Ann Bot (Lond) 111:895–904

Novak SJ, Mack RN (2005) Genetic bottlenecks in alien plant species. Influence of mating systems and introduction dynamics. In: Sax DF, Stachowicz JJ, Gaines SD (eds) Species invasions: insights into ecology, evolution and biogeography. Sinauer Associates, Sunderland, MA, USA, pp 201–228

Pellegrin D, Hauber DP (1999) Isozyme variation among populations of the clonal species, *Phragmites australis* (Cav.) Trin. ex Steudel. Aquatic Botany 63:241–259

Perdue RE (1958) *Arundo donax*: source of musical reeds and industrial cellulose. Economic Botany 12:368–404

Pilu R, Cassani E, Landoni M et al (2014) Genetic characterization of an Italian Giant Reed (*Arundo donax* L.) clones collection: exploiting clonal selection. Euphytica 196:169–181

Pleines T, Jakob SS, Blattner FR (2009) Application of non-coding DNA regions in intraspecific analyses. Plant Systematics and Evolution 282:281–294

Polunin O, Huxley A (1987) Flowers of the Mediterranean. Hogarth Press, London, UK

Prentis PJ, Wilson JRU, Dormontt EE et al (2008) Adaptive evolution in invasive species. Trends in Plant Sciences 13:288–294

Rejmánek M (2000) Invasive plants: approaches and predictions. Austral Ecology 25:497–506

Rice D, Rooth J, Stevenson JC (2000) Colonization and expansion of *Phragmites australis* in upper Chesapeake Bay tidal marshes. Wetlands 20:280–299

Sakai AK, Allendorf FW, Holt JS et al (2001) The population biology of invasive species. Annual Review of Ecology and Systematics 32:305–332

Saltonstall K (2002) Cryptic invasion by a non-native genotype of the common reed, *Phragmites australis*, into North America. Proc Natl Acad Sci USA 99:2445–2449

Saltonstall K (2003a) Genetic variation among North American Populations of *Phragmites australis*: implications for management. Estuaries 26:444–451

Saltonstall K (2003b) Microsatellite variation within and among North American lineages of *Phragmites australis*. Molecular Ecology 12:1689–1702

Saltonstall K, Hauber DP (2007) Notes in *Phragmites australis* (Poaceae: Arundinoidae) in North America. Journal of the Botanical Research Institute of Texas I(I):385–388

Saltonstall K, Peterson PM, Soreng RJ (2004) Recognition of *Phragmites australis* subsp. Americanus (Poaceae: Arundinoideae) in North America: evidence from morphological and genetic analyses. Sida 21:683–692

Saltonstall K, Stevenson JC (2007) The effect of nutrients on seedling growth of native and introduced *Phragmites australis*. Aquatic Botany 86:331–336

Sax DF, Stachowicz JJ, Brown JH et al (2007) Ecological and evolutionary insights from species invasions. Trends in Ecology and Evolution 22:465–471

SEMARNAP (1997) Programa de Conservación de la Vida Silvestre y Diversificación Productiva en el Sector Rural, 1997–2000. Secretaría de Medio Ambiente, Recursos Naturales y Pesca, México, DF, Mexico

Sexton JP, McKay JK, Sala A (2002) Plasticity and genetic diversity may allow salt cedar to invade cold climates in North America. Ecol Appl 12:1652–1660

Simberloff D (2009) The role of propagule pressure in biological invasions. Annu Rev Ecol Evol Syst 40:81–102

Simberloff D, Martin JL, Genovesi P et al (2013) Impacts of biological invasions: what's what and the way forward. Trends in Ecology and Evolution 28:58–66

Souza V, Escalante AE, Espinoza LE et al (2004) Cuatro Ciénegas: un laboratorio natural de astrobiología. Ciencias 75:4–12

Souza V, Espinosa-Asuar L, Escalante AE et al (2006) An endangered oasis of aquatic microbial biodiversity in the Chihuahuan desert. Proc Natl Acad Sci USA 103:6565–6570

Souza V, Siefert JL, Escalante AE et al (2012) The Cuatro Ciénegas Basin in Coahuila, Mexico: An astrobiological Precambrian Park. Astrobiology 12:641–647

Stein BA, Kutner LS, Adams JS (2000) Precious Heritage: The Status of Biodiversity in the United States. Oxford University Press, Oxford, UK

Sterling TM, Thompson DC, Abbott LB (2004) Implications of invasive plant variation for weed management. Weed Technology 18:1319–1324

Strauss SY, Webb CO, Salamin N (2006) Exotic taxa less related to native species are more invasive. Proc Natl Acad Sci USA103:5841–5845

Suarez AV, Tsutsui ND (2008) The evolutionary consequences of biological invasions. Molecular Ecology 17:351–360

Tarin DA, Pepper E, Goolsby JA et al (2013) Microsatellites Uncover Multiple Introductions of Clonal Giant Reed (*Arundo donax*). Invasive Plant Science and Management 6:328–338

Templeton AR (2006) Population genetics and microevolutionary theory. John Wiley, Hoboken, NJ, USA

Trewick SA, Morgan-Richards M, Chapman HM (2004) Chloroplast DNA diversity of *Hieracium pilosella* (ASTERACEAE) introduced to New Zealand: reticulation, hybridization, and invasion. American Journal of Botany 91:73–85

Vellend M, Harmon LJ, Lockwood JL et al (2007) Effects of exotic species on evolutionary diversification. Trends in Ecology and Evolution 22:481–488

Vilà M, Espinar JL, Hejda M et al (2011) Ecological impacts of invasive alien plants: a meta-analysis of their effects on species, communities and ecosystems. Ecol Lett 14:702–708

Ward DB (2010) North America has two species of *Phragmites* (Gramineae). Castanea 75:394–401

Ward SM, Gaskin JF, Wilson LM (2008) Ecological Genetics of Plant Invasion: What Do We Know? Invasive Plant Science and Management 1:98–109

Ward SM, Reid SD, Harrington J et al (2008) Genetic variation in invasive populations of yellow toadflax (*Linaria vulgaris*) in the western United States. Weed Sci 56:394–399

Williams CMJ, Biswas TK, Schrale G et al (2008) Use of saline land and wastewater for growing a potential biofuel crop (*Arundo donax* L.). In: Irrigation Australia 2008 Conference CD of Proceedings, Melbourne, Australia, 20–22 May. Available from: http://irrigation.org.au/publications-resources/2008-irrigation-australia-conference-papers

Zhao R, Cheng Z, Lu W et al (2006) Estimating genetic diversity and sampling strategy for a wild soybean (*Glycine soja*) population based on different molecular markers. Chinese Science Bulletin 51:1219–1227

Zhao J, Solis-Montero L, Lou A et al (2013) Population structure and genetic diversity of native and invasive populations of *Solanum rostratum* (Solanaceae). PLoS One 8:e79807

Zohary M (1962) Plant life of Palestiene: Israel and Jordan. Ronald Press, New York, NY, USA

Chapter 16
Disturbance and the Formation and Colonization of New Habitats in a Hydrological System in the Cuatro Ciénegas Basin

Irene Pisanty, Mariana Rodríguez-Sánchez, Polenka Torres Orozco, Luisa A. Granados-Hernández, Stéphanie Escobar, and María C. Mandujano

Abstract Disturbance can open new habitats that follow different plant colonization patterns. The Churince hydrological system from the Cuatro Ciénegas Basin has been desiccating progressively since 2003 because of the disturbance of the underground hydrological flow. The terminal lagoon and the last part of the Churince river have had no surface water since 2009, and the Intermediate Lagoon is now dry. The drying up of the system generated three new habitats: sinkholes, formed by the dispersion of soils in the South bank of the Churince river, the bed of the dry river, and the lacustrine bed of the desiccated Churince lagoon. Sinkholes, which started forming around 2003 at a high rate, are of different shapes, sizes, and depths. New sinkholes can be formed while others can be closed by the accumulation of sand and plants; others have persisted during at least nine years since their formation. Our aim was to record early vegetation stages in these three newly created habitats.

We found continuous changes in water content, as well as in plant species richness and plant cover. While water flowed towards the plain, many sinkholes had water, but today they are dry. Sinkholes are colonized mainly by riparian species, e.g., *Samolus ebracteatus* var. *coahuilensis* and *Flaveria chlorifolia*, which are also establishing in the dry river and lacustrine beds. Species composition and frequency in the sinkholes change as water availability diminishes, reducing the frequency of

I. Pisanty (✉) · M. Rodríguez-Sánchez · P. T. Orozco · L. A. Granados-Hernández · S. Escobar
Facultad de Ciencias, Departamento de Ecología y Recursos Naturales, Universidad Nacional Autónoma de México, UNAM, Mexico City, Mexico
e-mail: ipisanty@unam.mx

M. C. Mandujano
Departamento de Ecología de la Biodiversidad, Instituto de Ecología, Universidad Nacional Autónoma de México, UNAM, Mexico City, Mexico
e-mail: mcmandujano@iecologia.unam.mx

M. C. Mandujano et al. (eds.), *Plant Diversity and Ecology in the Chihuahuan Desert*, Cuatro Ciénegas Basin: An Endangered Hyperdiverse Oasis,
https://doi.org/10.1007/978-3-030-44963-6_16

the more hydrophilic species like *Schoenus nigricans* and *S. ebracteatus* var. *coahuilensis* and increasing that of less water demanding and more salt and gypsum tolerant plants, as *Sporobolus airoides* and *F. chlorifolia*. The same pattern is expected in the lake and river beds, where colonization started several years later than in the sinkholes. Vegetation patterns and species composition and cover were influenced by the availability of water, which diminished in time due to the desiccation of the hydrological system.

Keywords Disturbance · Desiccation · Colonization · Sinkholes · Succession · River and lake bed · Riparian species

Introduction

Disturbance is a relatively discrete event in time with disruptive effects on environmental physical and biotic features that can have structuring effects at different ecological scales affecting ecosystems, communities, and populations (Turner 1989). A common result of disturbance is the formation of new habitats because of the destruction and substitution of the original habitats, and the consequent effects on nearby places. Colonization processes usually follow this opening of new habitats (White and Jentsch 2001; White and Pickett 1985).

Desert springs and wetlands, as well as other isolated water bodies, are intensely used and overexploited all over the world. Several underground water systems have been recently reduced, or even have completely disappeared, thus deeply disturbing nearby and long-distance terrestrial ecosystems depending on them (Briggs 1996; Capon and Dowe 2007; Hendrickson et al. 2008; Unmack and Minckley 2008; Sivinski and Tonne 2011; Souza et al. 2018).

The disturbance of ground water systems in the Cuatro Ciénegas Basin (CCB) and in nearby valleys, like Ocampo and El Hundido, due to overexploitation of aquifers mainly for agriculture, is having huge ecological consequences (Evans 2005; Souza et al. 2006, 2018; Rodríguez et al. 2007; Wolaver et al. 2008; Pisanty et al. 2013), including the progressive desiccation of the Churince system (Fig. 16.1), which has shown alterations continuously since 2003 (Souza et al. 2006, 2018; Pisanty et al. 2013). The first parts to lose water in the Churince system were the furthest from the spring at *poza* Churince, located near the base of the San Marcos y Pinos *sierra*; the largest terminal lake of the basin (i.e., the Churince Lagoon) was the first to dry up in 2008, followed by the nearby river mouth (RM), the part of the river that drained in this lagoon, and by the Intermediate Lagoon, that dried recently, in 2018 (De Anda et al. 2018).

The first consequences of this major disturbance are the loss of surface water and, consequently, the change of the riparian habitats, followed by the opening of new disturbance-induced habitats represented by the dry beds of the Churince river

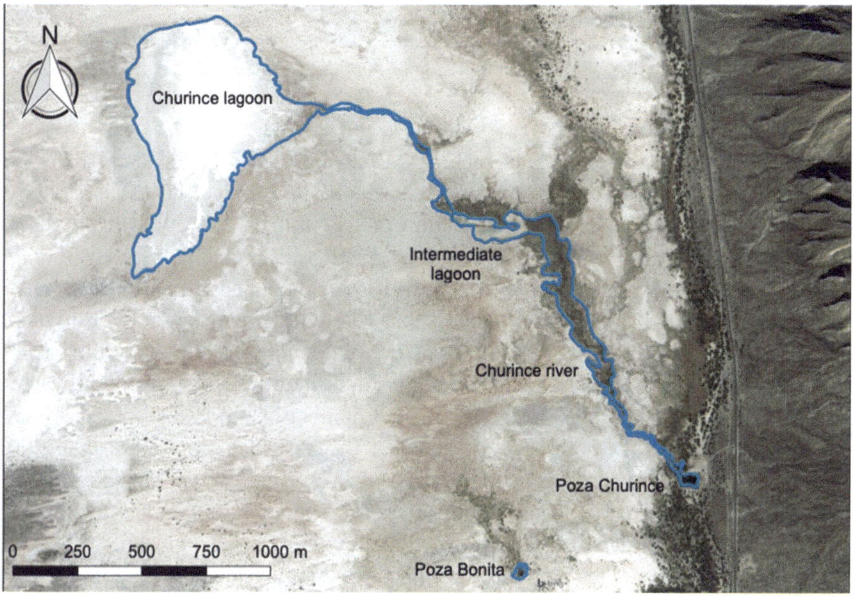

Fig. 16.1 The Churince System in the Cuatro Ciénegas Basin. Modified from Google Earth by S. Islas

(DRB) and Churince lagoon (DLB), the latter including the dry RM. The lagoon and the RM have been dry since 2009. As the lagoon dried, its bed first became marshy and then firmer, with less evidence of subsurface water. In the terminal part of the system, the river began to desiccate in 2009, and has been dry since 2012. However, during this study period there was still water in the last part before the Intermediate Lagoon, and today some water remains between the latter and the spring, implying a water gradient between the final and the initial part of the system. In 2015, seedlings and some mature plants of several species could be found in the part of the river where water was no longer flowing, along ca. 600 m from the RM. In the lake bed, one species (*Flaveria chlorifolia*) was the main colonizer. Additionally, a complex set of sinkholes (S) was formed starting in 2003 in the surroundings of the last part of the river and of the Intermediate Lagoon, due to the subsurface flow of the water that could not reach the RM (Pisanty et al. 2013).

In this chapter, using both community and population dynamics, we describe the plant colonization process of the different environments generated by the disturbance caused by the progressive drying of the Churince system, including sinkholes (S), the dry lake bed (DLB), and the dry river bed (DRB). These analyses allow us to understand the patterns of early successional stages in the three environments and the ecological behavior of the different plant species that colonize them, most of which characterized the original riparian habitat.

Methods

Study Site

Cuatro Ciénegas Basin is located at 27° 11′ 24″–26° 42′ 36" N and 102° 48′ 00″–101° 54′ 36" W, in the northern state of Coahuila, Mexico. It has an arid climate, with an annual average precipitation of 212 mm (CONANP 1999) which fluctuates considerably (Montiel-González et al. 2018). It is characterized by a high specific and ecosystem diversity (Minckley 1969; Souza and Eguiarte 2018) that includes aquatic and semiaquatic systems that belong to five different hydrological systems. Many *pozas* (spring fed pools) can be found in different parts of the basin. Pits and *pozas*, due to substrate characteristics and the flow of underground water, can be ephemeral or long lasting, so their exact number is impossible to determine, but Hendrickson et al. (2008) recognized that the basin contains more than 200 water bodies, including *pozas*, springs, and rivers, as well as two terminal lakes, and Meyer (1973) reported the formation of temporal, ephemeral pits. One of the systems, known as Churince System, is formed by the *poza* Churince, where the spring that feeds the system is located, the Churince river, an intermediate lagoon and the Churince Lagoon, the largest terminal (or desiccation) lake of the basin (Fig. 16.1).

Three different new environments were formed due to the desiccation process:

1. *Sinkholes (S)*. A set of soil depressions formed since 2003 in ca. 600 m^2, in a semi-parallel pattern along the South bank of the river in the terminal part of the system. These soil depressions are formed when water dissolves salt and gypsum particles, causing the loss of cohesion between them (Umesh et al. 2011). Sinkholes are relatively discrete environments, with either water or high humidity, where the extreme temperatures and strong winds of the desert are buffered (Fig. 16.2a). Originally, the flatland was recognized as a zone with no evident vegetation (Pinkava 1984), and only sparse, short stems of the halophyte *Distichlis spicata* (Poaceae) were found.
2. *Dry river bed (DRB)*. The last part of the river remains without surface water since the end of 2009. Many seedlings of riparian species can be found since the end of 2015 (Fig. 16.2b).

Fig. 16.2 Three disturbance-induced microenvironments formed due the desiccation of the terminal part of the Churince system, Cuatro Ciénegas, Coah: (**a**) sinkholes (2010, 2012); (**b**) dry river bed (2017); (**c**) dry lagoon bed (2017)

Fig. 16.2 (continued)

3. *Dry lake bed (DLB)*. Churince lagoon remains without surface water since 2009. The RM is included in this environment. *Flaveria chlorifolia* (Asteraceae), an herbaceous perennial plant associated with humid, saline and gypsic soils, started establishing in 2014, forming discrete patches (Fig. 16.2c).

Field Methods

1. *Sinkhole dynamics and colonization*

A census of the sinkholes was made from January 2008 to September 2017. In a total of 51 census, all open sinkholes were considered, and new ones were identified and incorporated as such. When a sinkhole closed, it was also registered and kept under observation because some re-opened. Re-opened sinkholes were not considered as new ones, and they were counted as open.

In every census, the presence or absence of water was registered, and specific composition, richness, and individual plant cover were determined in every sinkhole. Specific cover was determined adding individual cover, approximated to the area of an ellipse.

2. *Colonization of the river bed*

In June 2015, 13 plots were set every 50 m along the 600 m between the RM and the part of the river that still had surface water. Eight plots were 2 m × 10 m, and five, placed in the narrower part of the river, were 0.9 m × 10 m. Bimonthly, from June 2015 to January 2017, all plants in the plots were identified and plant cover was determined.

Frequency and plant cover of species colonizing the river bed from June 2015 to January 2017 were analyzed with an ANOVA. A Bray–Curtis similarity index was calculated for plant cover.

3. *Colonization of the lake bed*

After identifying *F. chlorifolia* as almost the only species establishing in the DLB, the demography of this species was chosen as the tool to analyze the colonization of this environment, contrasting with the community approach used in the other two. Six plots of 5 m × 5 m were set where patches of the species were found. In each plot, all individuals were numbered and tagged. Height and reproductive stage of each individual were registered bimonthly from September 2015 to September 2017. Established plants were classified in four height categories (1 = 0.1–4 cm; 2 = 5–15 cm; 3 = 16–30 cm; 4 ≥ 30 cm), and additional categories were included for the seed stage and for adult plants with dry aerial parts. Height was chosen to build categories because it is more accurate than cover due to the growth form and branching pattern of this species.

A Lefkovitch average transition probability matrix was built for each zone, followed by projection matrices, using the standard iterative procedure (Caswell 2001).

The finite growth rate (λ) was calculated once the stable stage distribution was obtained and elasticities (De Kroon et al. 1986; Caswell 2001) were calculated to evaluate the relative contribution of each process to the finite growth rate.

Results

1. Colonization of Sinkholes

A total of 247 sinkholes were censed between January 2008 and September 2017. In January 2008, when the census started, 126 sinkholes were identified. The maximum number of open sinkholes (157) was registered in June 2010, while the minimum (62) corresponds to August 2017 (Fig. 16.3). Sixty-two new sinkholes were formed between January 2008 and November 2011, while there was still water in the river, and another 61 opened since January 2012, when the latter was already dry. Contrastingly, only 22 sinkholes closed while the river had water and 162 did so when it was dry (Fig. 16.3).

Most sinkholes had plants throughout all the census period, but the presence of water was variable and, overall, the proportion of sinkholes with water decreased through time. In the first census, 50.79% of the sinkholes had water, but since 2012, when the river bed dried, only a few did. From May 2016 until September 2017 none had water (Fig. 16.3).

A total of 15 plant species colonized the sinkholes (Table 16.1). The average number of species in sinkholes with plants was 2.39, with a maximum of seven and a minimum of one species. Most of these species are riparian perennial herbs. Some sinkholes had no plants in some census. The most frequent species throughout the census period were *Samolus ebracteatus* var. *coahuilensis*, *Flaveria chlorifolia*,

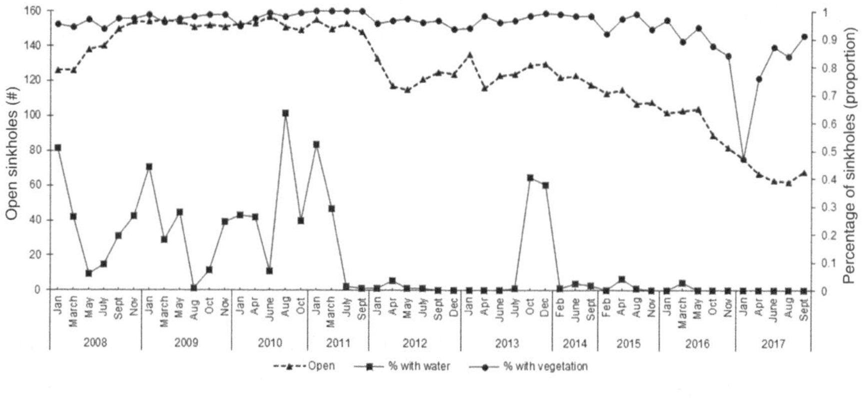

Fig. 16.3 Open sinkholes and proportion of sinkholes with water and with vegetation in the terminal part of the Churince system, Cuatro Ciénegas Basin, Coahuila. Census period: January 2008–September 2017

Table 16.1 Colonizing plant species in sinkholes and on the dry lagoon and river beds of the Churince system, Cuatro Ciénegas Basin

Species	Life cycle and growth habit	Environment
Cirsium coahuilense Ownbey & Pinkava; Asteraceae	Perennial/Erect Herbaceous	Si
Dasylirion sp.; Asparagaceae	Perennial/Shrub	Si
Distichlis spicata Greene; Poaceae	Perennial/Erect Herbaceous/ Clonal	Si/Rb
Erigeron cuatrocieneguensis G.L. Nesom; Asteraceae;	Annual/Erect Herbaceous/ Endemic	Si
Eustoma exaltatum L.; Gentianaceae	Annual or short lived perennial/ Erect Herbaceous	Si/Rb
Flaveria chlorifolia A. Gray; Asteraceae	Perennial/Erect Herbaceous or shrubby	Si/Rb/Lb
Haploesthes robusta I.M. Johnst; Asteraceae	Annual/Erect Herbaceous	Si/Rb
Nerysirenia incana Rollins; Brassicaceae	Perennial/Erect Herbaceous/ Gypsophyte	Si
Polygala parrasana Brandegee; Polygalaceae	Perennial/Herbaceous Sprawling	Si
Prosopis glandulosa Torr.; Fabaceae	Perennial/Erect Tree	Si/Rb
Sabatia tuberculata J.E. Williams; Gentianaceae;	Annual/Erect Herbaceous/ Endemic	Si/Rb
Samolus ebracteatus var. *coahuilensis* Henrickson; Primulaceae	Perennial/Herbaceous Sprawling/ clonal	Si/Rb
Schoenus nigricans L.; Cyperaceae	Perennial/Erect Herbaceous	Si/Rb
Sesuvium verrucosum Raf.; Aizoaceae	Annual/Erect Herbaceous	Si/Rb
Sporobolus airoides Torr.; Poaceae	Perennial/Erect Herbaceous/ Clonal	Si/Rb

Sinkhole (Si), River bed (Rb), Lake bed (Lb)

Distichlis spicata, *Schoenus nigricans*, and *Sporobolus airoides* (Table 16.1; Fig. 16.4). While water availability in the sinkholes was high, *S. nigricans* and *S. ebracteatus* var. *coahuilensis* were very frequent: however, as the river and the sinkholes lost their water, their frequency decreased (Fig. 16.4a, b). The frequency of *F. chlorifolia* (Fig. 16.4c) increased in 2015, and that of *D. spicata* remained relatively constant, although it periodically decreases in the cold season (Fig. 16.4d). *S. airoides* showed a low frequency during the first years, but since 2013 it increased until it was present in more than half of the sinkholes (Fig. 16.4e). Frequencies of the remaining species were much lower and are not shown in the figure.

Plant cover of the five most frequent species changed through time (Fig. 16.5). The loss of cover of *S. ebracteatus* var. *coahuilensis* coincides with the decrease of its frequency, while for *S. nigricans* frequency decreases more than cover, indicating the prevalence of large clumps in some sinkholes. The cover of *S. airoides* tends to increase with its frequency, while that of *D. spicata* remains low almost constantly, despite its high frequency.

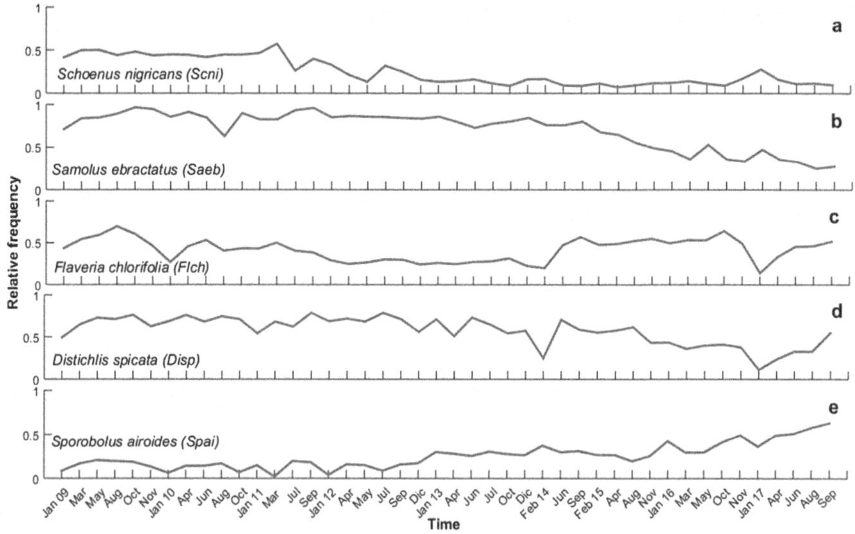

Fig. 16.4 Relative frequency of the most common colonizer plant species in the sinkholes at the terminal part of the Churince System, Cuatro Ciénegas, Coahuila. Census period: January 2009–September 2017. Number of census = 51

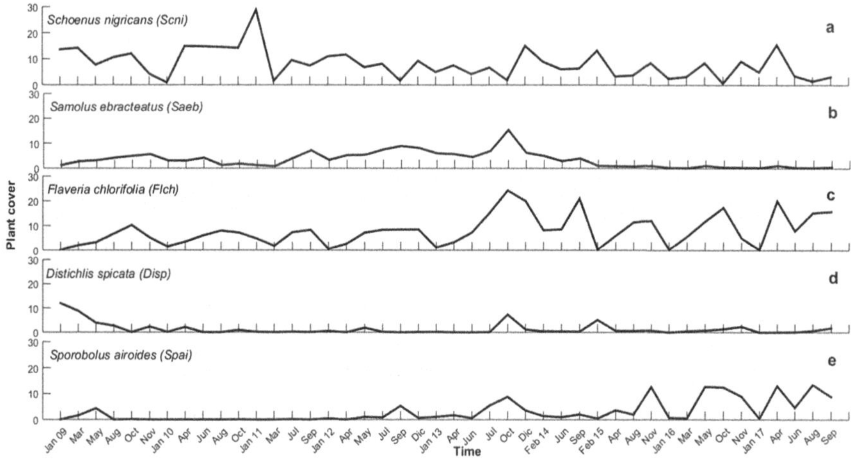

Fig. 16.5 Plant cover of colonizing species of sinkholes of the terminal part of the Churince System, Cuatro Ciénegas, Coahuila. Census period: January 2009–September 2017. Number of census = 51

2. Colonization of the River Bed

Ten different riparian species of plants colonized the DRB from June 2015 to June 2017 (Table 16.1). In this environment, three statistically significantly different groups based on species frequencies were identified with an ANOVA ($F_{8,12} = 14.92$, $p < 0.0001$): (1) *F. chlorifolia, S. airoides, D. spicata, S. ebracteatus* var. *coahuilensis,* and *Sesuvium verrucosum* (highest frequency); (2) *S. verrucosum* and *S. nigricans* (intermediate frequency); and (3) *Sabatia tuberculata, Eustoma exaltatum, Prosopis glanudulosa*, and *Haploesthes robusta* (lowest frequency).

Plant cover on the river bed changed in time ($F_{8,12} = 14.92$, $p < 0.0001$). The maximum cover values were registered on November 2015, and June and August 2016, while its minimum values were on January 2016 and 2017, showing a seasonal pattern. Lower plant cover values were registered in the coldest month (January) in both years. Specific plant cover varied between plots (Fig. 16.6a). The highest plant cover was in plots 7–13, and the lowest in plots 1–6. *F. chlorifolia, S. airoides*, and *D. spicata* were found in all plots (Fig. 16.6a); *S. ebracteatus* var. *coahuilenis* was not present in plots 3 and 13, and *S. nigricans* was found only from plot 6 to 12. All the other species were found in almost all plots but with a lower cover (Fig. 16.6b).

The Bray–Curtis similarity analysis shows three main clusters (Fig. 16.6b), one (S1) corresponding to the plots near the RM (1 to 6 plus 8), and two other (S2) closer to where water was still found, one including plots 7 and 9–11, and another formed by plots 12 and 13 (Fig. 16.6b).

3. Colonization of the Lake Bed

The DLB was the last disturbance-induced environment to be colonized. A total of 316 individuals of *F. chlorifolia* were monitored bimonthly. At the end of the study, only 20% of these individuals survived. Only two recruitments were observed.

During the coldest month (January) of 2016 and 2017, over 50% of the individuals entered the desiccated category but, after the cold period, meristems were activated and plants began to grow again, reaching the highest number of plants in the category of the tallest individuals (category 5) at the end of summer (Fig. 16.7).

The finite growth rate ($\lambda = 0.77$) indicates that the population is decreasing by 23%. The elasticity analysis showed that the two main processes that contribute the most to lambda are the stasis of the tallest individuals and of those that desiccated (0.13). The third most important process is the stasis in the category 3 (0.08) (Fig. 16.7).

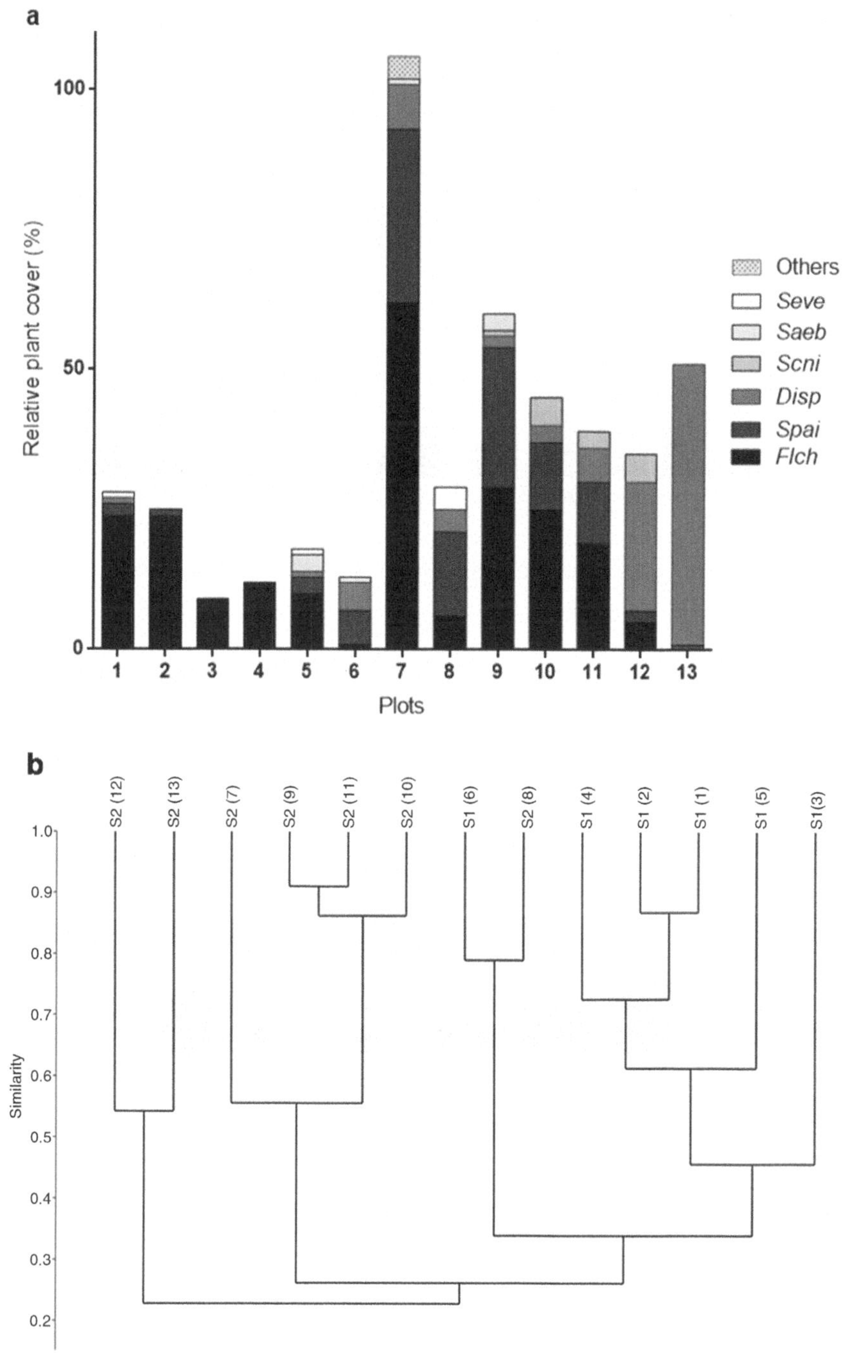

Fig. 16.6 Relative plant cover on the DRB: (**a**) mean plant cover in plots; (**b**) Bray–Curtis analysis. S1 = plots near to the dry RM. S2 = plots near to where the river had water

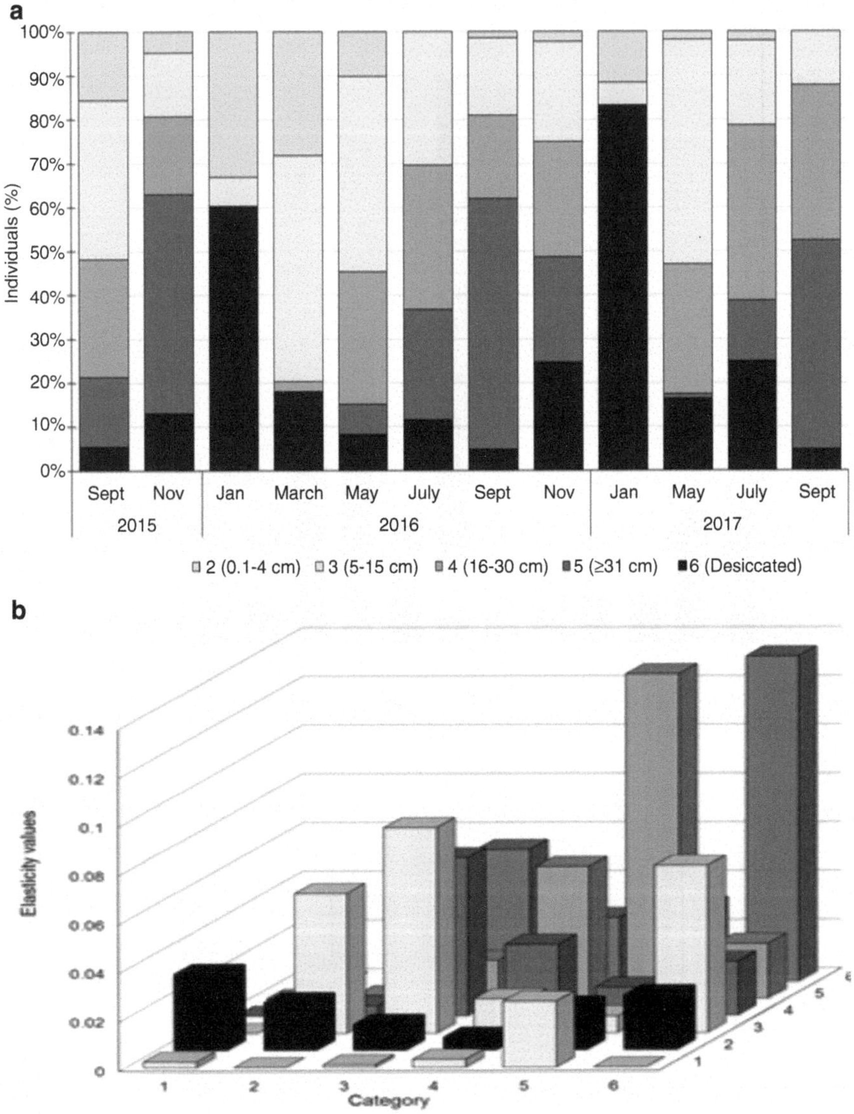

Fig. 16.7 Population dynamics of *Flaveria chlorifolia* on the dry lake bed: (**a**) population structure; (**b**) elasticities. Numbers in *x* and *y* axis represent each category in the matrix (1: seed; 2: 0.1–4 cm; 3: 5–15 cm; 4: 16–30 cm; 5: 31 cm; 6: desiccated)

Discussion

Disturbance in arid and semiarid ecosystems generates spatial heterogeneity, which plays an influential role in population and community dynamics, and explains, among other things, patchy plant distributions at both levels. Ecological succession, as a regional process, is seldom observed in these systems, because it is extremely slow; however, colonization of microenvironments or microhabitats, and differences in population or community behavior are common (Polis 1991; Mandujano et al. 1998; Vega and Montaña 2004). The same type of process is observed after disturbance both in the disturbed habitat and in its neighborhood.

Riparian habitats in deserts are more mesic than the surrounding areas (Briggs 1996). Species growing in different types of desert wetlands and springs face different selective pressures that include anoxic environments for the roots of plants growing on continuously or intermittent flooded sites, xeric conditions for aerial parts, and different types of saline or gypsic conditions (Briggs 1996; Ezcurra et al. 1998; Capon and Dowe 2007). The colonization potential of these species depends on their capacity to adapt to this complex set of selective forces. This capacity is evident in the riparian species, characteristic of the borders of undisturbed water bodies that colonize the environments created by disturbance as the S, DRB, and DLB analyzed here.

The three different habitats formed after the disturbance of the Churince system that we considered in this study have different characteristics and the periods in which colonization has been taking place vary among them; nonetheless, they are consistently and predominantly colonized by the same set of riparian species.

Sinkholes proved to be very dynamic since they can remain open, or close permanently or temporarily, and new ones can be formed. Formation of new sinkholes decelerated progressively as water from the river diminished together with the sublevel flow, for the dispersion of soils does not happen without the presence of water that dissolves salts and causes loss of particle cohesion (Umesh et al. 2011). Sinkholes closed more frequently after they stopped having water. Although the same five species remained as the most frequent through time, specific frequencies did change, together with cover. While the river had water, many sinkholes did also, and the most frequent species were the most hydrophilic, as *S. ebracteatus* var. *coahuilensis* and *S. nigricans*. As sinkholes became drier, species that are less water demanding, as *S. airoides*, and more tolerant to gypsum, as *F. chlorifolia*, became more frequent. Plant cover strongly depends on growth forms, but it reflects the success a species has in a given environment (Kent 2011). The loss through time of plant cover of some of the first colonizers, as *S. ebracteatus* and *S. nigricans*, reflects the change of conditions and of selective pressures as sinkholes become drier, as well as the limited responses some species have. Changes take place both seasonally and in a longer term. Thus, frequency reflects the colonizing ability of a species, while cover indicates its success after establishment. As underground and sublevel water continue to decrease, the more water demanding species, as *S. ebracteatus* var. *coahuilensis* and *S. nigricans*, are likely to continuously become less frequent and lose plant cover, and eventually disappear, while more resistant species as *F. chlorifolia* and *S. airoides* become dominant, at least temporarily. However, in a longer term, we can expect all riparian species will become locally extinct in the Churince system as desiccation continues. The concentration of salt and gypsum is probably changing due to the disturbance of the hydrological system, and some gypsophytes as *Nerisyrenia incana* and *Nama* spp. (see Ochoterena et al. 2020, this volume), formerly absent, can now be found on the flatland (pers. obs.).

The presence of water in the sinkholes depends on many factors, including sub-level availability of water and meteorological conditions, as seen in August 2010, when the number of sinkholes with water increased due to high precipitations caused by a hurricane. The peak of sinkholes with water in 2013 was caused because the main gate through which water is extracted from the basin through an artificial canal, that diverts it even beyond CCB, was closed for a short period, allowing the wetland and the river to recover water temporarily (Souza and Eguiarte 2018). As soon as the gate was opened again, water was lost and was never recovered.

The DRB was colonized much later than the sinkholes, because it only became an available habitat for terrestrial plants until the end of 2014. Despite the short period of colonization, the bed was quickly occupied by 10 riparian species, all found also in sinkholes with similar frequencies for that time period. Seeds allowing this colonization process were probably dispersed from recently formed fruits by plants on the sinkholes and from other cohorts represented in the seed bank. The DRB did not have the same conditions over space and time. The CCB has clear seasonal changes (De Anda et al. 2018) which explain the observed short term variations of plant cover on the DRB. Plant cover was higher in the rainy months than in the dry ones. The Bray–Curtis similarity analysis showed a spatially discontinuous establishment pattern along the DRB, determined by the distance between plots and the presence of nearby water. Two zones can be recognized: S1 was the zone of the DRB with drier soils, because it was the first to desiccate. In S2, since superficial water was intermittently present, water availability was higher, a fact confirmed by the presence of *S. nigricans*. The highest number of species was found in S2 apparently because the more tolerant plants, which can stand hydric stress produced by desiccation, were found in it as well as in S1. Plant cover of the more tolerant species was higher in S2 than in S1, confirming their hydrophilic character. Plant cover was relatively scarce except for plot 7, where it reached more than 100%, which could be explained by the higher humidity in this plot that enhanced the cover of *F. chlorifolia*.

Although the Churince lagoon was the first part to desiccate, it passed through a very swampy period before becoming suitable for plant establishment, which started later than in the sinkholes and on the DRB, probably due to its high gypsum content and low nutrient concentration (Escudero et al. 2014). Although scattered and scarce individuals of *S. ebracteatus* var *coahuilensis* and *D. spicata* could be found, *F. chlorifolia* was the main colonizer on the DLB, probably resulting of a massive, but punctual, germination and establishment event, associated to an unidentified environmental trigger and to its successful germination in natural conditions (Peralta-García et al. 2016). Establishment of *F. chlorifolia* was enhanced by its tolerance and relative affinity to gypsum soils (Powell 1978; Flores-Olvera et al. 2016). Despite its seasonal decline, its perennial character allows successful individuals to persist, an event relevant to its population dynamics, as observed in the elasticities values. The reactivation of meristems after the cold season is also crucial for the population, and it is the second major process of importance over the population growth rate. However, the high mortality rates in the DLB coincided with the death of most of the established shrubs in 2017 (pers. obs.), after a long and cold winter which, apparently, limited the survival of meristems.

Colonizers of the new habitats can show characteristics identified as favorable for the colonization of new microsites in deserts (Ezcurra et al. 2020, this volume), as high fecundity rates (Noy-Meir 1973, 1985), like in *S. ebracteatus* var. *coahuilensis* (Pisanty et al. 2013) and *F. chlorifolia* (Rodríguez-Sánchez 2018), and rapid germination responses, as those reported by Peralta-García et al. (2016) for them, *E. exaltatum,* and *S. tuberculata*, as well as for *S. nigricans* (Peralta-García et al. 2020, this volume). Clonality, present at least in *S. ebracteatus* var *coahuilensis*, *S. nigricans*, *S. airoides,* and *D. spicata*, has also been identified as a successful strategy for establishment in arid zones (Montaña et al. 1990; Mandujano 2007). All these characteristics play a relevant role in the way each species responds to disturbance.

Heterogeneity in arid and semiarid ecosystems plays an influential role in population and community dynamics, and explains, among other things, patchy distributions at both levels. Succession, as a regional process, is seldom observed in these systems, because it is extremely slow; however, colonization of microenvironments or microhabitats, and differences in population or vegetation dynamics are common (Polis 1991; Mandujano et al. 1998; Vega and Montaña 2004), although their effect at a larger scale are unknown.

The colonization of the three newly formed habitats reported in this chapter offers a unique possibility of observing succession after a disturbance on a semiarid zone in newly formed, different microenvironments that offer, at least temporarily, water availability, but are heterogeneous in space and in time. Changing conditions force the process to be dynamic in a relatively short time, and the fact that riparian species are the colonizers indicate the presence of some water that makes these habitats suitable for species that used to live close by.

The changing conditions, especially in sinkholes, are reflected by the change in frequency and cover of the main colonizing species. *S. airoides* can grow in alkaline and gypsic soils that offer water at least temporarily, and tolerate both drought and flooding once established. (USDA 2018), which explains the increase in frequency observed throughout this study. The increasing presence of *P. glandulosa* indicates there is still deep ground water available (Nilsen et al. 1983). Seeds of this species were spread by animals, mainly by horses that used to roam the study site grazing and looking for water.

The disappearance of the Gentianaceae *Sabatia tuberculata* from the borders of water bodies (pers. obs.) and its increasingly low frequency in the sinkholes is a cause of concern, because this is an endemic species (Williams 1982; Hendrickson et al. 2008; Vela Coiffier et al. 2015; Villarreal-Quintanilla et al. 2017) and it is not known whether it is endangered or not (SEMARNAT 2010; Pisanty and Rodríguez-Sánchez 2017).

The presence of riparian plants away from their natural habitats indicates the latter are being disturbed. When these species grow in areas where they could not establish before, they indicate the presence of water in areas that are usually dry, and their sole presence should be considered as an early warning of underground disturbance, even if this is not evident in the water bodies that seem undisturbed.

Acknowledgements This research was financed by programs UNAM/PAPIIT IN231811 and UNAM/PAPIIT IN217915, and supported by the Department of Ecology and Natural Resources (Facultad de Ciencias, UNAM) and the alliance WWF-Carlos Slim Foundation. MRS was supported by a scholarship from CONACyT. We thank Juan Carlos Ibarra, director of the Cuatro Ciénegas protected area, Valeria Souza, and Martín Carrillo for their constant support and commitment; Cristina Pérez y Sosa, and Erick Peña for their contributions, and Gabriel Gálvez and Jazmín Sánchez for their help. We thank the experts of the National Herbarium (UNAM), the Herbario de la Facultad de Ciencias (UNAM), and Patricia Dávila, Nelly Diego, Hilda Flores-Olvera, Manuel Villarreal-Quintanilla, and Manuel Rosas for the identification of species. We also acknowledge Israel Solano Zavaleta, Mariana Hernández-Apolinar, Pedro E. Mendoza-Hernández and Marco Romero Romero for their technical support, as well as the people of the Ejido Seis de Enero and Alma Zertuche. The advice of the late Salvador Contreras is fully appreciated. This chapter constitutes a partial fullfillment of the Doctorado en Ciencias (Biología) of the Universidad Nacional Autónoma de México (UNAM) of Irene Pisanty Baruch

References

Briggs MK (1996) Riparian ecosystem recovery in arid lands. The University of Arizona Press, Tucson, AZ, USA

Capon S, Dowe JL (2007) Diversity and dynamics of riparian vegetation. In: Lovett S, Price P (eds) Principles of riparian lands management. Land & Water Australia, Canberra, Australia, pp 13–32

Caswell H (2001) Matrix population models. Construction, analysis, and interpretation. Sinauer Associates Inc, Sunderland, MA, USA

Comisión Nacional de Áreas Naturales Protegidas (CONANP) (1999) Programa de manejo del Área de Protección de Flora y Fauna Cuatrociénegas. Instituto Nacional de Ecología SEMARNAP, México, DF, Mexico

De Anda V, Zapata-Peñasco I, Eguiarte LE, Souza V (2018) Towards a comprehensive understanding of environmental perturbations in microbial mats from the Cuatro Ciénegas Basin by network inference. In: García-Oliva F, Elser J, Souza V (eds) Ecosystem ecology and geochemistry of Cuatro Ciénegas. How to survive in an extremely oligotrophic site. Springer International, Cham, Switzerland, pp 85–98

De Kroon H, Plaisier A, van Groenendael J, Caswell H (1986) Elasticity: the relative contribution of demographic parameters to population growth rate. Ecology 67(5):1427–1431

Escudero A, Palacio S, Maestre FT, Luzuriaga AL (2014) Plant life in gypsum: a review of its multiple facets. Biol Rev 90(1):1–18

Evans S (2005) Using chemical data to define flow systems in Cuatro Cienegas. MSc thesis, University of Texas in Austin, Austin, TX, USA

Ezcurra E, Felger RS, Russell AD, Equihua M (1998) Freshwater island in a desert sand sea: the hydrology, flora and phytogeography of the Grand Desierto oasis of Northwestern Mexico. Desert Plants 9:35–63

Ezcurra E, Martínez-Berdeja A, Villanueva-Almanza L (2020) The evolution of North American deserts and the uniqueness of Cuatro Ciénegas. In: Mandujano MC, Pisanty I, Eguiarte LE (eds) Plant diversity and ecology in the Chihuahuan Desert. Springer International, Cham, Switzerland

Flores-Olvera H, Czaja A, Estrada JL, Méndez UR (2016) Floristic diversity of halophytic plants of Mexico. In: Khan M, Boër B, Öztürk M, Clüsener M, Gul B, Breckle SW (eds) Sabkha ecosystems. Tasks for vegetation science, vol 48. Springer, Cham, Switzerland

Hendrickson DA, Marks JC, Moline AB, Dinger EC, Cohen AE (2008) Combining ecological research and conservation: a case study in Cuatro Ciénegas, Mexico. In: Stevens L, Meretsky VJ (eds) Aridland springs in North America: ecology and conservation. University of Arizona Press, Tucson, AZ, USA, pp 127–157

Kent M (2011) Vegetation description and data analysis. Willey-Blackwell, Chichester, UK

Mandujano MC, Montaña C, Méndez I, Golubov J (1998) The relative contributions of sexual reproduction and clonal propagation in *Opuntia rastrera* from two habitats in the Chihuahuan Desert. J Ecol 86:911–921

Mandujano MC (2007) La clonalidad y sus efectos en la biología de poblaciones. In: Eguiarte LE, Souza V, Aguirre X (eds) Ecología Molecular. Secretaría de Medio Ambiente y Recursos Naturales, Instituto Nacional de Ecología, Universidad Nacional Autónoma de México, Comisión Nacional para el Conocimiento y Uso de la Biodiversidad, México, DF, Mexico, pp. 215–250

Meyer ER (1973) Late-quaternary paleoecology of the Cuatro Ciénegas Basin, Coahuila, México. Ecology 54(5):982–995

Minckley WL (1969) Environment of the Bolsón of Cuatro Ciénegas, Coahuila, Mexico: with special reference to the aquatic biota. Science Series, vol 2. Texas Western Press, El Paso, TX, USA

Montaña C, Lopez-Portillo J, Mauchamp A (1990) The response of two woody species to the conditions created by a shifting ecotone in an arid ecosystem. The Journal of Ecology 78(3):789

Montiel-González C, Bautista F, Delgado C, García-Oliva F (2018) The climate of Cuatro Ciénegas Basin: drivers and temporal patterns. In: Souza V, Olmedo-Álvarez G, Eguiarte LE (eds). Cuatro Ciénegas Ecology, Natural History and Microbiology. Springer International, Cham, Switzerland, pp 35–44

Nilsen ET, Sharifi MR, Rundel PW, Jarrell WM, Virginia RA (1983) Diurnal and seasonal water relations of the desert phreatophyte *Prosopis glandulosa* (honey mesquite) in the Sonoran desert of California. Ecology 64:1381–1393

Noy-Meir I (1973) Desert ecosystems: Environment and producers. Annu Rev Ecol Syst 4(1):25–51

Noy-Meir I (1985) Desert ecosystem structure and function. In Evanari M, Noy Meir I, Goodall DW (eds) Ecosystems of the World. Elsevier, Amsterdam, Netherlands

Ochoterena H, Flores-Olvera H, Gómez-Hinostrosa C, Moore MJ (2020) Gypsum and plant species: a marvel of Cuatro Ciénegas and the Chihuahuan Desert. In: Mandujano MC, Pisanty I, Eguiarte LE (eds) Plant diversity and ecology in the Chihuahuan Desert. Springer International, Cham, Switzerland

Peralta-García C, Sánchez-Coronado ME, Orozco-Segovia A, Orozco-Segovia S, Pisanty-Baruch I (2016) Germination of four riparian species in a disturbed semi-arid ecosystem. S Afr J Bot 106:110–118

Peralta-García C, Pisanty I, Orozco-Segovia A, Sánchez-Coronado ME, Rodríguez-Sánchez M (2020) Germination of riparian species in natural and experimental conditions. In Mandujano MC, Pisanty I, Eguiarte LE (eds.) Plant diversity and Ecology in the Chihuahuan desert. Springer International, Cham, Switzerland

Pinkava DJ (1984) Vegetation and Flora of the bolson of Cuatro Cienegas region Coahuila, Mexico: IV. Summary, endemism and corrected catalogue. J Ariz Nev Acad Sci 19(1):23–47

Pisanty I, Pérez y Sosa M, Gálvez G (2013) Agriculture, water mismanagement and ecosystem transformations in the Cuatrociénegas Valley in the Chihuahuan Desert, Mexico. In: Schwabe K, Albaic J, Connor JD, Hassan RM, Meza L (eds) Drought in arid and semi-arid regions, a multi-disciplinary and cross-country perspective. Springer, Dordrecht, Netherlands, pp 199–216

Pisanty I, Rodríguez-Sánchez M (2017) Amenazas a la biodiversidad en Cuatro Ciénegas. In: La biodiversidad de Coahuila. Estudio de Estado. Vol. I CONABIO/Gobierno del Estado de Coahuila de Zaragoza, Saltillo, Coahuila, Mexico, pp 485–498

Polis GA (1991) The ecology of desert communities. The University of Arizona Press, Tucson, AZ, USA

Powell AM (1978) Systematic of *Flaveria* (Flaveriinae-Asteraceae). Ann Mo Bot Gard 65(2):590–636

Rodríguez JM, Souza V, Arriaga L (2007) Effect of the overexploitation of the aquifer of the Hundido valley and the impact on the ecological reserve of Cuatrociénegas, Coahuila. Ciencia FIC. Universidad Autónoma de Nuevo León 1:32-38

Rodríguez-Sánchez M (2018) Dinámica poblacional, germinación, establecimiento y crecimiento de *Flaveria chorifolia* (Asteraceae) en un sistema perturbado en el Valle de Cuatrociénegas, Coahuila, México. Master Thesis, Facultad de Ciencias, Universidad Nacional Autónoma de México, Mexico City, Mexico

Secretaría de Medio Ambiente y Recursos Naturales (SEMARNAT) (2010) Norma Oficial Mexicana NOM-059-SEMARNAT-2010. Published December 30, 2010, México, DF, Mexico

Sivinski RC, Tonne P (2011) Survey and assessment of aridland spring ciénegas in the southwest region. Section 6, Segment 25, Progress Report submitted to: NM Forestry Diversion Santa Fe and USDI-Fish and Wildlife Service, Region 2, Albuqerque, NM, USA

Souza V, Eguiarte LE (2018) In the beginning, there was fire: Cuatro Ciénegas Basin (CCB) and the long history of life on earth. In: Souza V, Olmedo-Álvarez G, Eguiarte LE (eds). Springer International, Cham, Switzerland, pp 21–34

Souza V, Espinosa L, Escalante A, Eguiarte LE, Farmer J, Forney L, Lloret L, Rodríguez J, Soberón X, Dirzo R, Elser J (2006) An endangered oasis of aquatic microbial biodiversity in the Chihuahuan desert. Proc Natl Acad Sci USA 103(17):6565–6570

Souza V, Eguiarte LE, Elser J, Travisano M, Olmedo-Álvarez G (2018) A microbial saga: how to study an unexpected hot spot of microbial biodiversity from scratch? In: Souza V, Olmedo-Álvarez G, Eguiarte LE (eds) Cuatro Ciénegas ecology, natural history and microbiology. Springer International, Cham, Switzerland, pp 1–18

Turner MG (1989) Landscape ecology: the effect of pattern on process. Annu Rev Ecol Syst 20:171–197

Umesh T, Dinesh S, Sivapullaiah P (2011) Characterization of dispersive soils. Mater Sci Appl 2(6):629–633

United States Department of Agriculture (USDA) (2018) Natural Resources Conservation Service. https://plants.usda.gov/core/profile?symbol=SPAI. Accessed 4 Jan 2019

Unmack PJ, Minckley WL (2008) The demise of desert springs. Aridland springs in North America: ecology and conservation. University of Arizona Press, Tucson, AZ, USA, pp 11–34

Vega E, Montaña C (2004) Spatio-temporal variation in the demography of bunch grass in a patchy semiarid environment. Plant Ecol 175(1):107–120

Vela Coiffier MP, Vela Coiffier EL, Lozano García DF (2015) Evaluación del conocimiento actual del APFF de Cuatrociénegas, Coahuila: FASE 1. Instituto Tecnológico y de Estudios Superiores de Monterrey. Centro de Calidad Ambiental. Informe final SNIB-CONABIO, proyecto No. JN002. México, D.F., Mexico

Villarreal-Quintanilla JA, Bartolomé JA, Estrada E, Ramírez H, Martínez SJ (2017) El elemento endémico de la flora vascular del Desierto Chihuahuense. Acta Bot Mex 118:65–96

White PS, Jentsch A (2001) The search for generality in studies of disturbance and ecosystem dynamics. In: Esser K, Lüttge U, Kadereit JW, Beyschlag W (eds) Progress in Botany, vol 62. Springer, Berlin, Germany, pp 399–450

White PS, Pickett STA (1985) Natural disturbance and patch dynamics: an introduction. In: Pickett STA, White PS (eds) The ecology of natural disturbance and patch dynamics. Academic Press, New York, NY, USA, pp 3–13

Williams JE (1982) A new species of *Sabatia* (Gentianaceae) from the Chihuahuan Desert. Southwest Nat 27(4):379–382

Wolaver B, Sharp JM Jr, Rodríguez JM, Ibarra Flores JC (2008) Delineation of regional arid karstic aquifers: an integrative data approach. Ground Water 46(3):396–413

Chapter 17
An Unlikely Movable Feast in a Desert Hydrological System: Why Do Life Cycles Matter

Mariana Rodríguez-Sánchez, Irene Pisanty, María C. Mandujano, Hilda Flores-Olvera, and Ana Karen Almaguer

Abstract Riparian species colonize habitats induced by disturbance along the hydrologic Churince System in Cuatro Ciénegas (Coahuila, Mexico), like sinkholes and the river and lacustrine dry beds, the dry riverbank, and the surrounding flatland. Among the colonizers of newly opened habitats, dominant species are scarce. They include *Samolus ebracteatus* var. *coahuilensis*, a perennial sprawling, hydrophilous herb that can form clones; *Flaveria chlorifolia*, an erect gypsovage perennial herb that can also grow as a shrub; and *Eustoma exaltatum*, a herb that usually behaves as an annual, but can also behave as a short-lived perennial. The life cycles and population dynamics of these species reflect short- and long-term responses to these heterogeneous, unpredictable, and disturbed habitats. Two main strategies of the life cycle of *F. chlorifolia* and *E. exaltatum* are key for their survival, especially in bare zones where abiotic conditions are harder: deciduous leaves which are lost during the cold season after producing flowers and fruits, and regrowth of new branches and leaves from their basal stem. The capacity of *E. exaltatum* to shift from an annual life cycle to a short-lived perennial when conditions allow it depends on root development and regrowth. *S. ebracteatus* var. *coahuilensis* is not completely deciduous and perennates through the activation of under or above ground meristems, and, as *F. chlorifolia*, it can produce underground leaves that

M. Rodríguez-Sánchez (✉) · I. Pisanty (✉) · A. K. Almaguer
Departamento de Ecología y Recursos Naturales, Facultad de Ciencias, Universidad Nacional Autónoma de México, UNAM, Mexico City, Mexico
e-mail: mrs@ciencias.unam.mx; ipisanty@unam.mx

M. C. Mandujano
Departamento de Ecología de la Biodiversidad, Instituto de Ecología, Universidad Nacional Autónoma de México, UNAM, Mexico City, Mexico
e-mail: mcmandujano@iecologia.unam.mx

H. Flores-Olvera
Departamento de Botánica, Instituto de Biología, Universidad Nacional Autónoma de México, UNAM, Mexico City, Mexico

M. C. Mandujano et al. (eds.), *Plant Diversity and Ecology in the Chihuahuan Desert*, Cuatro Ciénegas Basin: An Endangered Hyperdiverse Oasis,
https://doi.org/10.1007/978-3-030-44963-6_17

emerge afterwards. The maintenance of living stems is crucial for the survival of individuals and permanence of these populations.

Keywords Life cycle · Perennation · Riparian species · Disturbance · Demography

Introduction

Riparian plants establish on the borders of water bodies and wetlands or on floodplains. Riparian systems are considered biological diversification centers, and their local flora contributes highly to worldwide biodiversity (Naiman et al. 1993; Tockner and Stanford 2002). These communities usually depend on groundwater reservoirs, flooding frequency and duration, and in arid and semiarid environments they are strongly influenced by streamflow (Stromberg et al. 2015).

Changes in water availability (i.e., water flow shifts or reductions) caused by natural or human disturbances may bring ecological consequences such as diversity loss, changes in the distribution of the species that can colonize new environments, variations in richness and composition of riparian communities, and the spreading of non-native species (Stromberg et al. 1996; Shafroth et al. 2002; Tockner and Stanford 2002; Capon and Dowe 2007; Poff and Zimmerman 2010). Due to their high sensibility to water availability, it is possible to identify a disturbance in riparian ecosystems by analyzing some population and community patterns. Plants that are adapted to tolerate these conditions will survive selective pressures and colonize the newly opened habitats, thus initiating a successional process.

The Cuatro Ciénegas Basin (CCB) is characterized by the presence of five hydrologic systems, formed by interconnected *pozas*, lagoons, and rivers, fed by springs which result from the emergence of groundwater to the surface (CONANP 1999). CCB is considered the most important wetland in the Chihuahuan Desert (Souza et al. 2004).

During a long time, the CCB has suffered from water overexploitation mainly for the cultivation of alfalfa (*Medicago sativa*), used for cattle raising. This disturbance has several impacts over the whole basin, including the Churince System (CS), which was one of the best preserved hydrological systems within CCB, where many riparian species thrived along the riverbank, surrounded by a flatland with almost no vegetation growing on it (Minckley 1969; Pinkava 1984; CONANP 1999). The terminal Churince Lagoon (the biggest in CCB) is now desiccated, along with the river mouth and the last part of the river, which are also dry nowadays. The disturbance of the groundwater reservoirs has also affected the soil structure of the surroundings of the river, because salts dissolve due to the abnormal sublevel flow of water from the river to the flatland, causing the dispersion of soils and the consequent formation of pits and sinkholes in the area. The latter represent alternative habitats for riparian species (Pisanty et al. 2013; Pisanty et al. 2020, this volume). The dry beds of the

Churince Lagoon and of the river and their surroundings also represent new habitats; they remained with no apparent vegetation during the desiccation period, but since 2014 some of the species previously reported on the riverbank and already established in the sinkholes began to colonize them, starting a successional process (Pisanty et al. 2020, this volume).

Life histories result from the evolution of vital parameters, while the variations in life cycles reflect short-term responses to punctual environmental conditions. The analysis of a life cycle and its variation in time through demographic dynamics allows the understanding of the effects of selective pressures and of the responses plants have to them in specific circumstances (Stearns 1992; Silvertown et al. 1993; Salguero-Gómez et al. 2016).

The aim of this chapter is to describe and analyze the life cycle traits that allowed the colonization of different microhabitats of three riparian species in a disturbed site in the CS. We used a demographic model to compare the dynamics of different populations and life cycle traits (Silvertown et al. 1993; Crone et al. 2011). This approach involves comparisons of finite population growth rates and vital parameters (birth, death, and growth rates, and survival or permanence) of three different species and populations, as well as the assessment of proportional contributions of specific processes or life cycle stages to the population growth rate (de Kroon et al. 1986; Silvertown et al. 1993; Caswell 2001). Two of the species are the perennials *Samolus ebracteatus* var. *coahuilensis* and *Flaveria chlorifolia*, which are among the most common first colonizers of the sinkholes. *Eustoma exaltatum* is reported mostly as an annual (Shinners 1957; Villarreal 2001) but can also behave as a short-lived perennial (Turner 2014) that forms dynamic, ephemeral patches.

Changes in life cycle and demographic parameters should be considered as early indicators of the changing environmental conditions these species endure, most of which derive from what seems to be an irreversible disturbance of the hydrological system.

Methods

Study Site

The CS is located within the CCB, at 26° 51′ N and 102° 08′ W; with a semiarid climate and an average annual precipitation of 220 mm. It is a hydrological system with a spring (*poza* Churince) that provides water to the Intermediate Lagoon, and to the terminal Churince Lagoon, interconnected by a creek (Churince River) (Fig. 17.1a). In this study we considered four specific zones of the terminal part of the CS:

1. Dry lagoon bed (DLB). Includes the Churince lagoon bed, the river mouth, and the contiguous southern bank (Fig. 17.1b).
2. Sinkholes (S). Soil depressions formed in the flatland in southern bank of the river, formed due to the loss of cohesion of soil particles and the subsequent col-

Fig. 17.1 Study site. (**a**) Churince System, Cuatro Ciénegas, Coahuila, (red rectangle size is 620 × 100 m). Specific zones of the study site are shown in red; (**b**) dry lagoon bed; (**c**) sinkhole; (**d**) flatland; (**e**) dry riverbank. Map modified from Google Earth

lapse of the soil surface; they are microenvironments that emulate the conditions of the riparian habitat (Pisanty et al. 2020, this volume) (Fig. 17.1c).

3. Flatland in the South bank of the river (F), previously reported as a zone with no apparent vegetation, now being colonized by riparian and other species (Fig. 17.1d).
4. Dry riverbank (RB). Originally highly saturated with water and exposed to flooding, this zone has now an irregular substrate and a scarce vegetation (Fig. 17.1e).

Species

Three hydrophilous species are considered in this chapter, due to their ability to colonize newly formed environments (Fig. 17.2).

1. *Flaveria chlorifolia* (Asteraceae) is common on the borders of water bodies and in S, but recently it is also establishing on DLB. It is an herbaceous perennial; with stems 0.3–2 m long, its leaves are connate-perfoliate; flowers aggregated into paniculate corymbs, yellow; fruit an achene. It is associated with saline sources of water and saline and gypsum soils. It is a gypsovage (plants that can grow both on gypsic and non-gypsic soils) native of North America (Powell 1978, Ochoterena et al. 2020, this volume).
2. *Eustoma exaltatum* (Gentianaceae) is the least frequent of the three species in S, but can be found in RB and especially in F. It is an herbaceous annual (Villarreal 2001) or short-lived perennial plant (SEINet 2019; Turner 2014); stems are 0.3–1 m long; leaves are opposite, sessile, glaucous; flowers paniculate, white to purple; fruits are capsules. It is associated with saline soils. It is found from south of the USA to Central and South America and the West Indies (Turner 2014).

Fig. 17.2 Riparian species in the terminal part of the Churince System, Cuatro Ciénegas, Coahuila, Mexico. From left to right *Eustoma exaltatum*, *Flaveria chlorifolia*, and *Samolus ebracteatus* var. *coahuilensis*

3. *Samolus ebracteatus* var. *coahuilensis* (Primulaceae), an endemic of the Chihuahuan Desert (Hendrickson 1983; Villarreal-Quintanilla et al. 2017) is a common riparian species. It is a short herbaceous perennial, with verticillate leaves, short stems (0.1–0.5 m), and small flowers aggregated in umbrellas. It is associated with aquatic and semiaquatic habitats with saline and gypsum soils (Hendrickson 1983, Ochoterena et al. this volume). It is one of the main colonizers in S and can also be found in F and DLB. Although it is still present in F, its frequency and cover have been decreasing as water becomes scarcer, and it is colonizing the DLB and the RB (Pisanty et al. 2020, this volume).

Field Work

Flaveria chlorifolia: Considering its patchy distribution, seven plots of 5 × 5 m were set on DLB, and all individuals within them were considered. The height of the longest stem was measured bimonthly for 2 years (September 2015–September 2017). Individuals were classified into four size categories, based on their height (2 = 0.5–4 cm; 3 = 5–15 cm; 4 = 16–30 cm; 5 = >30 cm), plus one category for seeds (category 1), and another one for individuals with desiccated aerial parts (category 6) that can either regrow or die. Categories are associated with size at the first reproduction (category 3). Recruitments of new individuals and mortality were registered, and the number of flower heads per plant was counted.

Eustoma exaltatum: six transects of 100 × 2 m, with a North–South direction, were set every 50 m starting from RB. Another transect, 100 × 5 m, was placed along RB. Individuals along transects were tagged and individual plant cover was measured bimonthly for 2 years (June 2015–June 2017) by approximating the shape of the plant to an ellipse (cm^2). Individuals were categorized by their cover into six size categories (2 = 1–60 cm^2; 3 = 61–130 cm^2; 4 = 131–250 cm^2; 5 = 251–450 cm^2; 6 = 451–1450 cm^2; 7 = >1450 cm^2), plus one for seeds (category 1), and another for desiccated plants (category 8). Recruitment of new individuals was recorded along transects. The number of flowers per individual was counted in all the plants. For *S. ebracteatus* var. *coahuilensis*, data from a previous demographic study (Pisanty et al. 2013) were used for comparison.

Average Lefkovitch matrices for each species and zone were calculated. The finite population growth rate (λ) (Caswell 2001), which indicates whether a population is growing ($\lambda > 1$), decreasing ($\lambda < 1$), or stable ($\lambda = 1$), was obtained, and general life cycles were described. Standard sensitivity (absolute contribution of each transition to λ) (Caswell and Trevisan 1994) and elasticity (proportional contribution of a transition to λ) (de Kroon et al. 1986) analysis were applied to identify the underlying processes that affect the dynamics of these populations.

Fecundity was estimated considering the average number of flowers, obtained from a random sample of 100 flowering individuals. Survival, growth, and shrinkage were estimated as transitions based on two consecutive demographic censuses (Caswell 2001).

Plant establishment, represented by the transition from seed to category 1 in matrix population models, was estimated through the recruitments of new individuals and the germination rate in natural conditions. The latter was obtained through the recording of the number of seeds that germinated in experimental units consisting of 30 randomly placed cloth bags around the plots of *F. chlorifolia* and 49 along the transects for *E. exaltatum*. Each bag contained 20 seeds. Bags were randomly collected every 2 months, and the number of germinated seeds was counted. Germination of *S. ebracteatus* var. *coahuilensis* was estimated elsewhere (Pisanty et al. 2013; Peralta-García et al. 2016).

Results

Life Cycle

Following germination and seedling establishment, plants enter a growth period after which they reproduce. The three species can reproduce since they are small (Category 3) (Table 17.1) and the number of flowers increases with size. When temperature decreases, the aerial parts dry and the leaves are lost, and when the cold season ends, regrowth starts with the emergence of new branches and leaves. Desiccation of the aerial parts and regrowth are characteristics of *F. chlorifolia* (Fig. 17.3) and *S. ebracteatus* var. *coahuilensis* (Fig. 17.3), but *E. exaltatum* does not follow the same pattern, as it usually behaves as an annual that dies after one flowering event, as shown in RB (Fig. 17.3).

Samolus ebracteatus var. *coahuilensis* has many stems and leaves that can die during the cold season, but the species is not completely deciduous, so even if the number of stems with leaves decreases, usually some leaves can be found. New branches and leaves can be formed from underground tissue and also from the basal part of stems with a damaged apex. In addition, this species clones through horizontal growth (Fig. 17.3).

Table 17.1 Population parameters of *Flaveria chlorifolia*, *Eustoma exaltatum*, and *Samolus ebracteatus* var. *coahuilensis* in six sites in the Churince System, Cuatro Ciénegas, Coahuila, Mexico

Species	*Flaveria chlorifolia*		*Eustoma exaltatum*		*Samolus ebracteatus*
	Dry lagoon bed	Sinkholes	Dry riverbank	Flatland	Sinkholes/ Flatland
Survival (%)	17	32	0	8	86
Size at first reproduction	5 cm	12 cm	4.7 cm^2	4.5 cm^2	1 stem (>10 leaves)
Minimal and maximum flowers per individual	1–471	1–113	1–93	1–230	195–550 (seeds)
λ	0.72	0.78	0.82	0.99	0.86
Field germination (%)	53	46	0	0	–

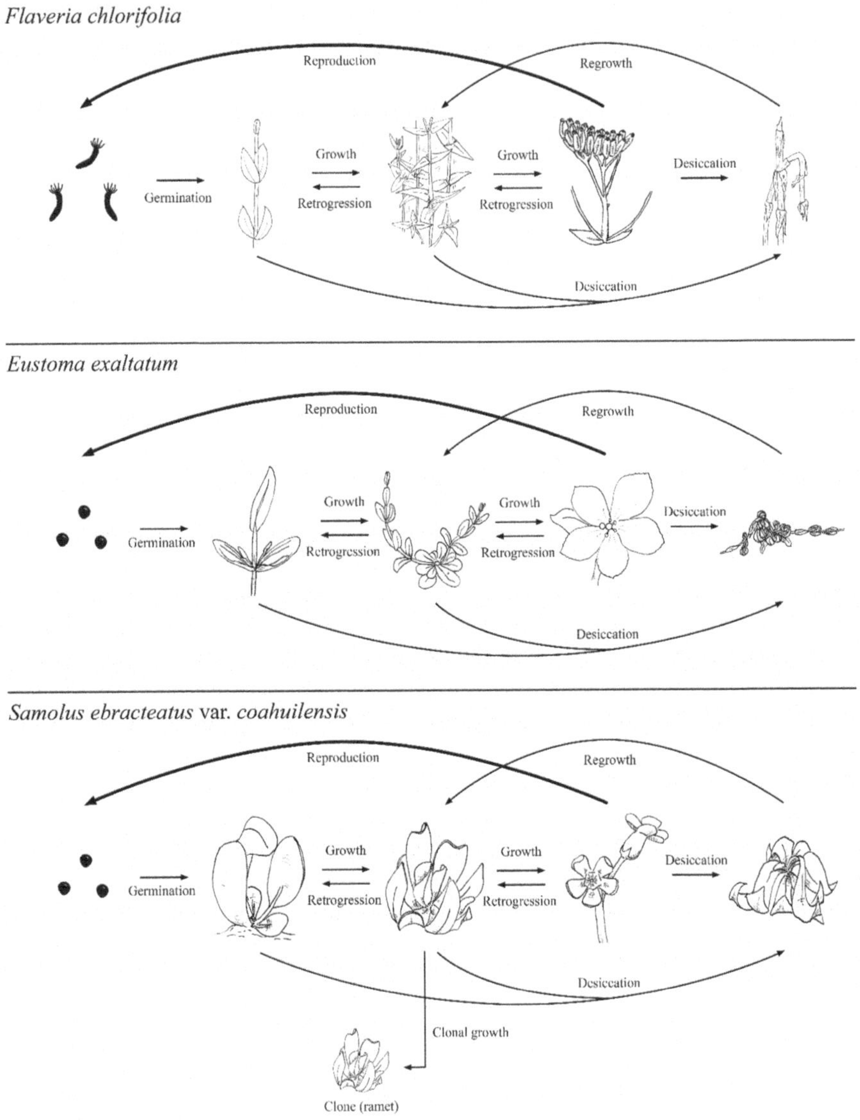

Fig. 17.3 General life cycle of three species in the Churince System, Cuatro Ciénegas, Coahuila, Mexico. From top to bottom *Flaveria chlorifolia, Eustoma exaltatum,* and *Samolus ebracteatus* var. *coahuilensis*. (Source: Illustration by D. Trujillo)

Flaveria chlorifolia had two reproductive events (November 2015 and September 2016) in both zones, while *E. exaltatum* had three (July 2015, July 2016, and June 2017) in F, but only one in RB (July 2015). Only three individuals, all in F, of this species survived until the third reproductive event. *S. ebracteatus var. coahuilensis* produced flowers from March until November, with a peak in spring (Cervantes-Campero et al. 2020, this volume). The number of flowers is shown in Table 17.1.

Population Dynamics

The finite population growth rate for the three species in all sites is less than 1 (Table 17.1), indicating that the population will disappear if conditions remain as those prevailing. Populations of *F. chlorifolia* (Fig. 17.4a), *E. exaltatum* (Fig. 17.4b), and *S. ebracteatus* var. *coahuilensis* decreased as no recruitments nor germination were recorded for *E. exaltatum*, and only two were registered in DLB for *F. chlorifolia* despite the high germination rate observed in the field experiment (Table 17.1). For *S. ebracteatus* var. *coahuilensis* recruitments were abundant while water was available both in S and in F sites (Pisanty et al. 2013), but they decreased through time.

Elasticities show that λ is sensitive to permanence or survival of adults with desiccated aerial parts in both sites for *E. exaltatum*, and in DLB for *F. chlorifolia*. In S, the highest elasticity value of *F. chlorifolia* corresponds to permanence of the tallest individuals, while for *S. ebracteatus* var. *coahuilensis* it corresponds to the permanence of individuals of intermediate size.

Discussion

Life Cycle

Flaveria chlorifolia and *Samolus ebracteatus* var. *coahuilensis* are perennials *sensu stricto*, while *Eustoma exaltatum* is usually reported as an annual species in the taxonomic literature (Shinners 1957; Villarreal 2001). However, recent studies (Turner 2014; SEINet 2019) indicate that it can also behave as a short-lived perennial, a behavior that is confirmed in this study for S and F, but not in RB, where all the individuals died after their first reproduction. In this now disturbed environment, individuals died faster and did not grow as much as in F, because in some parts the RB formerly behaved as a narrow flood plain, and *E. exaltatum* cannot endure this swampy condition as it is not strictly a riparian species. Probably remaining underground anoxic conditions limited its growth (Capon and Brock 2006), resulting in a short, less complex life cycle with a single reproductive event and a smaller number of flowers and seeds (Table 17.1). In F, where sublevel water became available due to disturbance, the substrate is less heterogeneous and has no history of flooding, many plants of *E. exaltatum* survived at least 1 more year than in the RB and were capable of maintaining living meristems that eventually initiate a new growth and reproduction cycle. Survival for more than 1 year depends completely on this

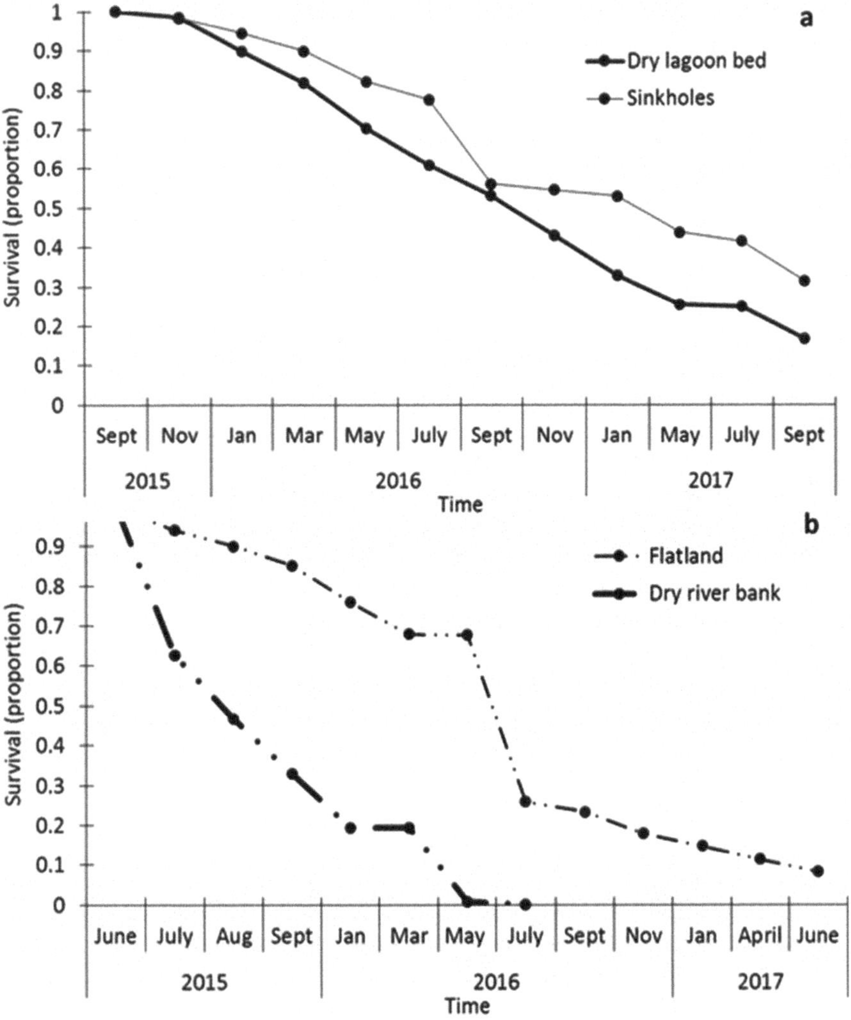

Fig. 17.4 Survival of *Flaveria chlorifolia* (**a**) in the dry lagoon bed and in the sinkholes, and *Eustoma exaltatum* (**b**) in the flatland and the dry riverbank in the Churince System, Cuatro Ciénegas, Coahuila, Mexico (June 2015–September 2017)

process, which has the highest elasticity values, and takes place when individuals have well-established roots.

Reproduction

The three species analyzed here showed early reproduction, and numerous flowers, fruits, and seeds, but the number of reproductive structures is affected, together with other traits, by microenvironmental conditions, as has been observed by other authors (e.g., Sultan 2000; Cervantes-Campero et al. 2020, this volume). The flowering season varies among the species. *F. chlorifolia* flowers in fall, *E. exaltatum* in early summer, and *S. ebracteatus* var. *coahuilensis* from March to November, with a clear spring peak. These differences in phenology, which can have important ecological consequences, indicate different responses to environmental cues (Park 2019), which still need to be identified in this case. Despite differences in their lifespan, the size at first reproduction of the three species was the same in all zones, and the number of flowers increased as the plant grew, but in *E. exaltatum* the production of flowers was higher in individuals with perennial behavior.

Seeds of the three species can remain on the parental plant, without being dispersed for some time. However, uprooted individuals of *E. exaltatum* are frequently tossed by the wind while seeds are still on the open fruits, thus dispersing them further away. Similar temporary retention of seeds has been reported for other desert species (Martínez-Berdeja et al. 2015; Ezcurra et al. 2020, this volume).

Growth

The three species showed a dynamic pattern of growth that includes permanence, fast, and slow transitions to subsequent or higher categories and retrogression. These diversified patterns should be considered part of the plasticity that these plants can have (Sultan 2000). Growth is related to prevailing and accumulated environmental conditions such as soil characteristics, temperature, and water availability (McAuliffe and Hamerlynck 2010), and since they are heterogeneous in time and space, differences within and between species are expected. Growth transition probabilities and elasticities are relatively high, but not enough to make this process an important relative contributor for the population growth rate. Growth results from the interaction of many factors, whose particular effects are difficult to identify, so further experimental work is needed to better explain the observed patterns in CCB.

Survival

The survival of *F. chlorifolia, E. exaltatum*, and *S. ebracteatus* var. *coahuilensis* depends strongly on the survival of meristems present in the lower parts of the stems after the desiccation of the aerial part. This seasonal process represents an alternative strategy to survive winter without an underground perennating structure such as bulbs or rhizomes, characteristic of geophytes (Dafni et al. 1981; Hervás 1999). The survival of *E. exaltatum* was higher in F which, as previously said, is not swampy and allows a more efficient plant development. The same pattern is observed for *F. chlorifolia* in DLB, which is now desiccated and, as F, does not get flooded but retains some sublevel and underground water. Both sites are highly disturbed, and plant cover and specific richness are lower than in S or RB, which might imply a less competitive environment that allows the survival of these hardy species.

Samolus ebracteatus var. *coahuilensis* had the highest survival rate of the three species, but conditions have changed since its demography was first analyzed, and less individuals are now found (M. Rodríguez-Sánchez and I. Pisanty, pers. obs.).

Population Dynamics

The low finite growth rates indicate that populations will continue to decrease and are likely to disappear if water loss continues, even if other factors, like temperature, are adequate. The fast colonization of the newly formed environments suggests that *F. chlorifolia* and *E. exaltatum* respond to very short, discrete periods of opportunity when abiotic factors are favorable. This is supported by the neglectable recruitment we observed, even for the case of *F. chlorifolia*, which showed high germination rates in the field experiment. Thus, we can expect a new colonization cycle to begin once favorable precipitation, water availability, and temperature conditions occur again, together with seed availability and adequate biotic conditions, especially regarding the persistence of seed banks and competition. The latter can be enhanced by the effects of the severe disturbance this zone is undergoing. On the other hand, the projection estimated with the population model for *S. ebracteatus* var. *coahuilensis* in 2008 (Pisanty et al. 2013) proved accurate, since fewer individuals were observed throughout the following years.

The life cycles and population dynamics of the three species exhibited a microenvironment-dependent behavior, which seems to be highly influenced by water availability as could be expected for riparian species and for arid lands (Stromberg et al. 1996; Shafroth et al. 2002; Tockner and Stanford 2002; Poff and Zimmerman 2010). Soil properties, as salt and gypsum concentration, can also be determinant (Escudero et al. 2015; Flores-Olvera et al. 2016; Ezcurra et al. 2020, this volume; Zavala and Jimenez 2020, this volume). This water-dependent behavior affected the distribution pattern of the riparian species through the colonization of new microenvironments.

Riparian species are able to colonize newly opened habitats, creating movable landscapes, but will not persist because these habitats are also ephemeral. These plants have a long biological history of adaptations to highly selective environments, but no life cycle can cope with such dramatic, short-term environmental changes as those CCB is suffering. Even feasts have limits.

Acknowledgments This research was financed by Project UNAM/PAPIIT IN217915 and by the Department of Ecology and Natural Resources, Facultad de Ciencias, UNAM. The first author had a scholarship from Consejo Nacional de Ciencia y Tecnología (CONACYT). We appreciate the support of the Alliance WWF-Fundación Carlos Slim and the constant support of Juan Carlos Ibarra, director of the Área de Protección de Flora y Fauna de Cuatrociénegas (CONANP), and of Martín Carrillo. We thank Valeria Souza for her constant support, Cynthia Peralta-García, Polenka Torres Orozco, and Stephanie Escobar for their help in the collection of data, and Diego Trujillo for the elaboration of Figure 17.3. We also acknowledge the technical support of Mariana Hernández-Apolinar, Pedro E. Mendoza-Hernández, Israel Solano-Zavaleta, and Marco Romero-Romero.

References

Capon SJ, Brock MA (2006) Flooding, soil seed bank dynamics and vegetation resilience of a hydrologically variable desert floodplain. Freshw Biol 51(2):206–223

Capon S, Dowe JL (2007) Diversity and dynamics of riparian vegetation. In: Lovett S, Price P (eds) Principles of riparian lands management. Land and Water Australia, Canberra, Australia, pp 13–32

Caswell H (2001) Matrix population models. Construction, analysis, and interpretation. Sinauer Associates Inc, Sunderland, MA, USA

Caswell H, Trevisan MC (1994) Sensitivity analysis of periodic matrix models. Ecology 75(5):1299–1303

Cervantes-Campero G, Pisanty I, Mandujano MC (2020) Morphological and phenological variation in *Samolus ebracteatus* var. *coahuilensis* in different environments in the churince system, Cuatro Ciénegas Basin (Coahuila). In: Mandujano MC, Pisanty I, Eguiarte LE (eds) Plant diversity and Ecology in the Chihuahuan desert. Springer International, Cham, Switzerland

Comisión Nacional de Áreas Naturales Protegidas (CONANP) (1999) Programa de manejo del Área de Protección de Flora y Fauna Cuatrociénegas. Instituto Nacional de Ecología SEMARNAP, México, DF, Mexico

Crone EE, Menges ES, Ellis MM, Bell T, Bierzychudek P, Ehrlén J, Kaye TN, Knight TM, Lesica P, Morris WF, Oostermeijer G, Quintana-Ascencio PF, Stanley A, Ticktin T, Valverde T, Williams JL (2011) How do plant ecologists use matrix population models? Ecol Lett 14(1):1–8

Dafni A, Cohen D, Noy-Mier I (1981) Life-cycle variation in geophytes. Ann Mo Bot Gard 68(4):652–660

de Kroon H, Plaiser A, van Geoenendael J, Caswell H (1986) Elasticity: the relative contribution of demographic parameters to population growth rate. Ecology 67(5):250–254

Escudero A, Palacio S, Maestre FT, Luzuriaga AL (2015) Plant life on gypsum: a review of its multiple facets. Biol Rev 90(1):1–18

Ezcurra E, Martínez-Berdeja A, Villanueva-Almanza L (2020) The evolution of North American deserts and the uniqueness of Cuatro Ciénegas. In: Mandujano MC, Pisanty I, Eguiarte LE (eds) Plant diversity and Ecology in the Chihuahuan desert. Springer International, Cham, Switzerland

Flores-Olvera H, Czaja A, Estrada-Rodríguez JL, Méndez UR (2016) Floristic diversity of halophytic plants of Mexico. In: Khan M, Boër B, Őzturk M, Clüsener-Godt M, Gul B, Breckle SW (eds) Sabkha ecosystems. Tasks for vegetation science, vol 48. Springer, Cham, Switzerland

Hendrickson J (1983) A revision of *Samolus ebracteatus* (*sensu lato*) (Primulaceae). Southwest Nat 28:303–314

Hervás JL (1999) Geófitos en el norte de Jaén. Aspectos de su ecología. Blancoana 16:65–69

Martínez-Berdeja A, Ezcurra E, Torres M (2015) Morphological variability in propagules of a desert annual as a function of rainfall patterns at different temporal and spatial scales. Funct Ecol 29(10):1260–1267

McAuliffe JR, Hamerlynck EP (2010) Perennial plant mortality in the Sonoran and Mojave deserts in response to severe, multi-year drought. J Arid Environ 74(8):885–896

Minckley WL (1969) Environment of the Bolsón of Cuatro Ciénegas, Coahuila, Mexico with Special Reference to the Aquatic Biota. The University of Texas at El Paso Science Series. Western Press, El Paso, TXT, USA

Naiman RJ, Decamps H, Pollock M (1993) The role of riparian corridors in maintaining regional biodiversity. Ecol Appl 3(2):209–212

Ochoterena H, Flores-Olvera H, Gómez-Hinostrosa C, Moore, MJ (2020) Gypsum and plant species: a marvel of Cuatro Ciénegas and the Chihuahuan Desert. In: Mandujano MC, Pisanty I, Eguiarte LE (eds) Plant diversity and Ecology in the Chihuahuan desert. Springer International, Cham, Switzerland

Park JS (2019) Cyclical environments drive variation in life-history strategies: a general theory of cyclical phenology. Proc R Soc B 286(1898):20190214

Peralta-García C, Sánchez-Coronado ME, Orozco-Segovia A, Orozco-Segovia S, Pisanty-Baruch I (2016) Germination of four riparian species in a disturbed semi-arid ecosystem. S Afr J Bot 106:110–118

Pinkava D (1984) Vegetation and flora of the Bolson of Cuatro Ciénegas region Coahuila, México: IV. Summary, endemism and corrected catalogue. J Ariz Nev Acad Sci 19(1):23–47

Pisanty I, Pérez y Sosa C, Gálvez G (2013) Agriculture, water mismanagement and ecosystem transformations in the Cuatrociénegas Valley in the Chihuahuan Desert, Drought. In: Schwabe K, Albiac J, Connor JD, Hassan RM, Meza LG (eds) Arid and semi-arid regions: a multidisciplinary and cross-country perspective. Springer, Dordrecht, Netherlands, pp 199–216

Pisanty I, Rodríguez-Sánchez M, Torres Orozco P, Granados-Hernández LA, Escobar S, María C. Mandujano MC (2020) Disturbance and the formation and colonization of new habitats in a hydrological system in the Cuatro Ciénegas Basin. In: Mandujano MC, Pisanty I, Eguiarte LE (eds) Plant diversity and Ecology in the Chihuahuan desert. Springer International, Cham, Switzerland

Poff NL, Zimmerman JK (2010) Ecological responses to altered flow regimes: a literature review to inform the science and management of environmental flows. Freshw Biol 55(1):194–205

Powell M (1978) Systematics of *Flaveria* (Flaveriinae-Asteraceae). Ann Mo Bot Gard 65(2):590–636

Salguero-Gómez R, Jones OR, Jongejans E, Blomberg SP, Hodgson DJ, Mbeau-Ache C, Zuidema PA, de Kroon H, Buckley YM (2016) Fast–slow continuum and reproductive strategies structure plant life-history variation worldwide. Proc Natl Acad Sci 113(1):230–235

SEINet (2019). http://swbiodiversity.org/seinet/taxa/index.php?taxon=Eustoma+exaltatum&formsubmit=Search+Terms#

Shafroth PB, Stromberg JC, Patten DT (2002) Riparian vegetation response to altered disturbance and stress regimes. Ecol Appl 12(1):107–123

Shinners LH (1957) Synopsis of the genus *Eustoma* (Gentianaceae). Southwest Nat 2(1):38–43

Silvertown J, Franco M, Pisanty I, Mendoza A (1993) Comparative plant demography—relative importance of life-cycle components to the finite rate of increase in woody and herbaceous perennials. J Ecol 81:465–476

Souza V, Escalante AE, Espinoza L, Valera A, Cruz A, Eguiarte LE, Pichel FG, Elser J (2004) Cuatro Ciénegas: un laboratorio natural de astrobiología. Revista Ciencias 75:4–12

Stearns SC (1992) The evolution of life histories. Oxford University Press, New York, NY, USA

Stromberg JC, Tiller R, Richter B (1996) Effects of groundwater decline on riparian vegetation of semiarid regions: the San Pedro, Arizona. Ecol Appl 6(1):113–131

Stromberg JC, Gallo E, Lohse K, Meixner T, Moody E, Sabo J, Setaro DL (2015) Structure and function of ephemeral streams in the arid and semiarid southwest: implications for conservation and management. Arizona State University. Tempe, AZ, USA

Sultan SE (2000) Phenotypic plasticity for plant development, function and life history. Trends Plant Sci 5(12):537–542

Tockner K, Stanford JA (2002) Riverine flood plains: present state and future trends. Environ Conserv 29(3):308–330

Turner BL (2014) Taxonomic overview of *Eustoma* (Gentianaceae). Phytologia 96(1):7–11

Villarreal QL (2001) Gentianaceae. Fasciculo 21. In: Sosa V, Cabrera RL, Escamilla M, Moreno NP, Mejía-Saulés MT, Nee M, Nevling L, Rzedowski J (eds) Flora de Veracruz. University of Riverside, Instituto de Ecología, AC, Xalapa, Veracruz, Mexico

Villarreal-Quintanilla JA, Bartolomé JA, Estrada E, Ramírez H, Martínez SJ (2017) El elemento endémico de la flora vascular del Desierto Chihuahuense. Acta Botánica Mexicana 118:65–96

Chapter 18
Morphological and Phenological Variation in *Samolus ebracteatus* var. *coahuilensis* in Different Environments in the Churince System, Cuatro Ciénegas Basin (Coahuila)

Gabriel Cervantes-Campero, Irene Pisanty, and María C. Mandujano

Abstract *Samolus ebracteatus* var. *coahuilensis* (Primulaceae) is a perennial creeping herb common in riparian habitats in the Cuatro Ciénegas Basin. It grows on the border of water bodies and in zones where water becomes available due to disturbance. Its distribution in the terminal part of the Churince system expanded to newly formed sinkholes and to the surrounding flatlands. Variation in height and leaf area, as well as in the number of stems and reproductive structures was evaluated as a short-term response to the conditions prevailing in three environments (riverbank, flatlands, and sinkholes formed due to subsurface water flow) in the terminal part of the Churince River. Bimonthly measurements of these parameters were taken from September 2012 to November 2013. Statistical analyses were performed for each dependent variable. The factors were time (months) and the three microenvironments. The analyzed traits showed significant differences between the latter and through time, indicating a plastic, short-term response to environmental conditions. The tallest plants were found in the sinkholes, while the shortest were in the flatlands. Plants with the highest number of stems were found on the riverbank, while those with the lowest number were found on the flatland. Leaf area proved to be smaller in the flatland than in the other microenvironments, and the highest number of plants with inflorescences was found on the riverbank.

Keywords Microenvironments · Riparian species · Disturbance · Phenotypic plasticity · Phenology

G. Cervantes-Campero · I. Pisanty (✉)
Departamento de Ecología y Recursos Naturales, Facultad de Ciencias, Universidad Nacional Autónoma de México, UNAM, Mexico City, Mexico
e-mail: ipisanty@unam.mx

M. C. Mandujano
Departamento de Ecología de la Biodiversidad, Instituto de Ecología, Universidad Nacional Autónoma de México, UNAM, Mexico City, Mexico
e-mail: mcmandujano@iecologia.unam.mx

M. C. Mandujano et al. (eds.), *Plant Diversity and Ecology in the Chihuahuan Desert*, Cuatro Ciénegas Basin: An Endangered Hyperdiverse Oasis,
https://doi.org/10.1007/978-3-030-44963-6_18

Introduction

Morphological changes in plants, such as variation among leaves, flowers, and fruits, as well as phenological ones, are due to various endogenous and exogenous variables, including water availability (Hutchings and de Kroon 1994, Valladares et al. 2007; Meier and Leuschner 2008; Sultan 2000; Furet et al. 2014). They have been extensively studied due to their ecological and economic importance (Gianoli 2004). Different features of the life history of plants, as well as the expression of sex and phenology, also respond to environmental factors, contributing to fitness (Sultan 2000).

The Cuatro Ciénegas valley faces a severe loss of water, due to the overexploitation of groundwater for agriculture. The Churince system has been badly affected by this disturbance and endures a progressive and probably irreversible desiccation process. The terminal part, which includes what was the largest lake in the basin, the Churince lagoon, is currently completely dry, as well as the river between the latter and the Intermediate lagoon, which is also dry nowadays (Souza et al. 2006; Pisanty et al. 2013; Souza and Eguiarte 2018; Pisanty et al. 2020, this volume).

The desiccation process included a subsurface flow of water from the South bank of the river, which caused the rapid formation of numerous small sinkholes at an abnormally high rate (S. Contreras, A. Contreras, J.C. Ibarra, pers. com.; Pisanty et al. 2013; Pisanty et al. 2020, this volume). These structures are common in dispersive soils of arid and semi-arid zones (Umesh et al. 2011).

Sinkholes are colonized mainly by riparian species (Pisanty et al. 2013; Pisanty et al. 2020, this volume; Rodríguez-Sánchez et al. 2020, this volume), among which *Samolus ebracteatus* var. *coahuilensis* (Primulaceae), a perennial reclining-spreading herb (Henrickson 1983) is frequent (Pisanty et al. 2013) (Fig. 18.1). This species forms small clumps and can clone through horizontal underground growth and by the separation of clumps in independent parts (pers. obs.).

Fig. 18.1 *Samolus ebracteatus* var. *coahuilensis*, Primulaceae, in Cuatro Ciénegas, Coahuila

Samolus ebracteatus var. *coahuilensis* is a hydrophilic species (Pinkava 1984), distributed in the Southern United States, Mexico, Belize, and the Antilles. The *coahuilensis* variety of this species has been recorded only in the Chihuahuan Desert (Henrickson 1983; GBIF 2018), growing near water bodies on salty and gypsum soils (Villarreal-Quintanilla and Encina-Domínguez 2005). Pinkava (1984) registered this plant in Cuatro Ciénegas within aquatic habitats and permanently flooded sites. It is frequent and abundant in the riparian zones of the valley (Henrickson 1983; Pinkava 1984; Villarreal-Quintanilla and Encina-Domínguez 2005).

Recently, due to the alteration of the Churince system, moisture became available and favored the establishment of *S. ebracteatus* var. *cohuilensis* in newly formed microenvironments near the mouth of the river and Lake Churince, as well as in the sinkholes in the South bank of the river, and on the flatland that surrounds them. It is one of the first species to colonize sinkholes (Pisanty et al. 2013, Pisanty et al. 2020, this volume; Rodríguez-Sánchez et al. 2020, this volume), and it is an effective colonizer because of its good germination and growth responses (Pisanty et al. 2013; Peralta-García et al. 2016). Establishment and growth in the original riparian habitat and in the new microenvironments can induce variations in some plant characters, depending both on the plasticity of the plants and on environmental conditions.

Here we analyze the morphological variations of *S. ebracteatus* var. *coahuilensis* in the different environmental conditions where it can be found, to better understand the mechanisms of colonization and survival that this plant has. We identified differences in morphological variation in *Samolus ebracteatus* var. *coahuilensis* among three microenvironments (riverbank, sinkholes, and flatland) in the terminal part of the Churince system of the Cuatro Ciénegas Basin, Coahuila. We expected that plants growing in the different environments would vary in height, leaf area, and the number of stems and of reproductive structures.

Methods

Study Site

Cuatro Ciénegas Basin is part of the Chihuahuan Desert in Northwestern Mexico. The Churince system is located in the western part of the basin. This study was made in the terminal part of the system (26° 51′ 07.61″ N, 102° 08′ 50.76″ W) (Fig. 18.2), which is deeply disturbed due to the desiccation process.

Samolus ebracteatus var. *coahuilensis* grows in three easily identified environments (Table 18.1). To explain the morphological variations found among them, moisture, light, and nutrients were considered as variables, because they can affect the morphology of plants. The availability of these variables in the three microenvironments was qualitatively assessed (Table 18.1), and plant responses were

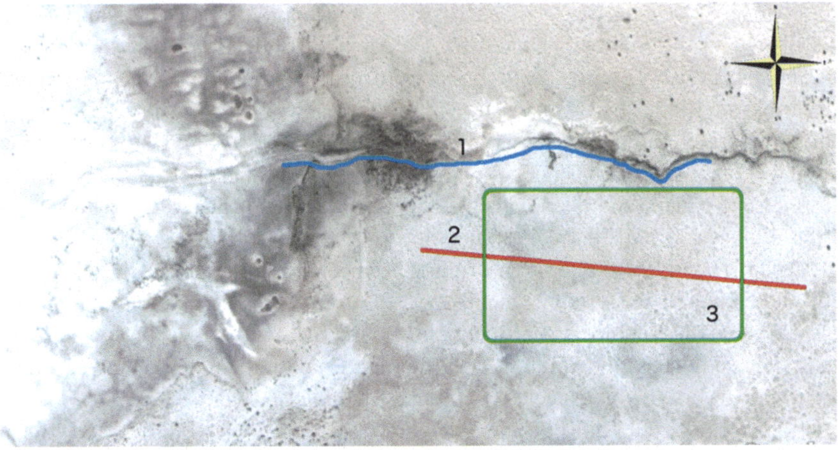

Fig. 18.2 Microenvironments and sampling zones of the terminal part of the Churince System, in the Cuatro Ciénegas Basin. 1. Transect along the riverbank (blue line, 550 m). 2. Transect on the flatland (red line). 3. Sinkhole zone (green box, ca 600 m^2). Modified from Google Earth, 2018. 26°51′07, 61″ N 102°08′50,76″ O

Table 18.1 Qualitative characteristics of microenvironments studied along the Churince system, in the Cuatro Ciénegas Basin

	Humidity	Irradiation	Nutrients
Riverbank	++	+	++
Flatland	–	++	–
Sinkholes	+	––	+

compared between them. Figure 18.2 shows the three environments, which can be described as follows:

1. The riverbank, which is the original habitat of this species. Water was readily available in it, and flooding was relatively frequent. It is slightly more shaded than the flatland, due to the irregularity of the substrate and to the established plants, but it has less shade than most sinkholes. Due to the decomposition of the plants that die at the riverbank, and to the deposition of fluvial sediment, it is rich in nutrients.
2. The flatland, which is the driest microenvironment. Organic matter is scarce, and irradiation is very high; it is affected by strong seasonal winds.
3. Sinkholes, which are shaded, humid microenvironments, where water could accumulate while the river had water. They are richer in organic matter than the flatland, due to the presence of water and plants that provide organic matter. They result from the dispersion of soil particles and emulate the conditions of the riparian habitat (Pisanty et al. 2013; Pisanty et al. 2020, this volume; Rodríguez-Sánchez et al.2020, this volume) (Fig. 18.2).

One hundred fifty *S. ebracteatus* var. *coahuilensis* plants were randomly selected on the flatland, another 150 plants in the riverbank, and 110 plants in the sinkholes. For each selected individual, height (Fig. 18.3a), number of stems with leaves (Fig. 18.3b), and the number of plants with reproductive structures (Fig. 18.3c and d) were bimonthly determined from September 2012 to November 2013.

Additionally, from January 2013 to January 2014, leaves were collected to determine variations in leaf area. In each microenvironment, 120 leaves were collected from randomly chosen plants, different from those that were being measured, and their leaf area (cm^2) was measured with the LI-3000C Portable Leaf Area Meter.

Data Analysis

In each microenvironment, the annual average of height, number of stems, and leaf area were obtained. Subsequently, because some of the selected plants died, a two-way variance analysis was performed with STATISTICA 8 for each variable. The factors considered in this analysis were time (months) and microenvironments (river, flatland, and sinkholes).

Results

Plants of *S. ebracteatus* var. *coahuilensis* were significantly different in height, leaf area, and number of stems among environments (Fig. 18.3). Height varied among months ($F = 74.02$, $d.f. = 7$, $P < 0.01$) and environments, as was the interaction between them ($F = 3.28$, $d.f. = 14$, $P < 0.01$). Plants growing in the sinkholes were taller (13.3 ± 0.3 cm) (Mean ± Standard Error), while the plants from the flatland were shorter (7.63 ± 0.1 cm) and intermediate in the riverbank (11.27 ± 0.17 cm) (Fig. 18.4a).

There were also significant differences in the number of stems among months ($F = 8.77$, $d.f. = 7$, $P < 0.01$) and microenvironments ($F = 142.43$, $d.f. = 2$, $P < 0.01$), and in the environment–month interaction ($F = 2.011$, $d.f. = 14$, $P < 0.01$). The average

Fig. 18.3 *Samolus ebracteatus* var. *coahuilensis*. Measured characters: (**a**) height; (**b**) stems; (**c**) inflorescence; (**d**) infructescence

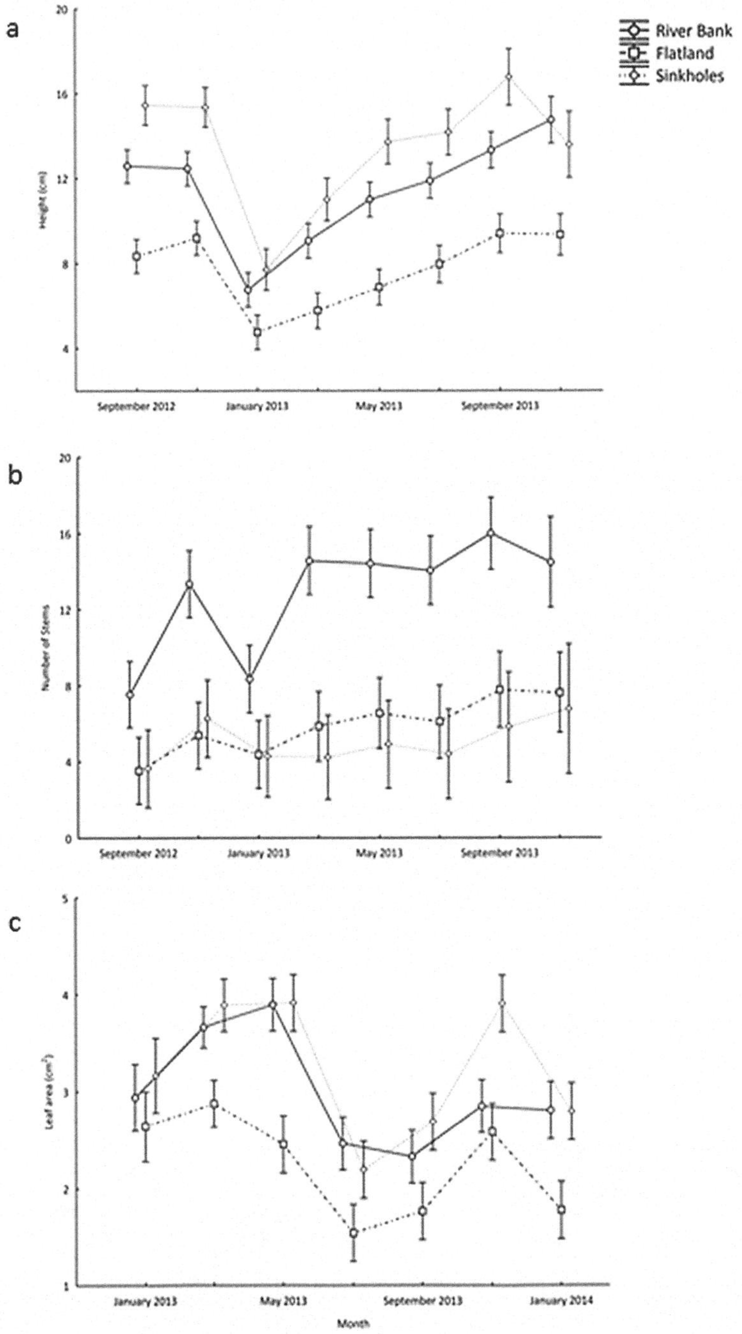

Fig. 18.4 Averages of measured characters of *Samolus ebracteatus* var. *coahuilensis* in each microenvironment (river, sinkholes, and flatland). (**a**) Height. (**b**) Number of stems. (**c**) Leaf area

number of stems of the plants on the riverbank (12.67 ± 0.5 stems) was higher than in the plants of the flatland (5.75 ± 0.17 stems) and of the sinkholes (4.85 ± 0.24 stems). There were no significant differences between plants established in the sinkholes and on the flatland (Fig. 18.4b). On the riverbank, the average number of stems during January 2013 (8.34 ± 0.71 stems) was significantly lower than in the rest of the year (P <0.01, *d.f.*= 14). In the other two areas, no significant differences were observed (Fig. 18.4b).

Significant differences were also found in the average leaf area of the three environments (F = 80.62, P <0.01) (Fig. 18.4c). Plants growing on the flatland had smaller leaf areas (2.26 ± 0.5 cm^2) than the plants in the other microenvironments (river: 3.06 ± 0.6 cm^2; sinkholes: 3.24 ± 0.07 cm^2). Significant differences were observed within each microenvironment over time (F = 47.98, *d.f.* = 6, P <0.01); leaves recorded in July and September had the smallest areas in the three microenvironments. The interaction month × environment also had a significant effect on the leaf area (F = 3.46, *d. f.* = 12, P <0.01).

The proportion of individuals with reproductive structures (inflorescences and infructescences) was higher on the riverbank (0.92) than on the flatland (0.8) and the sinkholes (0.65). The highest percentage of plants with one or more inflorescences was observed in March, with flowers in the pre-anthesis and anthesis stages. After this flowering peak, the presence of inflorescences diminishes until November 2013, when very few individuals exhibited them and all were senile (Fig. 18.5a). When sampling started, many plants had infructescences with senile fruits from the previous reproductive period. In May a new cohort was formed, and young or mature fruits were found, so the largest number of these structures was registered in this month. On the riverbank and the flatland, a proportion of 0.91 of the plants had infructescences, and in the sinkholes this proportion was 0.87, but these differences were not significant. On the riverbank, infructescence production continued until July, when 0.96 of the plants had at least one infructescence. The proportion of plants with infructescences reached its lowest values in November 2013 (0.18 in the riverbank, 0.63 in the flatland, and 0.20 in the sinkholes) (Fig. 18.5b).

Discussion

There was evident morphological variation among individuals of *Samolus ebracteatus* var. *coahuilensis* growing in each of the environments in the Churince system. Higher plants were found in the sinkholes, while the shortest were found in the flatland; the average number of stems in the river bank was significantly higher than that of the sinkholes and the flatland, which were not different from each other; the leaf area in the flatland was significantly lower than that observed in the other microenvironments. The conditions of the riverbank were more favorable for the production of reproductive structures, while the sinkholes were the least favorable environment.

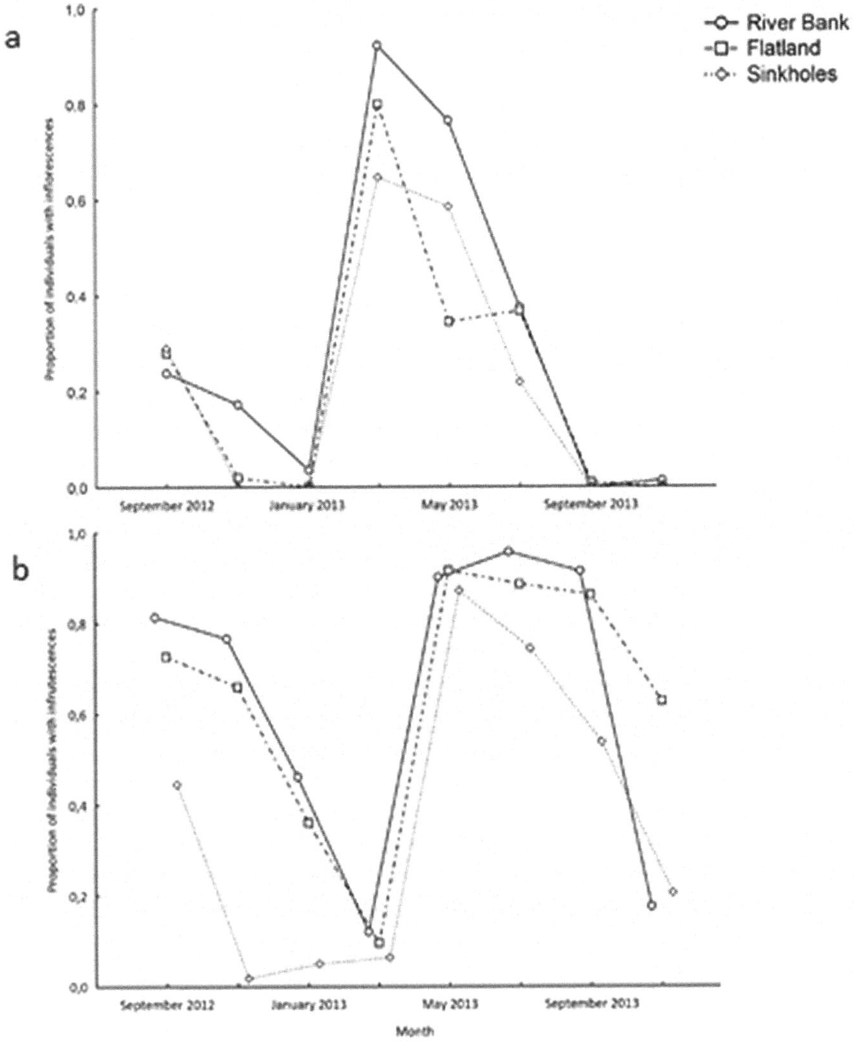

Fig. 18.5 Proportion of individuals of *Samolus ebracteatus* var. *coahuilensis* with reproductive structures in each microenvironment (river, sinkholes, and flatland). (**a**) Inflorescences. (**b**) Infructescences

Differences in height among of *S. ebractatus* var. *coahuilensis* plants can be explained by the incidence of light on plants, which is lower in the sinkholes, thus forcing stems to elongate (Bidwell 1993; Hutchings and de Kroon 1994). Although other factors, as temperature, can have a seasonal influence, making plants always shorter during the cold months. Plants with intermediate height were found in the riverbank, where the incidence of light is high, and nutrients are relatively abundant.

The proliferation of branches in the plants growing on the riverbank was probably due to higher water and nutrient availability, as both promote their development

(Daniels 1986; Hutchings and de Kroon 1994; Lortie and Aarssen 1997; Furet et al. 2014). Plants tend to have fewer branches when they develop in shaded conditions than when they are established in full light. In the sinkholes, fewer stems were consistently observed, coinciding with the typical responses of plants to low light availability, including the increase in leaf area and thinning and elongation of stems (Daniels 1986; Hutchings and de Kroon 1994; Huber 1996; González and Gianoli 2004). The small number of stems recorded in the flatland and in the sinkholes contrasts with the numerous stems in the riverbank and indicates that plants grow better in their original habitat than in the newly opened ones, despite their ability to colonize them.

Lower nutrients and water availability, as well as higher gypsum content in the flatland, possibly had a negative effect on leaf growth (Pedrol et al. 2000; Meier and Leuschner 2008). Givnish (1984) proposed a model for optimal leaf size, considering transpiration as a cost and photosynthesis as a benefit. Thus, the area of the leaves and the transpiration surface that it represents tends to decrease in dry environments and to increase in humid ones. The smaller size of the leaves in the flatland possibly corresponded to an optimal size that avoids excess perspiration and uses the scarce resources efficiently, while the greater size of the leaves in the riverbank and in the sinkholes points to more favorable conditions, especially regarding water and nutrient availability.

In plants exposed to light, leaves are usually smaller, contain less chlorophyll, and are thicker than plants in shaded environments (Valladares et al. 2004; Gratani et al. 2006). Leaves can respond to shade by increasing leaf surface and consequently improving the capture of light in environments where it is limited (Gianoli 2004; Valladares et al. 2004), which may explain the smaller size of leaves in the flatland and the highest values in the sinkholes. Nutrient shortage can also cause the reduction of leaf area, and this can be happening in the flatland (Knops and Reinhart 2000; Meier and Leuschner 2008). Givnish (1984) suggested that the shortage of nutrients produces small leaves, because the cost of absorbing nutrients from the soil is greater and restricts the production of photosynthetic cells and enzymes. However, this pattern is not always observed and the response of this trait to nutrient scarcity may vary between species and be influenced by other environmental conditions that exert pressure at the same time (Meier and Leuschner 2008). On the riverbank, plants had the largest leaf areas that, in general, did not differ significantly from those established in sinkholes, possibly due to the availability of nutrients. However, it is still necessary to identify which nutrients could be specifically causing a reduction in leaf area.

Seasonal and phenological cycles also influence leaf area, as can be seen in the average leaf area in July 2013, when the lowest values were observed in the three microenvironments, due to an increase in the production of new leaves during the beginning of summer.

Flowering showed a clearly seasonal pattern in the plants established in the three environments. The flowering of *S. ebracteatus* var. *coahuilensis* is reported to happen between Spring and Autumn (SEINet 2018), and this pattern was followed in the three microenvironments. The percentages of plants with inflorescences showed differences between microenvironments during March and May, when these structures were more abundant. In these months, most of the flowers were active, while in the rest of the months senile flowers predominated, without significant differences among microenvironments.

In March, the production of inflorescences responded to the conditions of each of the microenvironments. Plants in the sinkholes had the lowest inflorescence production of the three microenvironments, possibly due to the lower quantity and quality of light in them (Puentes et al. 1993; Arboleda 2011) and to the low branch production. Plants of the flatland had an intermediate proportion of inflorescences; however, in May this proportion decreased, until it was the lowest, maybe due to the less favorable conditions and to the exposure to external factors as the strong winds that can break an erect inflorescence. Due to the high number of stems, and to the effects of the higher availability of water and nutrients (Inouye et al. 2003; García et al. 2008), plants growing on the riverbank had numerous reproductive structures.

A clear seasonal pattern was also observed in the formation of infructescences in the three microenvironments, where fruit production took place primarily in May. The presence of infructescences decreased with time, and since July mainly senile fruits without seeds were observed. The maximum proportion of plants with infructescences observed in May is a consequence of the maximum proportion of plants with inflorescences 2 months before. This increase in the number of infructescences prior to the rainy season allows seeds to germinate during the most favorable season of the year (Mott and McComb 1975; Simpson and Dean 2002), since seeds are quiescent, and it is water availability that determines germination of this species (Peralta-García et al. 2016; Peralta-García et al. 2020, this volume). During most of this study, the proportion of plants with infructescences was significantly lower in the sinkholes than in the other two microenvironments, which is consistent with the lower production of inflorescences.

Samolus ebracteatus var. *coahuilensis* shows a relevant phenotypic plasticity that includes growth and reproductive parameters, which probably affect individual survivorship. Variation of these parameters can contribute to the fitness of this species in heterogeneous and unstable environments (Bazzaz 1996). This study provides a first approach for the analysis of the morphological behavior and phenology of the species, but further studies are needed to determine which specific factors (e.g., specific nutrients) are responsible for the morphological differences between microenvironments.

The water systems of the Cuatro Ciénegas valley are currently changing, as evidenced by the drying of the river and the lagoons, and in the opening of sinkholes. These changes have modified the original habitat of several species that developed in aquatic, underwater, and riparian environments. Pinkava (1984) located *S. ebracteatus* var. *coahuilensis* in aquatic and permanently flooded environments, but its distribution expanded successfully, as it colonized, at least temporarily, new areas such as the sinkholes and the flatland that surrounds them.

It is important to mention that the conditions in the study area, the Churince system from the Cuatro Ciénegas Basin, are changing, and currently the sinkholes and the plain are dry (Pisanty et al. 2020, this volume), so the permanence of this plant in these microenvironments is uncertain. Nevertheless, *S. ebracteatus* var. *coahuilensis* is an indicator of water availability, and changes in its original and present distribution should be considered an early alarm of groundwater disturbance even before it becomes evident in the surface of the water bodies of the Basin.

Acknowledgements This project was financed by UNAM/ PAPIIT IN 231811 "Formación, dinámica y colonización de hundimientos diferenciales (abras) del Valle de Cuatrociénegas, Coahuila," and supported by the Alliance WWF- Slim Foundation. We appreciate the help of Ma. Esther Sánchez-Coronado, Mariana Rodríguez-Sánchez, and Jazmín Sánchez Rosales, and the support of Mariana Hernández Apolinar, Pedro Mendoza Hernández, and Marco Romero Romero. Special thanks to Juan Carlos Ibarra, director of the Cuatro Ciénegas protected area, and to Valeria Souza, for their continuous support.

References

Arboleda ME (2011) Efecto de la irradiancia en el crecimiento y desarrollo de *Aptenia cordifolia* (Lf) Schwantes como cobertura ornamental. Bioagro 23:175–184

Bazzaz FA (1996) Plants in changing environments: linking physiological, population, and community ecology. Cambridge University Press, Cambridge, UK

Bidwell RGS (1993) Fisiología vegetal. AGT Editor, México, DF, Mexico, 784 pp

Daniels RE (1986) Studies in the growth of *Pteridium aquilinum* (L.) Kuhn (bracken). 2. Effects of shading and nutrient application. Weed Res 26:121–126

Furet PM, Lothier J, Demotes-Mainard S, Travier S, Henry C, Guérin V, Vian A (2014) Light and nitrogen nutrition regulate apical control in *Rosa hybrida* L. J Plant Physiol 171:7–13

García OJ, Dueñez EY, Fischer G, Chaves B, Quintero O (2008) Efecto del nitrato de potasio, fosfato de potasio y ethephon en la inducción floral de la feijoa o goiabeira serrana (*Acca sellowiana* [O. Berg] Burret). Rev Bras Frutic 30:577–584

GBIF (2018) GBIF Home Page. https://www.gbif.org. Accessed Nov 2018

Gianoli E (2004) Plasticidad fenotípica adaptativa en plantas. In: Cabrera HM (ed) Fisiología Ecológica en Plantas. Editorial de la Universidad Católica de Valparaíso, Valparaíso, Chile, pp 13–26

Givnish TJ (1984) Leaf and canopy adaptations in tropical forests. In: Medina E, Mooney HA, Vázquez-Yánes C (eds) Physiological ecology of plants of the wet tropics. Springer, The Hague, Netherlands, pp 51–85

González AV, Gianoli E (2004) Morphological plasticity in response to shading in three *Convolvulus* species of different ecological breadth. Acta Oecol 26:185–190

Gratani L, Covone F, Larcher W (2006) Leaf plasticity in response to light of three evergreen species of the Mediterranean maquis. Trees 20:549–558

Henrickson J (1983) A revision of *Samolus ebracteatus* (*sensu lato*) (Primulaceae). Southwest Natural 28:303–314

Huber H (1996) Plasticity of internodes and petioles in prostrate and erect *Potentilla* Species. Funct Ecol 10:401–409

Hutchings MJ, de Kroon H (1994) Foraging in plants: the role of morphological plasticity in resource acquisition. Adv Ecol Res 25:159–238

Inouye DW, Saavedra F, Lee-Yang W (2003) Environmental influences on the phenology and abundance of flowering by *Androsace septentrionalis* (Primulaceae). Am J Bot 90:905–910

Knops JM, Reinhart K (2000) Specific leaf area along a nitrogen fertilization gradient. Am Midl Nat 144:265–272

Lortie CJ, Aarssen LW (1997) Apical dominance as an adaptation in *Verbascum thapsus*: effects of water and nutrients on branching. Int J Plant Sci 158:461–464

Meier IC, Leuschner C (2008) Leaf size and leaf area index in *Fagus sylvatica* forests: competing effects of precipitation, temperature, and nitrogen availability. Ecosystems 11:655–669

Mott JJ, McComb AJ (1975) The role of photoperiod and temperature in controlling the phenology of three annual species from an arid region of Western Australia. J Ecol 63:633–641
Pedrol N, Ramos P, Reigosa MJ (2000) Phenotypic plasticity and acclimation to water deficits in velvet-grass: a long-term greenhouse experiment. Changes in leaf morphology, photosynthesis and stress-induced metabolites. J Plant Physiol 157:383–393
Peralta-García C, Sánchez-Coronado ME, Orozco-Segovia A, Orozco-Segovia S, Pisanty-Baruch I (2016) Germination of four riparian species in a disturbed semi-arid ecosystem. S Afr J Bot 106:110–118
Peralta-García C, Pisanty I, Orozco-Segovia A, Sánchez-Coronado ME, Rodríguez-Sánchez M (2020) Germination of riparian plants in natural and experimental conditions. In: Mandujano MC, Pisanty I, Eguiarte LE (eds) Plant Diversity and Ecology in the Chihuahuan Desert. Springer International, Cham, Switzerland
Pinkava DJ (1984) Vegetation and flora of the Bolson of Cuatro Cienegas Region, Coahuila, Mexico: IV, summary, endemism and corrected catalogue. J Ariz Nev Acad Sci 19:23–47
Pisanty I, Pérez y Sosa C, Gálvez G (2013) Agriculture, water mismanagement and ecosystem transformations in the Cuatrociénegas Valley in the Chihuahuan Desert. In: Schwabe K, Albiac J, Conor JD, Hassan RM, Meza González L (eds) Drought in arid and semiarid regions. A multidisciplinary and Cross-Country Perspective. Springer Science Business Media, Dordrecht, Netherlands, pp 199–216
Pisanty I, Rodríguez-Sánchez M, Granados-Hernández L, Escobar S, Mandujano MC (2020) Disturbance and the formation and colonization of new habitats in the terminal part of the Churince System in the Cuatro Ciénegas Basin. In: Mandujano MC, Pisanty I, Eguiarte L E (eds) Plant diversity and ecology in the Chihuahuan Desert. Springer International, Cham, Switzerland
Puentes DA, de Armas JH, Almirall AL (1993) Fenología y estructura floral de *Trichilia havanensis* Jacq. (Meliaceae). Ann Mo Bot Gard 80:862–869
Rodríguez-Sánchez M, Pisanty I, Mandujano MC, Almaguer K (2020) An unlikely moveable feast in a desert hydrological system: why do life cycles matter. In: Mandujano MC, Pisanty I, Eguiarte L (eds) Plant Diversity and Ecology in the Chihuahuan Desert. Springer International, Cham, Switzerland
SEINet (Southwest Environmental Information Network) (2018) Flora of North America. *Samolus ebracteatus* Kunth. http://swbiodiversity.org/seinet/taxa/index.php?taxon=18678. Accessed Nov 2018
Simpson GG, Dean C (2002) *Arabidopsis*, the Rosetta stone of flowering time? Science 296:285–289
Souza V, Eguiarte L (2018) In the beginning, there was fire: Cuatro Ciénegas Basin (CCB) and the Long History of Life on Earth. In: Souza V, Olmedo-Álvarez G, Eguiarte LE (eds) Cuatro Ciénegas ecology, natural history and microbiology. Springer International, Cham, Switzerland, pp 21–34
Souza V, Espinosa L, Escalante A, Eguiarte LE, Farmer J, Forney L, Lloret L, Rodríguez J, Soberón X, Dirzo R, Elser J (2006) An endangered oasis of aquatic microbial biodiversity in the Chihuahuan Desert. Proc Natl Acad Sci USA 103(17):6565–6570
Sultan SE (2000) Phenotypic plasticity for plant development, function and life history. Trends Plant Sci 5:537–542
Umesh T, Dinesh S, Puvvadi V. Sivapullaiah P (2011) Characterization of Dispersive Soils. Materials Sciences and Applications 2(6):629–633
Valladares F, Aranda I, Sánchez-Gómez D (2004) La luz como factor ecológico y evolutivo para las plantas y su interacción con el agua. In: Valladares F (ed) Ecología del bosque mediterráaneo en un mundo cambiante. Organismo Autónomo de Parques Nacionales. Ministerio de Medio Ambiente, Madrid, Spain, pp 335–370
Valladares F, Gianoli E, Gómez JM (2007) Ecological limits to plant phenotypic plasticity. New Phytologist. Tansley Rev 176:749–763
Villarreal-Quintanilla JA, Encina-Domínguez J (2005) Plantas vasculares endémicas de Coahuila y algunas áreas adyacentes, México. Acta Bot Mexican 70:1–46

Chapter 19
Germination of Riparian Species in Natural and Experimental Conditions

Cynthia Peralta-García, Irene Pisanty, Alma Orozco-Segovia, Ma. Esther Sánchez-Coronado, and Mariana Rodríguez-Sánchez

Abstract In altered or newly opened habitats, germination is a crucial event for the natural regeneration and for the colonization of environments. In the Cuatro Ciénegas Basin, the Churince hydrological system has been deeply disturbed and numerous sinkholes were formed in the surrounding flatland, which can provide favorable conditions for the establishment of riparian plants, whose original habitat is disappearing. The effect of light and storage conditions and time on the germination of five riparian species from the Churince system (*Eustoma exaltatum*, *Flaveria chlorifolia*, *Sabatia tuberculata*, *Samolus ebracteatus* var. *coahuilensis*, and *Schoenus nigricans*) that are colonizing the sinkholes were analyzed in field and controlled conditions. Additionally, in a greenhouse, we analyzed germination on different substrates. All the species were indifferent to light. Germination in field conditions was low, particularly for *Schoenus nigricans*. Germination of control seeds and of those recovered from the field experiment was high in controlled conditions. *Samolus ebracteatus* var. *coahuilensis* and *F. chlorifolia* germinated as soon as hydrated and had germination proportions close to 0.9. In the greenhouse, the highest percentage of germination was on black soil, except for *Schoenus nigricans*, followed by silica sand. Germination was low on substrates from the collection site, especially for *Schoenus nigricans*. The seeds may be conditionally dormant, allowing the formation of a soil seed bank. The different germinative responses suggest an important functional diversity in the disturbed Churince system.

Keywords Desert riparian ecosystems · Riparian species · Disturbance · Germination · Sinkholes

C. Peralta-García · I. Pisanty (✉) · M. Rodríguez-Sánchez
Departamento de Ecología y Recursos Naturales, Facultad de Ciencias, Universidad Nacional Autónoma de México, UNAM, Mexico City, Mexico
e-mail: ipisanty@unam.mx

A. Orozco-Segovia · M. E. Sánchez-Coronado
Departamento de Ecología Funcional, Instituto de Ecología, Universidad Nacional Autónoma de México, UNAM, Ciudad Universitaria, Mexico City, Mexico

M. C. Mandujano et al. (eds.), *Plant Diversity and Ecology in the Chihuahuan Desert*, Cuatro Ciénegas Basin: An Endangered Hyperdiverse Oasis,
https://doi.org/10.1007/978-3-030-44963-6_19

Introduction

Riparian ecosystems have a wide distribution and can be found in a diversity of environments (Zaimes 2007). In arid and semiarid zones, water bodies have high concentrations of salts and fluctuating water regimes, and they can form isolated oasis (Ezcurra et al. 1988). Desert springs and wetlands, as well as the riparian habitats associated with them, contribute to the diversity of arid lands because they are unique and usually discrete habitats, frequently rich in endemic species, which contribute to make them diversity hotspots (Hubbard 1977; Ezcurra et al. 1988; Briggs 1996; Tiner 2003). Additionally, they include habitats for different resident and migratory animal species, like birds and insects (Skagen et al. 1998; Naiman and Décamps 1997; Capon and Dowe 2007). This type of ecosystems has been altered by humans for a long time and has suffered overexploitation for agriculture and cattle raising; thus many aquatic, sub-aquatic, and riparian ecosystems have been severely degraded or completely lost (Patten 1998; Naiman et al. 2005; Poff et al. 2011).

Overexploitation of ground and surface water, mainly for alfalfa cultivation, has deeply affected the major hydrological systems in the Cuatro Ciénegas Basin (CCB). The Churince system, one of them, located in the southern part of the protected area of the basin, has been drastically disturbed during the last 10 years by this process, which has caused its progressive desiccation (Souza et al. 2006; Pisanty et al. 2013). The terminal Churince lagoon and the Intermediate lagoon, as well as the in-between part of the Churince River are presently dry. The flatland surrounding the Churince river was classified by Pinkava (1984) as an area with no evident vegetation, but it has been occupied by those riparian species that are more tolerant to changing conditions (Pisanty et al. 2013; Pisanty et al. 2020, this volume). The edaphic structure of this zone has been altered as the progressive desiccation of the river caused a subsurface flow of water in the South bank, due to which numerous sinkholes have been formed since ca. 2003 (Rodríguez et al. 2005; Pisanty et al. 2013; Pisanty et al. 2020, this volume). Sinkholes are soils depressions of different sizes, depths, and shapes, formed by the loss of cohesion between particles and their dispersion, that cause the collapse of the soil surface (Heinzen and Arulanandan 1977). Sinkholes proved to be safe sites for germination (Peralta-García et al. 2016) and establishment of riparian species (Pisanty et al. 2013; Pisanty et al. 2020, this volume).

Riparian plant species growing in arid and semiarid lands face a heterogeneous environment with different selective pressures, like underground anoxic conditions due to flooding, desiccation due to extreme temperatures, and poor soils with high pH values, due to considerable concentrations of salt and gypsum (Capon and Dowe 2007). Light and temperature fluctuations are also important cues that allow seeds to respond when conditions can be favorable for seedling establishment, allowing them to break dormancy or quiescence and germinate (Thompson and Grime 1983; Baskin et al. 1989). This is frequent in species from these environments (Thompson and Grime 1983; Baskin et al. 1989). Germination responses to these conditions have proven to be both complex and variable (Shipley and Parent 1991; Leck and Brock 2000; Peralta-García et al. 2016), and very little is known about them. Leck and Brock (2000) and Capon (2007) report clear effects of flooding on germination and seedling emergence patterns in different arid wetlands, and the former authors recognize that water regimes, seed bank dispersal, dormancy germination strategies, life history patterns, and functional groups affect germination and establishment in these heterogeneous (both in time and space) environments.

In this chapter, we analyze the germination responses of five riparian species to different conditions that include light or darkness, field and controlled conditions (growth chambers), and different substrates in semi-controlled conditions (greenhouse) in order to explain how they may establish in different environments. Because changes in hydrological systems can alter the vegetation related to them (Patten et al. 2008), the study of germination patterns in the sinkholes formed due to disturbance of the hydrological system is crucial to understand how these sites are colonized, depending on the ability of seeds to respond to heterogeneous and relatively unpredictable environments with increasing stress that, eventually, can cause riparian species to disappear in this area, as has been observed in other places (e.g., Patten et al. 2008).

Methods

Study Area

CCB is located at 27°11′–26°42′ N and 102°48′–101°54′ W, in the Chihuahuan Desert, Coahuila, Mexico, 740 m above sea level (Minckley 1969; INE-SEMARNAP 1999). Climate is arid with an average annual precipitation of 212 mm. Average temperature is 30 °C, but variation is high (0–50 °C) (INE-SEMARNAP 1999). The Churince system (26°51′N and 102°08′W) is formed by *Poza* Bonita, *Poza* Churince, Churince River, Intermediate (or *Los Güeros*) lagoon, and Churince (or *Grande*) Lagoon. This study took place in the surroundings of the terminal part of the river and of the Intermediate lagoon, where seeds of five riparian species and soil samples were collected, and germination experiments in the sinkholes, under natural conditions, were performed.

Species

Five riparian species were included in this study: *Samolus ebracteatus* HBK *var. coahuilensis* Henrickson (Primulaceae), *Flaveria chlorifolia* A. Gray (Asteraceae), *Eustoma exaltatum* (L.) Salisb. Ex G. Don (Gentianaceae), *Sabatia tuberculata* J. E. Williams (Gentianaceae), and *Schoenus nigricans* L. (Cyperaceae), which are among the most common riparian species at CCB (Pinkava 1984; INE-SEMARNAP 1999) as well as colonizers of the sinkholes (Pisanty et al. 2013; Pisanty et al. 2020, this volume). *Samolus ebracteatus* var. *coahuilensis, F. chlorifolia,* and *Schoenus nigricans* are perennials; *E. exaltatum* is an annual or a short-lived perennial (Turner 2014, Rodríguez-Sánchez et al. 2020, this volume), and *Sabatia tuberculata* is an annual.

Effect of Light on Germination

Most seeds were collected in 2008–2009, except for those of *Schoenus nigricans* that were collected in 2016. Seeds of all the species were sown in agar plates (0.8%) in Petri dishes and exposed to light or darkness. Experimental design for each species was: two light conditions × 5 replicates (5 Petri dishes) × 10 seeds in each Petri dish. For *Schoenus nigricans* 20 seeds were sown per replicate. Petri dishes were placed inside a growth chamber provided with white light lamps at a photoperiod 12/12 h and a thermoperiod 18/32°C (18/6 h) based in previous observations. Darkness condition was obtained by wrapping the dishes with two layers of aluminum foil. Due to the lack of normality and/or homoscedasticity, a logistic regression was used for each species (Hosmer and Lemeshow 2000) to calculate the probability of germination in light or darkness, and differences in final probabilities of germination between light conditions were contrasted. This statistical test was made using JMP ver. 10 software (SAS Institute Inc. Cary, NC, USA).

Germination in Natural and Controlled Conditions

In August 2009, 50 seeds of *Samolus ebracteatus* var. *coahuilensis*, *F. chlorifolia*, *E. exaltatum,* and *Sabatia tuberculata* were placed in nylon mesh bags (36 bags per species) and placed in randomly chosen sinkholes of the Intermediate lagoon under natural conditions. Control seeds were stored in the laboratory. Six randomly chosen bags per species were recovered every 2 months, for 2 years. Germinated seeds in the bags were counted, and non-germinated seeds (recovered seeds hereafter) were sown on agar plates in the previously described germination chamber. Control seeds (laboratory stored) per species were simultaneously sown and placed in the germination chamber. For *Schoenus nigricans*, 300 bags with 20 seeds were placed in the sinkholes in October 2016; germination in the field and in recovered and control seeds was treated as described above. For each species, logistic regressions were used to calculate germination probabilities and to test the effect of storage condition (laboratory or field), storage time, and the interaction of both factors.

Germination in Greenhouse Conditions

A greenhouse experiment was performed using six different substrates to determine their effects on the germination of *Samolus ebracteatus* var. *coahuilensis*, *F. chlorifolia*, *Schoenus nigricans,* and *E. exaltatum*. Nutrient-rich commercial black soil, silica sand, and substrates from the sinkholes, the river mouth, the river bed, and the river bank were used. Soils were collected from the surface to 30 cm depth. Fifty plastic pots of 9.5 cm of diameter were filled with each of the substrates, and five seeds of each species were sown per pot. Pots were placed in a greenhouse (average temperature = 19.55 °C; maximum = 37.8 °C; minimum = 10.3 °C; relative humidity = 52.8%); during the experiment, temperature and humidity were measured continuously with a data logger (HOBO Pro V2, Onset Computer Corporation, Pocasset, MA, USA). Pots were kept humid, and seedling emergence was registered every 3 days. The experiment ended after 4 continuous weeks without emergence. Differences between final germination percentages were compared using Kruskal–Wallis test (Zar 2010).

Results

Effect of Light on Germination

Seeds of all the species germinated with light and in darkness, in proportions higher than 0.5; nevertheless, the germination response to light differed between species. *Samolus ebracteatus* var. *coahuilensis* attained a germination proportion of 1 both in light and in darkness, so it can be considered truly indifferent to light. *F. chlorifolia* had a small but significantly higher germination proportion in darkness than in light (0.98 and 0.86, respectively; $\chi^2 = 5.45$, *d.f.* = 1, $p = 0.02$). On the contrary, germination proportions were significantly higher in light for *Schoenus nigricans* (0.86), *E. exaltatum* (0.82), and *Sabatia tuberculata* (0.98) ($\chi^2 = 13.25$, *d.f.* = 1, $p < 0.05$; $\chi^2 = 7.00$, *d.f.* = 1, $p < 0.05$; $\chi^2 = 4.23$, *d.f.* = 1, $p = 0.04$, respectively) than in darkness (0.64, 0.58 and 0.88, respectively).

Germination in Natural and Controlled Conditions

Storage type, storage time, and their interaction had effect on the germination proportions (Table 19.1), which were significantly higher for the control seeds (from 0.6 to 1; Fig 19.1a, d, g, j, and m) than for the recovered seeds and for those that germinated in field conditions, which were the lowest (Fig. 19.1b, e, h, c, k, and n). In this last condition no germination was observed in *E. exaltatum* (in September and November 2009 and in May 2010), *F. chlorifolia* (in January 2010), *Sabatia*

Table 19.1 Results of the logistic regression analysis for the effect of storage type (ST), storage time (st), and the interaction between both factors (ST × st) on the germination of seeds stored in the field, recovered seeds, and control seeds of five riparian species from Cuatro Ciénegas Basin, Coahuila, Mexico

Species	Source of variation	*Wald* χ^2	*d.f.*	*p*
Eustoma exaltatum	Storage type (ST)	515.23	2	<0.0001
	Storage time (st)	72.08	5	<0.0001
	ST × st	282.13	10	<0.0001
Flaveria chlorifolia	ST	672.52	2	<0.0001
	st	18.82	5	0.0021
	ST × st	110.81	10	<0.0001
Sabatia tuberculata	ST	675.51	2	<0.0001
	st	117.72	5	<0.0001
	ST × st	160.65	10	<0.0001
Samolus ebracteatus var. *coahuilensis*	ST	1241.94	2	<0.0001
	st	44.22	5	<0.0001
	ST × st	429.31	10	<0.0001
Schoenus nigricans	ST	3309.20	2	<0.0001
	st	64.38	5	<0.0001
	ST × st	39.30	10	<0.0001

tuberculata (in November 2009 and January and May 2010), and *Samolus ebracteatus* var. *coahuilensis* (in May 2010). The maximum field germination proportion (0.6) was observed in *Samolus ebracteatus* var. *coahuilensis* in July 2010. The field germination proportions of *Schoenus nigricans* were close to zero (Fig. 19.1n). On the other hand, minimum proportions of germination of recovered seeds were higher than 0.4, and maximum proportions reached at least 0.7 (Fig. 19.1c, f, i, l, and o).

Germination in Greenhouse Conditions

Seed germination of the four species included in this experiment was significantly and differentially affected by the substrates. Substrates from the study site induced germination percentages lower than 50% in all the species, and higher percentages were found at least in one of the commercial substrates (black soil and silica sand) (Fig. 19.2). *E. exaltatum* and *Samolus ebracteatus* var. *coahuilensis* showed the significantly higher germination percentages on black soil (46 ± 1.76%, mean ± s.d.) and 57 ± 1.39%, respectively) and on silica sand (43 ± 1.41% and 70 ± 1.65%, respectively; Fig. 19.2a and c; H = 51.83, $p < 0.05$, H = 168.59, $p < 0.05$, respectively). The lowest germination percentage was registered on soil from the river bank for both species (10 ± 1.09% and 1 ± 0.23%, respectively), nevertheless differences between maximum and minimum values were more contrasting for *Samolus ebracteatus* var. *coahuilensis* (Fig. 19.2c). *F. chlorifolia* showed germination percentages close to 40% in almost all the substrates, except in black soil, where a

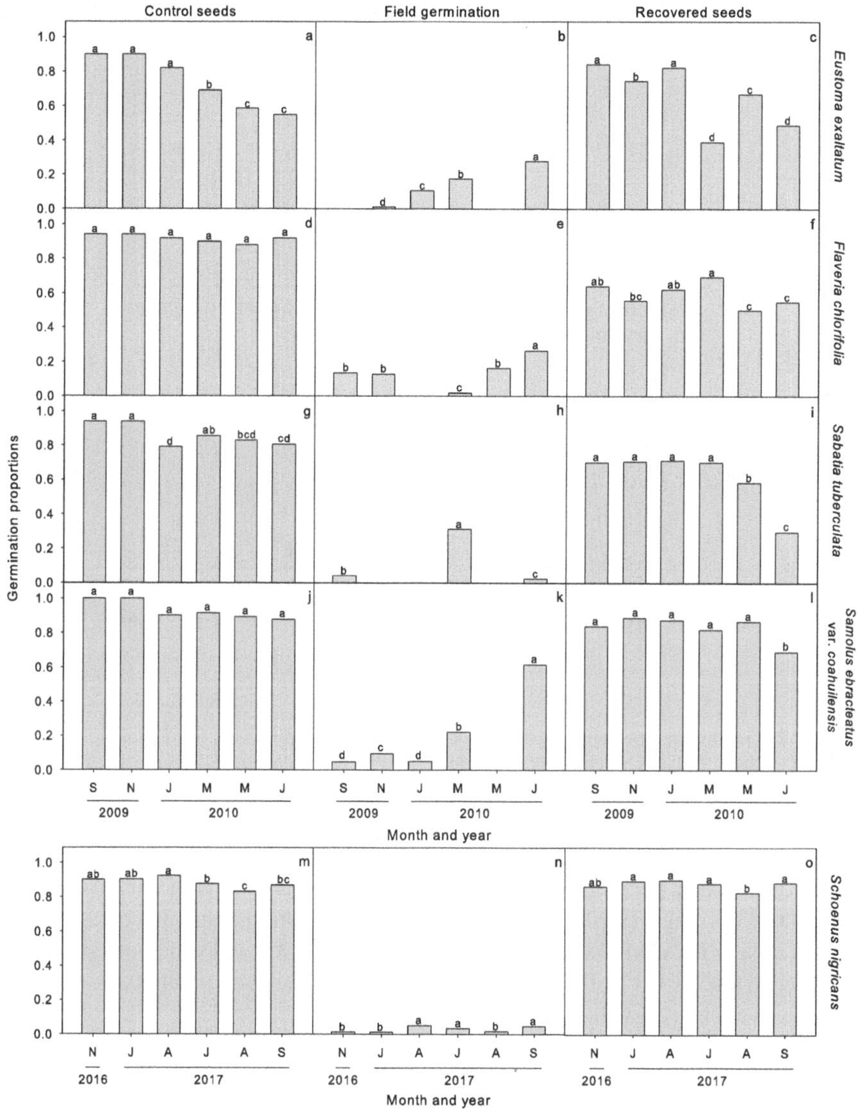

Fig. 19.1 Germination proportions of control seeds, field germination, and recovered seeds (field storage) of *Eustoma exaltatum (a,b,c)*, *Flaveria chlorifolia (d,e,f)*, *Sabatia tuberculata (g,h,i)*, and *Samolus ebracteatus* var. *coahuilensis (j,k,l)* in 2009–2010 and *Schoenus nigricans (m,n,o)* in 2016–2017. Seeds were collected in Cuatro Ciénegas Basin, Coahuila, Mexico. Letters indicate groups with significant differences within each species

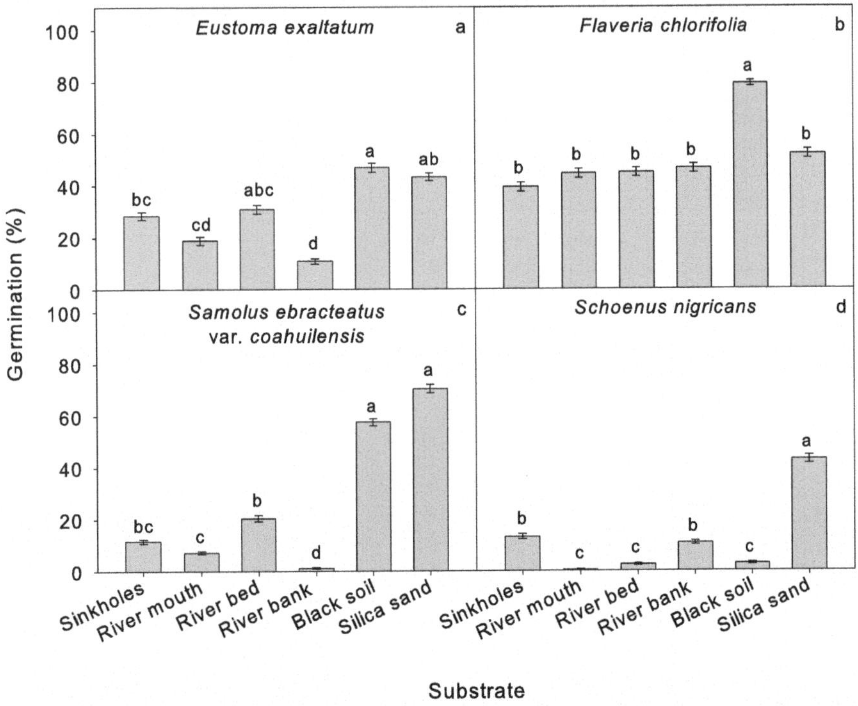

Fig. 19.2 Germination percentages of the seeds of four riparian species from Cuatro Ciénegas Basin, Coahuila, Mexico, sown in six different substrates in a greenhouse. Mean ± s.d. Different letters indicate differences within each species

significantly higher value was attained (79 ± 1.07%; Fig. 19.2b; H = 42.42, $p < 0.05$). *Schoenus nigricans* showed the lowest germination percentages in this experiment (<43%, Fig. 19.2d). Its significant highest germination percentage was observed in silica sand (43 ± 1.57%, H = 116.11, $p < 0.05$), while the lowest corresponded to black soil (2 ± 0.35%), and substrates from the river mouth (0.4 ± 0.14%) and the river bed (2 ± 0.38%).

Discussion

The ability to germinate in different light environments might be an advantage for the colonization of the river bank and the sinkholes, which vary in light availability and quality, temperature, and humidity, due to the differences in size and plant cover (Pisanty et al. 2013). Despite the fact that all species have light insensitive seeds, as germination was observed both with and without light (Vázquez-Yanes and Orozco-Segovia 1993), there were differences in the response to light availability among the five species, suggesting specific functional diversity (Orozco-Segovia and

Sánchez-Coronado 2009) that allows varied roles in succession and ecosystem regeneration, as well as in colonization. The capacity of buried seeds to germinate is crucial for the recruitment of new individuals and non-photoblastic seeds can do it. *Samolus ebracteatus* var. *coahuilensis* did not show different germination proportions with or without light, thus proving it is an indifferent species. Differences in *F. chlorifolia*, *Schoenus nigricans,* and *Sabatia tuberculata* are significant but minimal. Besides germinating in darkness, *F. chlorifolia* and *Samolus ebracteatus* var. *coahuilensis* can also produce leaves if meristems are covered with soil (Pisanty et al. 2013; Rodríguez-Sánchez 2018), implying that tolerance to darkness involves both seeds and photosynthetic structures, which can be advantageous for species growing on heterogeneous and unstable substrates both in their original habitat and in the disturbance induced ones. On the contrary, the proportion of seeds of *E. exaltatum* that do not germinate in darkness is considerable, probably allowing the formation of a soil seed bank through which it spreads risks in time and avoids adverse conditions for seedlings (Vázquez-Yanes and Orozco-Segovia 1996; Fenner and Thompson 2005; de Jong and Klinkhamer 2005).

Contrasting with germination of control and recovered seeds, field germination proportions were from low to intermediate. This could also indicate a conditional dormancy (*sensu* Baskin and Baskin 1985; Baskin et al. 1989), as suggested by the increased germination observed in *E. exaltatum*, *F. chlorifolia,* and *Samolus ebracteatus* var. *coahuilensis* in July 2010, related with the intense precipitations originated by a hurricane in 2010 (Peralta-García et al. 2016) and by *E. exaltatum*, *Sabatia tuberculata*, and *Samolus ebracteatus* var. *coahuilensis* in March 2010, at the end of winter, when conditions may be favorable, in less fluctuating temperature and water availability in the sinkholes. The poor germination response to precipitation observed in *Sabatia tuberculata* and *Schoenus nigricans* may be due to their long lag times (Peralta-García et al. 2016; Peralta-García pers. observations) which could prevent these species from taking advantage of random and unpredictable increases in water availability, before drying out again. Seeds in the seed bank thus avoid desiccation risk for slow growing seedlings (Peralta-García pers. observations). Contrastingly, the control seeds of *E. exaltatum, F. chlorifolia,* and *Samolus ebracteatus* var. *coahuilensis* germinated after a few days with high germination rates (Peralta-García et al. 2016) as a quick response to water availability.

In all species, conditional dormancy in field conditions and high germination responses of recovered seeds suggest that seeds can remain viable in the soil seed bank at least 12 months. This is important for the species growing in this disturbed environment, where the conditions are continuously changing in time and space. In species with short life cycles, like *E. exaltatum*, an annual or short-lived perennial (Turner 2014), and *S. tuberculata*, an annual endemic species (Williams 1982; Villarreal-Quintanilla and Encina-Domínguez 2005), the seed bank is the only way to remain as part of the plant community.

In the greenhouse, the differential germination response on the different substrates could be related to the osmotic potential of substrate more than with the presence of gypsum, for Escudero et al. (1997) demonstrated that gypsum concentrations did not affect the germination responses of 11 species growing in gypsum rich sites in Spain. In the greenhouse experiment, pots were watered every 3 days and solutes might have accumulated in different forms in the soil solution afterwards, affecting the response of the four species. The high germination percentages of *F. chlorifolia* on black soil and silica sand proved that this species, frequently associated with gypsum, is a gypsovage more than a gypsophyte (Powell 1978; Flores-Olvera et al. 2016). Better germination responses in growth chambers, with high water availability and stable conditions (light, water, and temperature fluctuating range) might explain the differences with germination in field conditions and in the greenhouse. Sporadic massive germination, such as the one observed after the 2010 hurricane, and peak establishment events allow colonization periods determined by germination and, afterwards, survival and growth of seedlings (Rodríguez-Sánchez et al. 2020, this volume). Different responses in similar environments reinforce the importance of functional diversity for the fitness of this species.

Changes in water availability due to disturbance are expected to affect germination, establishment, growth, and survival following different space and time patterns than those expected when conditions are more stable, even if they are as stressful as they can be in a desert. Unmack and Minckley (2008) recognized the early response of riparian vegetation to changes in water flow and identified plants around springs (e.g., sedges and grasses) that root in waterlogged conditions, as stable for long periods. However, when water flow is altered and water levels decline, compromising sedges, grasses, and other riparian species, which can be substituted by shrubs, forbs, and trees. All the species considered in this chapter are being affected by the disturbance of the hydrological systems on CCB. Environmental changes induced by disturbance have changed the intensity of selective forces, especially of those related with water availability and unless perturbation is reversed and the system is allowed to recover, at least partially, riparian habitats and species will soon disappear, with the concomitant loss of species and ecosystem diversity as well as of the environmental services provided by this area.

Acknowledgments This study was financed by projects UNAM/ PAPIIT IN231811, UNAM/ PAPIIT IN217915, UNAM /PAPIIT 205715, CONACyT 221015, and the WWF-Carlos Slim Alliance. We especially thank Juan Carlos Ibarra, director of the Área de Protección de Flora y Fauna de Cuatrociénegas for his help and outstanding efforts, Valeria Souza for her constant support, and Maria C. Mandujano for her contribution to this project. We appreciate the contribution of Ana Karen Almaguer and the support of Beatriz Zúñiga Ruiz, Manuel Hernández Quiroz, Laura P. Olguín Santos, María Eugenia Muñiz Díaz de León, J. Luis Vigosa Mercado, Pedro E. Mendoza-Hernández, Mariana Hernández-Apolinar, Israel Solano-Zavaleta, and Marco Romero- Romero. We fully appreciate the help of the people from Cuatro Ciénegas and the *ejido* Seis de enero.

References

Baskin JM, Baskin CC (1985) The annual dormancy cycle in buried weed seeds: a continuum. Bioscience 35:492–498

Baskin JM, Baskin CC, Spooner DM (1989) Role of temperature, light and date: seeds were exhumed from soil on germination of four wetland perennials. Aquat Bot 35:387–394

Briggs MK (1996) Riparian ecosystem recovery in arid lands: strategies and references. The University of Arizona Press, Tucson, AZ, USA

Capon SJ (2007) Effects of flooding on seedling emergence from the soil seed bank of a large desert floodplain. Wetlands 27(4):904–914

Capon SJ, Dowe JL (2007) Diversity and dynamics of riparian vegetation. In: Lovett SS, Price P (eds) Principles for riparian lands management. Land and Water Australia, Canberra, Australia, pp 13–32

de Jong T, Klinkhamer P (2005) Evolutionary ecology of plant reproductive strategies. Cambridge University Press, Cambridge, UK

Escudero A, Carnes LF, Pérez-García F (1997) Seed germination of gypsophytes and gypsovags in semi-arid central Spain. J Arid Environ 36:487–497

Ezcurra E, Felger RS, Russell AD, Equihua M (1988) Freshwater islands in a desert sand sea: the hydrology, flora, and phytogeography of the Gran Desierto Oasis of Northwestern Mexico. The University of Arizona Press, Tucson, AZ, USA

Fenner M, Thompson K (2005) The ecology of seeds. Cambridge University Press, New York, NY, USA

Flores-Olvera H, Czaja A, Estrada-Rodríguez JL, Romero-Méndez U (2016) Floristic diversity of halophyte plants of Mexico. In: Khan MA (ed) Sabkha Ecosystems Volume V: The Americas, Tasks for Vegetation Science 48. Springer International Publishing, Cham, Switzerland

Heinzen RT, Arulanandan K (1977) Factors influencing dispersive clays and methods of identification. ASTM Spec Tech Publ 623:202–217

Hosmer DW, Lemeshow S (2000) Applied logistic regression. John Wiley & Sons, New York, NY, USA

Hubbard JP (1977) Importance of riparian ecosystems: biotic considerations. In: Johnson RR, Jones DA (eds) Importance, preservation and management of riparian habitat: a symposium, USDA Forest Service General Technical Report RM-43. US Department of Agriculture, Rocky Mountain Forest and Range Experimental Station, Fort Collins, CO, USA, pp 14–18

Instituto Nacional de Ecología (INE), Secretaría de Medio Ambiente, Recursos Naturales y Pesca (SEMARNAP) (1999) Programa de manejo del área de Protección de flora y fauna Cuatrociénegas. Instituto Nacional de Ecología, México, DF, Mexico

Leck MA, Brock MA (2000) Ecological and evolutionary trends in wetlands: evidence from seeds and seed banks in New South Wales, Australia and New Jersey, USA. Plant Species Biol 15:97–112

Minckley WL (1969) Environments of the bolson of Cuatro Ciénegas, Coahuila, Mexico, with special reference to the aquatic biota, Science Series, vol 2. Texas Western Press, The University of Texas, El Paso, TX, USA, pp 1–65

Naiman RJ, Décamps H (1997) The ecology of interfaces: riparian zones. Annu Rev Ecol Syst 28:621–658

Naiman RJ, Décamps H, McClain M (2005) Disturbance and agents of change. In: Naiman RJ, Decamps H, McClain M (eds) Riparian: ecology, conservation and management of streamside communities. Academic Press, Cambridge, UK, pp 189–232

Orozco-Segovia A, Sánchez-Coronado ME (2009) Functional diversity in seeds and its implications for ecosystem functionality and restoration ecology. In: Gamboa-de Buen A, Orozco-Segovia A, Cruz-García F (eds) Functional diversity of plant reproduction. Research Signpost, Kerala, India, pp 195–236

Patten DT (1998) Riparian ecosystems of semi-arid North America: diversity and human impacts. Wetlands 18(4):498–512

Patten DT, Rouse L, Stromberg J (2008) Vegetation dynamics of Great Basin Springs: potential effects of groundwater withdrawal. In: Stevens LE, Meretsky VJ (eds) Aridland springs in North America: ecology and conservation. The University of Arizona Press, Tucson, AZ, USA, pp 279–289

Peralta-García C, Pisanty I, Sánchez-Coronado ME, Orozco-Segovia A et al (2016) Germination of four riparian species in a disturbed semi-arid ecosystem. S Afr J Bot 106:110–118

Pinkava DJ (1984) Vegetation and flora of the Bolson of Cuatro Cienegas region, Coahuila, Mexico: IV, summary, endemism and corrected catalogue. J Arizona-Nevada Acad Sci 19:23–47
Pisanty I, Pérez y Sosa MC, Gálvez G (2013) Agriculture, water mismanagement and ecosystem transformations in the Cuatrociénegas Valley in the Chihuahuan Desert, Mexico. In: Schwabe K, Albiac J, Connor JD et al (eds) Drought in arid and semi-arid regions: a multi-disciplinary and cross-country perspective. Springer Publishing, Dordrecht, Holland, pp 148–181
Pisanty I, Rodríguez-Sánchez M, Torres Orozco P, Granados-Hernández L, Escobar S, Mandujano MC (2020) Disturbance and the formation and colonization of new habitats in a hydrological system in the Cuatro Ciénegas Basin. In: Mandujano MC, Pisanty I, Eguiarte LE (eds) Plant Diversity and Ecology in the Chihuahuan Desert. Springer International, Cham, Switzerland
Poff B, Koestner KA, Neary DG, Henderson V (2011) Threats to riparian ecosystems in western North America: an analysis of existing literature. J Am Water Resour Assoc 47(6):1241–1254
Powell MA (1978) Systematics of *Flaveria* (Flaveriniae Asteraceae). Ann Missouri Bot Gard 65(2):590–636
Rodríguez JM, Souza V, Díaz de León LE (2005) The overexploitation of the aquifer of El Hundido Valley and the corresponding effects on the reserve Cuatro Ciénegas Valley of Coahuila, Mexico. In: Stevanović Z, Milanović P (eds) Water resources and environmental problems in karst. Proceedings of the International conference and field seminars. National Committee of the International Association of Hydrogeologists of Serbia and Montenegro, Belgrade, Serbia, pp 303–310
Rodríguez-Sánchez M (2018) Dinámica poblacional, germinación, establecimiento y crecimiento de *Flaveria chorifolia* (Asteraceae) en un sistema perturbado en el Valle de Cuatrociénegas, Coahuila, México. Master Thesis, Facultad de Ciencias, UNAM, Mexico city, Mexico.
Rodríguez-Sánchez M, Pisanty I, Mandujano MC, Flores-Olvera, H, Almaguer, K (2020) An unlikely moveable feast in a desert hydrological system: why do life cycles matter. In Mandujano MC, Pisanty I, Eguiarte LE (eds) Plant Diversity and Ecology in the Chihuahuan Desert. Springer International, Cham, Switzerland
Shipley B, Parent M (1991) Germination responses of 64 wetland species in relation to seed size, minimum time to reproduction and seedling relative growth rate. Funct Ecol 5(1):111–118
Skagen S, Melcher CP, Howe WH, Knopf FL (1998) Comparative use of riparian corridors and oasis by migrating birds in Southeast Arizona. Conserv Biol 12(4):896–909
Souza V, Espinosa L, Escalante A, Eguiarte LE, Farmer J, Forney L, Lloret L, Rodríguez J, Soberón X, Dirzo R, Elser J (2006) An endangered oasis of aquatic microbial biodiversity in the Chihuahuan Desert. Proc Natl Acad Sci USA 103(17):6565–6570
Thompson K, Grime JP (1983) A comparative study of germination responses to diurnally-fluctuating temperatures. J Appl Ecol 20:141–156
Tiner RW (2003) Geographically isolated wetlands of the United States. Wetlands 23(3):494–516
Turner BL (2014) Taxonomic overview of *Eustoma* (Gentianaceae). Phytologia 96(1):7–11
Unmack PJ, Minckley WL (2008) The demise of desert springs. In: Stevens LE, Meretsky VJ (eds) Aridland springs in North America. Ecology and conservation. The University of Arizona Press, Tucson, AZ, USA, pp 11–35
Vázquez-Yanes C, Orozco-Segovia A (1993) Patterns of seed longevity and germination in the tropical rainforest. Annu Rev Ecol Syst 24:69–87
Vázquez-Yanes C, Orozco-Segovia A (1996) Physiological ecology of seed dormancy and longevity. In: Mulkey SS, Chazdon RL, Smith AP (eds) Tropical forest plant ecophysiology. Springer, Boston, MA, USA, pp 535–558
Villarreal-Quintanilla JÁ, Encina-Domínguez JA (2005) Plantas vasculares endémicas de Coahuila y algunas áreas adyacentes, México. Acta Bot Mex 70:1–46
Williams JE (1982) A new species of *Sabatia* (Gentianaceae) from the Chihuahuan Desert. Southwest Nat 27(4):379–382
Zaimes G (2007) Defining Arizona's riparian areas and their importance to the Landscape. In: Zaimes G (ed) Understanding Arizona's riparian areas. The University of Arizona, College of Agriculture and Life Sciences, Tucson, AZ, USA, pp 1–13
Zar JH (2010) Biostatistical analysis, 5th edn. Pearson Prentice-Hall, Pearson, NJ, USA

Index

M. C. Mandujano et al. (eds.), *Plant Diversity and Ecology in the Chihuahuan Desert*, Cuatro Ciénegas Basin: An Endangered Hyperdiverse Oasis,
https://doi.org/10.1007/978-3-030-44963-6

The manufacturer's authorised representative in the EU is Springer Nature Customer Service Centre GmbH, Europaplatz 3, 69115 Heidelberg, Germany. If you have any concerns regarding our products, please contact ProductSafety@springernature.com

Printed and bound by CPI Group (UK) Ltd, Croydon, CR0 4YY
30/09/2026
02225503-0010